The Race to the Moon Chronicled Postcards, and Postmarks
A Story of Puffery vs. the Pragmatic

Umberto Cavallaro

The Race to the Moon Chronicled in Stamps, Postcards, and Postmarks

A Story of Puffery vs. the Pragmatic

Published in association with
Praxis Publishing
Chichester, UK

Springer

Umberto Cavallaro
Italian Astrophilately Society
Villarbasse, Italy

SPRINGER-PRAXIS BOOKS IN SPACE EXPLORATION

Original Italian edition published by Impremix, Edizioni Visual Grafika, Torino, 2011
Springer Praxis Books
ISBN 978-3-319-92152-5   ISBN 978-3-319-92153-2   (eBook)
https://doi.org/10.1007/978-3-319-92153-2

Library of Congress Control Number: 2018954650

© Springer Nature Switzerland AG 2018
This work is subject to copyright. All rights are reserved by the Publisher, whether the whole or part of the material is concerned, specifically the rights of translation, reprinting, reuse of illustrations, recitation, broadcasting, reproduction on microfilms or in any other physical way, and transmission or information storage and retrieval, electronic adaptation, computer software, or by similar or dissimilar methodology now known or hereafter developed.
The use of general descriptive names, registered names, trademarks, service marks, etc. in this publication does not imply, even in the absence of a specific statement, that such names are exempt from the relevant protective laws and regulations and therefore free for general use.
The publisher, the authors, and the editors are safe to assume that the advice and information in this book are believed to be true and accurate at the date of publication. Neither the publisher nor the authors or the editors give a warranty, express or implied, with respect to the material contained herein or for any errors or omissions that may have been made. The publisher remains neutral with regard to jurisdictional claims in published maps and institutional affiliations.

Cover design: Jim Wilkie
Project Editor: Michael D. Shayler

This Springer imprint is published by the registered company Springer Nature Switzerland AG
The registered company address is: Gewerbestrasse 11, 6330 Cham, Switzerland

Contents

Acknowledgments viii
Dedication ix
Foreword by Walt Cunningham x

1 Sputnik Triggers the USSR–USA Competition 1
   USSR-USA Space Race: Ignited in Italy 1
   Sputnik: The Opening Shot of the Space Race 7
   Sputnik 2: A Rocket Six Times More Powerful in Four Weeks! 23
   Explorer 1: One of the Main Discoveries of the IGY 27
   Vanguard 1: The Most Ancient Satellite in Orbit 30
   Sputnik 3: The Emblem of the Soviet Satellites 31
   Rivalry and Inefficiency in the Soviet Space Program 36
   NASA Starts its Adventure 39
   Pioneer and Lunik in a Race to the Moon 42
   Luna 1: The Year Begins with a New Soviet Record 46
   Luna 2: The First Man-Made Object on the Moon 48
   Luna 3: A New Soviet Triumph 49
   Corona: Eye in the Sky 55
   Zenit: The Soviet Corona 61
   Kennedy: The Space Program Leads to a Winning Presidential Campaign 62

2 Man in Space 65
   USA in the Running to Put a Man in Space Before the Soviets 65
   USSR: Determined to Keep Pre-Eminence in Space at Any Cost 70
   The First Accidents and Casualties 72
   Humans in Space: Two Different Paths 76
   1960: Bad Luck for the Soviet Program 78
   The Nedelin Catastrophe 79

Further Year-End Misadventures 84
The Americans Regain Confidence 86
How the Americans Lost the Race to Put a Man in Space First 89
A Soviet is the First Man in Space! 90
"Baikonur" and Soviet Lies 102

3 The Space Race changes direction 107
Kennedy and America's Prestige 107
Alan Shepard: The High-Tech Cold War Gladiator in a Silver Space Suit 108
Kennedy Lays Down the Gauntlet. The Soviet Reaction. 111
Gus Grissom: America Puts the Third Man in Space 114
A New Russian First: An Entire Day in Orbit 116
It is Time to Think About the Moon 120
Soviet Lunar Program 123
John Glenn: The First American in Earth Orbit. 124
Scott Carpenter's Flight Equals the Soviets. 136
The First Soviet "Group Flight" 138
Wally Schirra: A "Textbook" Flight. 141
Gordon Cooper: A New American Record for Space Endurance 142
Valery Bykovsky Pulverizes Every Solo Space Flight Record 143
Valentina Tereshkova: The First Woman in Space. 145
Soviet Supremacy Confirmed 149

4 The astronaut sits in the driver's seat 153
The Gemini Program Announced 153
Gemini Slowly Takes Shape 155
NASA Gets Back on Track 159
The Dyna-Soar Becomes Extinct. 159
The Astronauts of Group 3: Fresh Forces Arrive. 161
The Gemini Program Lifts Off. 161
Voskhod 1: The Most Absurd Adventure in Space Ever 163
GT-2: Halted by a Lightning Strike 169
The Last Dramatic Chance to Beat the Americans 170
Voskhod 2: The First Spacewalk 171
GT-3: The First Spacecraft 'Flown' in Space Under Pilot Control 176
GT-4: The Americans Also Walk in Space. 181
GT-5: The USA Surpasses Soviet Records 186
GT-6 & 7: A New Endurance Record and the First Rendezvous in Space 187
GT-8: An Emergency in Space. 191
Voskhod 3: In Search of an Unlikely New Soviet Space Spectacular 192
GT-9 Lifts-Off with the Backup Crew. 195
GT-10: A Rendezvous with Two Satellites 198
GT-11: An Unprecedented Record Altitude Flight 199
MOL: The Secret American Military Space Station 202
GT-12: The Grand Finale. 205

5	**Two tragedies block the race in space**	209
	NASA: Strong Direction to Achieve Success	209
	The Soviet Lunar Program	212
	The Sudden Passing of Korolev	215
	The Soviets Take the First Pictures from the Moon	217
	The American Lander is Behind Schedule	220
	Apollo on the Launch Pad	223
	"We've Got a Fire in the Cockpit!"	225
	Soyuz 1: An Eminently Predictable Tragedy	229
6	**The final leap**	241
	A New Beginning at NASA	241
	Five Giant Steps	242
	The Russians are Coming!	243
	The Mysterious Death of Yuri Gagarin	247
	NASA Speeds up and Modifies its Programs	249
	Zond 5: A New Soviet Record	250
	Apollo 7: The Flight of the Phoenix	252
	Soyuz 3: The Soviet Reply	257
	Zond 6: Officially "A Complete Success"	258
	The Final Sprint	259
	The Soviets Intensify Testing and the Success of Soyuz 4 and 5	265
	A New Attempt at a Circumlunar Trip	275
	Apollo 9: Flight Test of the Lunar Module	277
	Apollo 10: Dress Rehearsal of the Moon Landing	279
	The Final Shot of the Soviets: A Flurry of Unfortunate Failures	280
	Apollo 11: American Footprints on the Lunar Surface	285
	The Race is Over	287
7	**First Lunar Landing: The Philatelic Side of Apollo 11**	291
	Insurance Covers: Far More Than Plain Collectors' Covers	291
	The First Lunar Post Office: The Moon Letter	297
	The Apollo 11 Flown Covers	302
	'First Man on the Moon': The Greatest Philatelic Success Ever	307

Appendix A: Apollo 15: "The Problem We Brought Back From the Moon" ... 315

Appendix B: Timeline of Main Space Events: 1957–1969 ... 328

Bibliography ... 331

Index ... 335

Acknowledgments

Among the many friends who have supported me in this effort, I must especially thank Jeff Dugdale for his patient first revision of this manuscript and for his precious comments and suggestions.

Thanks also to the many collectors who helped me to document this history by making available some unique items from their collections: Dave Ball (USA); Chris Calle (USA); Pietro Della Maddalena (Italy); Dennis Dillman (USA); Steve Durst (USA); Don Hillger (USA); Peter Hoffman (USA); Walter Hopferwieser (Austria); the late Viacheslav Klochko (Russia); Jaromir Matejka (Austria); Stefano Matteassi (Italy); Renzo Monateri (Italy); Renato Rega (Italy); Jim Reichman (USA); and Antoni Rigo (Spain).

A special thanks to former NASA astronaut Walter Cunningham for his contributions to this book, especially the Foreword. I was also honored to have him attend the presentation of the first version of this manuscript when it appeared in Italian some years ago.

A final thanks to Mike Shayler, who added value to the text with his professional editing and his friendly advice.

Unless otherwise indicated, the pictures of collectibles (covers, stamps, cards, posters, mission patches, manuals, etc.) are from the collection of the author. Despite extensive research, the author has been unable to trace the exact origins of some of the images used in this title and would welcome any assistance that would enable him to credit the appropriate sources.

Throughout this text, there are additional sections aimed at clarifying special philatelic aspects of the story. They are not part of the main flow of the text and can be found in boxes highlighted with a different background color.

To Anna Yukiko
Who will eye-witness
What we only dared to dream.

Foreword

Reading this book brought me back to the wonderful and intense years I spent at NASA when the events described by Umberto were taking place. It reminded me of the enthusiasm, the emotions, anxiety, expectations, frustration, and elation of those "golden years" when I was living them.

The book presents the two different worldviews confronting each other at the turn of the 1950s, when two superpowers came together in a head-to-head competition. Fortunately, the rivalry extended beyond our planet and slid into a more peaceful battlefield. Sputnik had changed the direction of American science and had touched all our lives, but the awakening that triggered a heated space race was, in the United States, the launch into space of Yuri Gagarin on Vostok 1 in April 1961. A few weeks later, American President John F. Kennedy announced that America was going to land a man on the Moon within 10 years.

We ran all the way keeping an eye on the Soviets, feeling their breath on our necks. Still, by the summer of 1967 we believed that, in the unproclaimed race to the Moon, the Russians were ahead of the United States. When Russia launched Zond 5 in September 1968, we worried that it was a prelude to an imminent Russian manned flight around the Moon and, eventually, a lunar landing. The Soviets had upstaged us so often that NASA was concerned that they would attempt their own manned circumlunar mission prior to the United States.

Only years later, when the United States and Russia began to move closer together with the joint Apollo/Soyuz Program – and to an even greater degree when American astronauts participated in the Russian Mir space station program – did we begin to understand the limitations of the Russian space program and the differences between the way our two countries were exploring space. As the book captures well, there was a radical difference in the human approach to space exploration. Early astronauts successfully fought for more human involvement and manual control of the spacecraft. Early cosmonauts were basically "passengers" on missions where operational activities were almost exclusively handled by ground controllers. Cosmonauts were sometimes perceived as a troublesome substitute for an onboard sequencer. The Soviets never fully trusted the cosmonaut crews as opposed to their ground "collective" support. The philosophy of central control led to the pecking order of ground over crews.

An automatic orbit and recovery was no more appealing to an astronaut in 1960 than automatic landings are today to a passenger on a commercial airliner. This attitude may have been a factor in one of the most obvious differences between the American and the Russian manned space programs: The Soviets returned from manned space missions on land, while in America we splashed down in the ocean. As a result, the Soviets developed relatively simple hardware and flew relatively simple missions. It was a real eye-opener to us in the Astronaut Office when we learned that so many Russian "firsts" had been accomplished, and they had gained such worldwide prestige, with such simple hardware. (It was one of the few areas where we could learn from them.)

An interesting side point is that philately – that at the time was a sort of national pastime, both in the USSR and in the USA – was heavily used by Soviets for propaganda purposes. Secrecy and propaganda, used with fantasy, or – if you want – with creativity, helped to mask the differences, and the limits, for years.

With the Apollo 50th anniversary approaching, thank you, Umberto, for doing this history!

Walter Cunningham, Apollo VII

1

Sputnik Triggers the USSR–USA Competition

USSR-USA SPACE RACE: IGNITED IN ITALY

In September 1956, for the first time ever, an artificial satellite was featured on a postal stamp. The Italian stamp, designed by Corrado Mancioli, was issued to mark the 7th International Astronautical Congress (IAC), which was hosted between September 17 and 22 that year in the Italian capital, Rome.

The IAC is organized by the International Astronautical Federation (IAF), a non-governmental international organization, with the first IAC held in Paris back in 1950. The 7th such Congress (which has no permanent home) in 1956 was hosted by the Italian Rocket Association (Associazione Italiana Razzi), headed by Professor General Gaetano Crocco. The chief topic of IAC-7 was the artificial unmanned satellite, heralded by the newspapers as the *"first step towards sidereal space."* The Congress was attended by almost 400 delegates, coming from the 20 national astronautical societies that were members of the IAF. The Soviets were also invited to attend, for the second time, with 'observer' status.

A year earlier, during the previous IAC-6 Congress, held in Copenhagen, Denmark, the Soviet delegation had held a press conference in their hotel, during which they announced a plan to launch a man-made object into space during the International Geophysical Year (IGY). The IGY would run from July 1, 1957 to December 31, 1958, to correspond with the maximum activity of the Sun's eleven-year cycle.

This would be the Soviet contribution in response to the resolution adopted in October 1954 by the *Comité Speciale de l'Année Geophysique Internationale* (CSAGI), during its meeting held in Rome. That resolution had indeed called for the launch of artificial satellites during the IGY, to contribute to the mapping of the

2   Sputnik Triggers the USSR–USA Competition

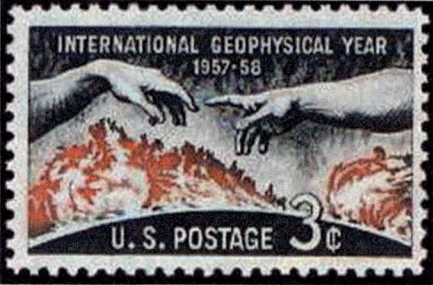

Figure 1.1: (top left) The first postage stamp ever to feature an artificial satellite. (top right) U.S. stamp issued to commemorate the International Geophysical Year – IGY 1957-8. (main) Cover commemorating the first International Astronautical Congress, Paris 1950.

Earth's surface. Quite coincidentally, the American president had issued a similar statement few weeks earlier, announcing the launch of Vanguard, the first American satellite.

During the IAC-7 Congress in Rome, a half-dozen American scientists circulated to illustrate the American plan in greater detail. It turned out that the UK, France, the Netherlands, and the USSR were all preparing their own satellites, unveiling a quiet scientific competition that until then had been played out in the greatest secrecy. However, no one gave much credence to the vague pronouncements of a possible launch by Leonid Sedov, the head of the Soviet delegation,

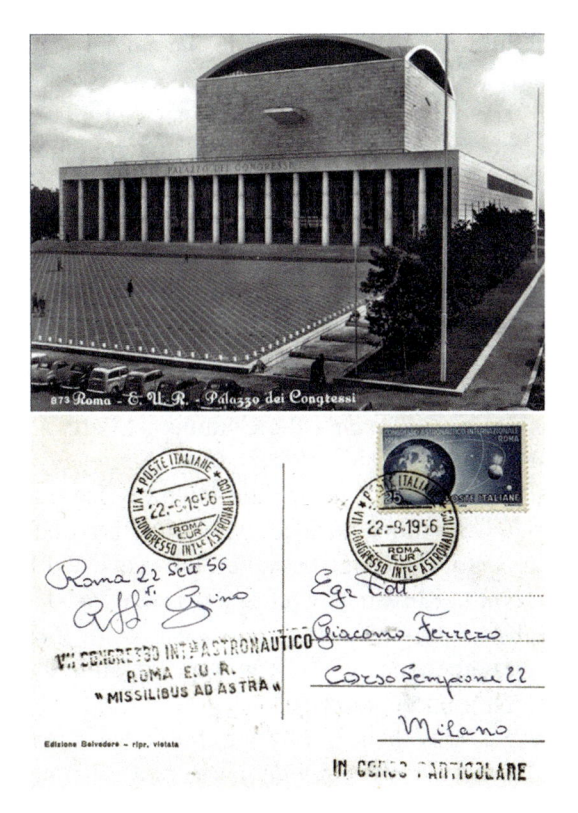

Figure 1.2: Postcard commemorating the 7th IAC in Rome, 1956.

whose statement was virtually under-valued and all but ignored[1]. Everybody *knew* that the United States would launch the world's first satellite![2]

Leonid Sedov, a university professor and Member of the Soviet Academy of Sciences, with no direct connection to the space program, would be destined to achieve great notoriety as a figurehead, presented to the Western media as a guiding force of the Soviet space program. In fact, the fledgling Soviet satellite program was controlled with an iron fist by the military, and Sedov's Commission had little real authority and virtually no contact with it.

---

[1] The same would happen the following year on September 30, 1957, just a week before the launch of Sputnik, when Sergei M. Poloskov, the Soviet speaker at the CSAGI Conference in Washington, announced that the Soviet launch was imminent, but the expression he used could not be literally translated.

[2] The name 'Vanguard' reflected American confidence that their satellite would be the first in the world, as Nikita Khrushchev ironically pointed out during his speech for the 40th anniversary of the Revolution, on November 6, 1957.

Sedov, who was allowed to travel outside the Soviet state to represent the USSR, would be undeservedly credited with the successes achieved with Sputnik, Lunik and Vostok by the mysterious 'Chief Designer' Sergei Korolev, whose identity remained a State secret until after his death[3]. Though Sputnik's launch in 1957 had become the talking point of the entire world, no one had a clue as to the identity of its chief designer. His deliberate anonymity would later be confirmed in an interview by Sergei Khrushchev, son of Soviet Premier Nikita: *"At that time, nobody knew the name 'Sergei Korolev'; it was classified."* [1]

Korolev was never allowed to travel abroad, nor to meet foreign scientists at home in Russia, or at international congresses on space matters. As the sole concession, in recognition of the key role he played, he was allowed to write articles in the important publication *Pravda* – the Communist Party's daily newspaper – but only under the pseudonym of either 'Professor K. Sergeyev', or 'Konstantinov'. Khrushchev was always careful to keep Korolev away from the spotlight. Even when the Nobel Prize Committee decided, without polling the world's scientists, to give an award to Sputnik's Chief Designer and requested his name from the Soviet government, Nikita Khrushchev refused to reveal his identity, claiming that in order to ensure the country's security, and the lives of these scientists, engineers, technicians and other specialists, it was not possible to make their names known or to publish their photographs.

According to Sergei Khrushchev, however, his father's real concern was not confidentiality [2]. *"The KGB knew that there was really no need to keep his name secret, but, as KGB chief Ivan Serov told me, the enemy's resources were limited, so [we] let them waste their efforts trying to uncover 'non-secret' secrets. As for real secrets, the enemy's arms were too short to reach them."* In fact, Nikita Khrushchev's main concern was that Korolev was the head of the council of chief designers, in charge of all space projects. Khrushchev knew that the other designers harbored their own ambitions and considered themselves no less significant. They would all have been madly jealous if Korolev alone had received such publicity. After the launch of Sputnik, all of the designers (including Korolev, Glushko, Chelomey, Tikhonravov, Keldysh, Mishin, Voskresensky, Chertok, etc.) had been

---

[3] The most objective biography of Sergei Pavlovich Korolev is "Королев. Факты и мифы" (*Korolev: fakty i mify* – in English "Korolev: Facts and Myths"), issued in 1994 by the Russian writer and journalist Yaroslav Golovanov. Between 1965 and 1966, Golovanov was one of the team of three journalists who were unofficial cosmonaut candidates. The team was disbanded after Korolev's death. Golovanov became a space correspondent of the daily newspaper *Komsomolskaya Pravda* for almost 30 years and worked on the biography of Korolev by interviewing about 300 people who personally knew him. It is noteworthy that, in 1964, Korolev was able to persuade the Kremlin to let him co-opt trustworthy newspaper reporters into his cosmonaut corps, in the hope that the ensuing publicity would inspire greater support for space exploration. This was decades before NASA realized the public-relations value of sending schoolteachers and senators into space.

jointly awarded the Lenin Prize and other Soviet honors. "*If the Nobel prize went only to Korolev,*" Sergei Khrushchev explained, "*my father thought the [other] members would get upset and that the team would simply disintegrate, and with it, the hopes of [the] Soviet Union's future space research and missile design. As my father saw it, you could order scientists to work together, but you couldn't force them to create.*"

Perhaps, as Anatoli Fedoseyev observed however, there were also other, more subtle reasons: "*There is another reason for the secrecy, especially as it applies to the leading scientists upon whom the level of science and technology in the Soviet Union really depends. It is not the fear of their being kidnapped which prompts the Soviet authorities to keep them incognito. It is rather because, if such people were known to the public, they might acquire sufficient fame and influence to represent a powerful and possibly dangerous opposition to the political leaders.*" [3]

In his reply to the Nobel committee, therefore, Premier Khrushchev stated that all the Soviet people had contributed to the project and that every Soviet citizen would deserve the reward… and the Nobel prize went elsewhere. This concept of collective achievement became one of the main recurring themes of the visual art of Soviet propaganda, designed to give all Soviet citizens a sense of pride and of belonging.[4] The sentiment was frequently expressed through many well-known posters, which were widely circulated at the time (see Figure 1.3).

However, Khrushchev had deprived Sputnik's creator of the highest honor in the field of science and, of course, Sergei Korolev felt deeply hurt. The price of technological success in the Soviet Union of the 1950s and 1960s was to disappear from public view. Korolev's daughter, Natasha Koroleva, recalled in a book that the veil of secrecy had vexed her father throughout his life: "*We are like miners – we work underground,*" she recalled him saying. "*No-one sees or hears us.*" [4]

The man who could pick up the phone to call Nikita Khrushchev and who would ultimately humiliate the mighty United States of America in the early years of the Space Race was condemned to be a faceless nonentity. The rest of the Soviet Union, and the world, would only learn of Korolev's name following his death in 1966.

---

[4] Visual art took on a very important role during the October proletarian revolution in Russia and the subsequent civil war. Very few newspapers existed in those days, so such posters often replaced the tabloids. Millions of posters were reproduced and circulated, posted on walls in cities and villages, where they were widely accessible to the less literate masses. The simple, emphatic, vibrantly colored designs they depicted were easily understood by everyone, while the short and energetic slogans with powerful propaganda messages that accompanied them stuck in the viewer's mind as a rallying call for action. Soviet posters continued to keep pace with the times. During the 'Space Era', their unique laconic, expressive and straightforward style delivered vigorous and effective slogans, glorifying the Soviet Union's technological prowess and importance in the world (and in the universe) and focused on the role that the workers played in the Space Race. They helped to inform, educate and instill pride in the average citizen about the achievements of the space program and Mother Russia's accomplishments.

6 Sputnik Triggers the USSR–USA Competition

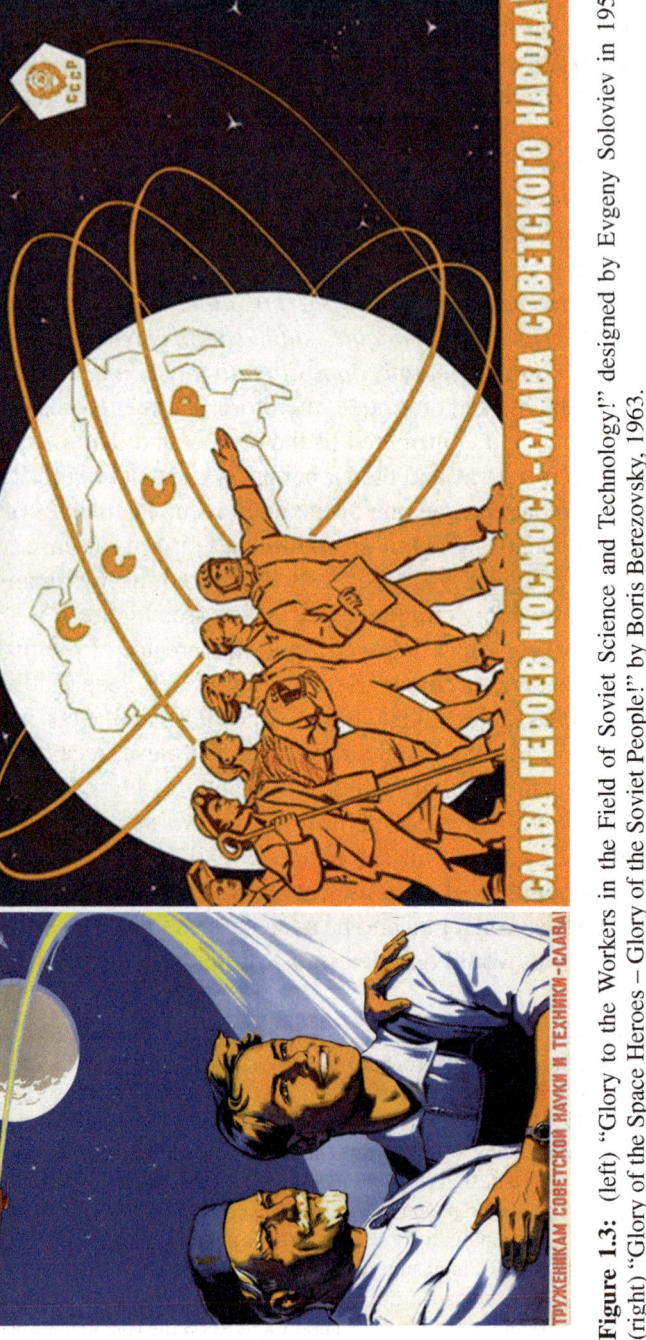

Figure 1.3: (left) "Glory to the Workers in the Field of Soviet Science and Technology!" designed by Evgeny Soloviev in 1959. (right) "Glory of the Space Heroes – Glory of the Soviet People!" by Boris Berezovsky, 1963.

Figure 1.4: Sergei Korolev, featured on a 1986 stamp.

SPUTNIK: THE OPENING SHOT OF THE SPACE RACE

When he returned from the IAC, Sedov reported back with the details of the announced American Vanguard program. Sergei Pavlovich Korolev – the genial and mysterious *Deus ex Machina* of the Soviet space program, with an innate initiative, drive and energy – soon suggested the ambitious project to launch the first artificial satellite to Khrushchev. [5] The Premier was excited about the idea of being able to "*overtake America*[5]."

Korolev had first raised the idea of space exploration with his government as far back as a meeting on April 30, 1955, but nothing had come of it. In an interview, the text of which was published after his death, Korolev recalled: "*We had followed closely the reports of preparations going on in the United States of America to launch a sputnik called, significantly, Vanguard. It seemed to some people at the time that it would be the first satellite in space. So, we then reckoned up what we were in a position to do, and we came to the conclusion that we could lift a good 100 kilograms (220 pounds) into orbit. We then put the idea to the Central*

---

[5] Korolev is often described as a man who favored a cautious, step-by-step approach to space exploration, but who was pressured by Khrushchev into staging space spectaculars to beat the Americans. Although the pressure from the Kremlin should certainly not be underestimated (starting after the launch of Sputnik 1, when Khrushchev realized the propaganda effect of space), Golovanov describes Korolev himself as a man almost obsessed with clinching space firsts. At one point, he even quotes Khrushchev's son as saying that the Soviet leader was somewhat vexed at Korolev's excessive urge to set space records. It was Korolev, not Khrushchev, who masterminded, up to a certain point, most of the spectacular space firsts.

Committee of the Party, where the reaction was: 'It's a very tempting idea. But we shall have to think it over.' In the summer of 1957, I was summoned to the Central Committee offices. The 'OK' had been given. That was how the first Sputnik was born." [6]

Unfortunately, the go-ahead came too late for what Korolev originally had in mind, because Mikhail Tikhonravov's satellite, Object-D[6], carrying many scientific instruments in the spirit of the IGY, was behind schedule. Now, a desperate race against time would begin. The R-7 rocket[7], capable of reaching orbital velocity, was almost ready, even though five out of its first six launch attempts had failed, but the same was not true of the heavy-duty satellite carrying several scientific instruments that Korolev and Tikhonravov were unofficially working on. However, sending *any* object into orbit would serve the political propaganda goals of the Soviet leadership, as long as it could announce its presence to the whole world. For this reason, and to save as much time as possible, Korolev decided to simplify the Sputnik down to basics, so that it would contain only a radio transmitter with sufficient power for even amateur radio enthusiasts to be able to track it. With the excellent collaboration of the equally brilliant Leonid Voskresensky, Korolev devised the new satellite configuration for an object that would simply be known as 'P.S.' (standing for 'Prosteishy Sputnik', or 'the Simplest Satellite'). The launch was scheduled for October 6, 1957.

When the program was announced for the 8th IAC, to be hosted in Barcelona, Spain, beginning on October 6, Korolev perceived that the Americans were about to launch their own satellite. He immediately cancelled some last-minute tests and moved up the launch of Sputnik by two days, to October 4.

The successful launch on that date saw the first man-made object accompany the Earth in its orbit around the Sun. The Soviet 'Sputnik' transmitted the first signals from orbit. The era of 'cosmonautics', as the Soviets called it, was inaugurated, and Sputnik became the first of a series of humiliations for the Americans in the early years of the space program.

Ironically, the great scientific cooperation that was called for to coordinate efforts to understand the mysteries of our world in the spirit of the International Geophysical Year (IGY) of 1957–58 was precisely what had triggered the political-technological rivalry between the two superpowers. Both were resolutely engaged in demonstrating to their citizens, allies and opponents that theirs was the most technologically advanced and militarily powerful nation. Just as

---

[6] 'Object D' (or D-1) was so named because it would be the fifth type of payload to be carried on an R-7 rocket. Objects A, B, V and G were designations for different nuclear warhead containers.

[7] An evolution of the ICBM developed in a forced cooperation with Valentin Glushko, for whom Korolev held a long-standing antipathy.

Sputnik: The Opening Shot of the Space Race 9

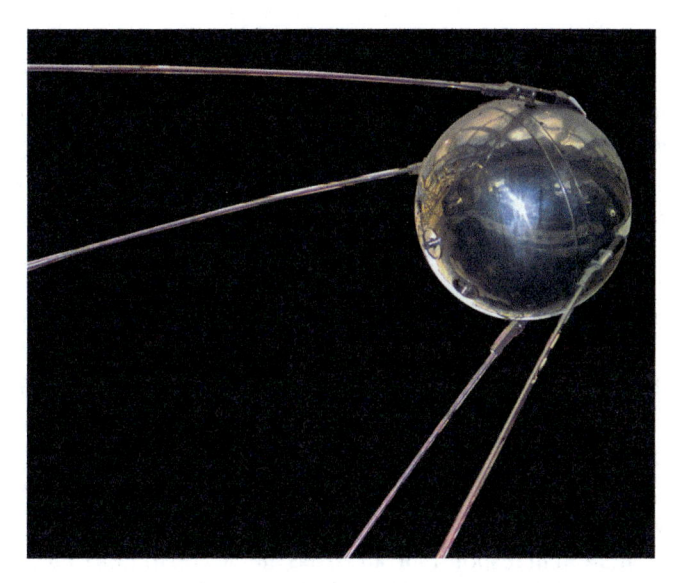

Figure 1.5: Sputnik (meaning 'traveling companion' in Russian) was a polished metal sphere with four long antennas. It was about 22 inches (56 cm) in diameter and weighed 184 pounds (83.6 kg), Circling the Earth every 98 minutes, it used a radio beacon that was able to pinpoint spots on the Earth's surface. Image © NASA-NSSDC

ironically, this momentous launch of the first artificial satellite in history, far from being the result of a well-planned strategy to demonstrate communist superiority over the West, was instead a spur-of-the-moment gamble, driven by the dream of one visionary scientist and iron-willed manager, who pressed the Kremlin to enter into an adventure which nobody desired and for which nobody felt the need.

After the successful completion of Sputnik's first orbit, Korolev called Soviet leader Nikita Khrushchev, who was in the Ukraine on military business, and reported the satellite's success. But nobody immediately grasped the importance of this event, which would mark a turning point in history. Khrushchev's son, Sergei, was in the Ukraine alongside his father at the time. He would recall later that they listened to the satellite's 'beep-beep' signal and went to bed. *"Sputnik's launch made the front page of Pravda but without banner headlines or enthusiastic comments,"* Sergei Khrushchev said in an interview in 2007. *"The story occupied the same amount of space as a report on Zhukov's visit to Yugoslavia but ran in a less prestigious position on the page. The reason was simple. My father and all the Soviet people thought that Sputnik's success was natural, and that, step-by-step, we were getting ahead of the Americans. After all, we – not the Americans – had opened the world's first nuclear power plant; our MiG jets set world records in the '50s, and the Soviet Tu-104 was the most efficient airliner of its class*[8]. *So Sputnik did not surprise us. All of us saw that as just another accomplishment, showing that the Soviet economy and science were on the right track. A lot of popular books had been published in the Soviet Union about future space stations and flights to the Moon and Mars. Space travel seemed quite feasible, and the readers of those books – including me – looked forward to it. We just couldn't understand why the engineers were taking so long.*" [1]

The article in *Pravda* on October 5 was, indeed, positioned modestly in a right-hand column part way down on the first page. Titled routinely 'TASS Report', it succinctly detailed the facts of the launch in a few paragraphs, plainly explaining to readers what 'satellite' meant. Two days later, *Pravda* led with a banner headline quoting the global furor.

The unexpected launch of Sputnik had surprised the whole world. It surprised the incredulous Khrushchev, who had only dreamed of success in outpacing the Americans with the satellite, and certainly didn't expect its powerful effect and the Western consternation in response. It surprised the Soviet military and political leaders, who had always fought against the "*useless satellites*", fearing that such 'toys' would interfere with the major intercontinental missile projects and slow down the development of the R-7 ICBM. It took them several days to understand the extent of what had happened. It surprised the U.S. experts, who had always believed that this was a competition which the Americans would win hands down and were now disappointed by the perceived 'missile gap', when Intelligence reports had always claimed that American missile technology was far more advanced than that of the Soviets. Apparently, it also surprised the Eisenhower Administration, which had so far considered these activities as mere scientific experiments[9].

---

[8] Experts and historians point out that the first nuclear power plant (Chicago Plant-1 or CP-1) was assembled and designed in the USA by Enrico Fermi on December 2, 1942. MiG jets were powered by unlicensed copies of the Rolls-Royce Nene engine which had been supplied by Great Britain. The De Havilland Comet, making its maiden flight in 1949, and the Boeing 707, the first widely-used jet airline, were the movers and shakers, not the Tu-104. The main reason the Tu-104 was the most efficient airliner of its class was because it was the *only* member of its class, a twin-engine airliner powered by Rolls-Royce Nene clones. (*theguardian.com* - accessed in February 2018)

[9] The Sputnik crisis depicted President Eisenhower as passive and unconcerned. This led to bitter accusations of complacency and contributed to the election of John F. Kennedy, who emphasized the space gap and the role of the Eisenhower Administration in creating it. According to some historians, however, Eisenhower knew far more than he could publicly admit about the status of the Russian missile programs. On the basis of the secret U-2 surveillance intelligence, he knew that there *was* no missile gap, and had strategic reasons to support his 'Open Skies' policy. (see the following section on "*Explorer I: one of the main discoveries of IGY*", p. 27).

The Soviet satellite was now a nagging irritation in American heads, repeating its incessant 'beep-beep' signal – resembling the soundtrack of an early Mickey Mouse movie – and reminding the world of the USSR's accomplishment. [7] It created a perception of American weakness and a wider sense of insecurity and apprehension.

Pravda also published a description of Sputnik's orbit and the frequencies of the satellite's radio transmitters, like a kind of 'train timetable', to help people watch and hear it pass. The article failed to mention that the light seen moving across the sky was not the tiny orbiter, whose size meant it was invisible to the naked eye, but was in fact the huge second stage of the booster rocket which was in roughly the same orbit. Other than this article, information remained scarce. No technical details, no name of the location from which it was launched, and no interviews with the people involved.

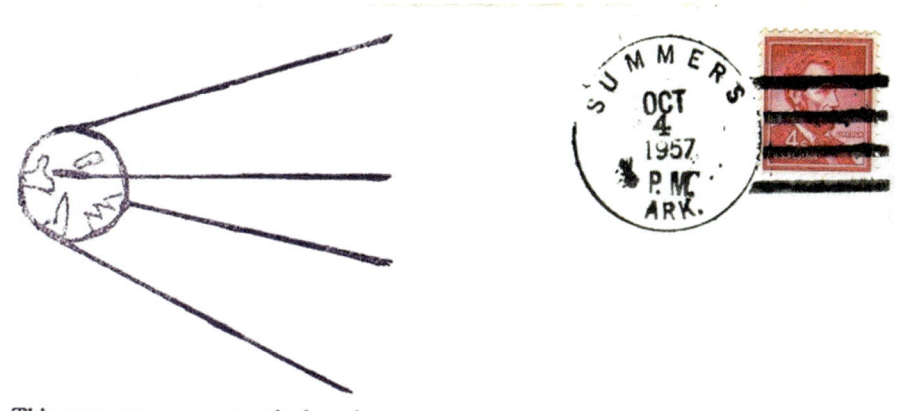

This cover commemorates the launching of the world's first satellite, called the **Sputnik**, by the Soviet Union, on October 4, 1957. The picture above was copied from a TV news feature.

Figure 1.6: The impact of Sputnik's passage over the United States sky caused reactions ranging from amazement and anger to panic. Newspapers gave the times that Sputnik would be passing overhead and instructions on how to locate it in the sky. TV also emphasized the event, as evidenced by the cover shown in Figure 1.7. Postmarked in Summers, Arkansas. (From the collection of Steve Durst, USA)

After Sputnik 1, the world would no longer be the same. Its impact on the United States and on the wider world was enormous and unprecedented. On the morning of October 5, the *New York Times* printed an unusual three-line head in half-inch capital letters, running full length across the front page (see Figure 1.7). In Great Britain, the *London Daily Mirror* proclaimed the birth of the "*Space Age*" in huge headlines. Almost immediately, two new phrases entered the language: 'pre-Sputnik' and 'post-Sputnik'. [8]

12   Sputnik Triggers the USSR–USA Competition

Figure 1.7: Unusual three-line head from the *New York Times* announcing the flight of Sputnik 1.

The Western world realized that the Soviet success was due to a modified intercontinental ballistic missile, and this was enough to assume Soviet military superiority and their lead in missile technology, as well as speculating apocalyptically on what the Russians might now do with their perceived capability of hitting any chosen target anywhere in the world.

The French daily *Le Figaro* led with the banner headline "*MYTH HAS BECOME REALITY: EARTH'S GRAVITY CONQUERED*" and went on to report the "*disillusion and bitter reflections of the Americans, (who) have had little experience with humiliation on the technical domain.*" [9] In West Germany, a new name was coined for America's still unorbited Vanguard. They called it 'Spätnik', *spät* being the German word for 'late'. [10] The idea of a Soviet-made object orbiting the skies above continental America terrified ordinary Americans, who feared that, with this kind of technology, the next thing the 'Ivans' would be doing would be "*dropping nuclear bombs on them like rocks from a highway overpass*[10]." Building a backyard bomb shelter quickly became a cottage industry.

Western panic over the nuclear threat (covered by the most stringent secrecy, which only served to increase the level of panic) and the collective feeling of being at the mercy of powerful Soviet missiles and the target of direct nuclear attacks, led to that fascinating and, in some ways, worrying chapter of our recent history known

---

[10] On the other hand, the same fear was true for the Soviets. Shannon Lucid reported long conversations she had with Yuri Onufriyenko and Yuri Usachev during her record-setting expedition on the MIR in 1996 and concluded: "*After a while we realized we had all grown up with the same fear: an atomic war between our two countries. I had spent my grade school years living in terror of the Soviet Union. We practiced bomb drills in our classes, all of us crouching under our desks, never questioning why. Similarly, Onufriyenko and Usachev had grown up with the knowledge that U.S. bombers or missiles might zero in on their villages. After talking about our childhoods some more, we marveled at what an unlikely scenario had unfolded. Here we were, from countries that were sworn enemies a few years earlier. I was living on a Russian space station, working and socializing with a Russian air force officer and a Russian engineer. Just 10 years ago, such a plot line would have been deemed too implausible for anything but a science-fiction novel.*" (Cavallaro [2017] p. 70).

as 'The Space Race.' As Neil Armstrong called it, *"The most elaborate non-military competition in history. It is unlikely that the space race was the diversion which prevented war."* Nevertheless, it *was* a diversion and provided an outlet to replace the 'brinkmanship' of the early 1950s that might well have led to armed conflict. [11] We can say that the 'Space Race' *sublimated* the Cold War and moved the competition between the two superpowers beyond our planet, to a crossroads where technology, armaments, science and fantasy crossed each other[11].

> **Soviet philately and propaganda**
>
> The launch of Sputnik initiated a clever new way of using stamps, for propaganda purposes. Once the Soviet Union grasped the importance of what had happened with the launch of Sputnik 1, space exploration became one of the favorite topics in Soviet philately for several years. During the late Fifties and the Sixties, the USSR issued more than 160 stamps with space topics, compared to just five stamps issued during the same period in the USA.
>
> Because of how it is used, the postage stamp is widely circulated and goes from hand to hand and from town to town, reaching the farthest corners and provinces of a country, or indeed the world. The fact that it does not convey an obvious message enhances its peculiar effectiveness and makes it an ideal means for subliminally influencing public opinion. [12]
>
> Philatelic propaganda reaches not only the recipient of the letter, but also everyone who handles that letter, starting with the individual who sends it. The envelope passes through many hands in the different postal offices and goes through many cities – and often through many countries – before reaching its final destination.
>
> Advertising through the use of stamps is now an established practice, an effective and cheap way to spread a message far and wide. At some time, every nation has utilized its stamps to promote domestic products, vacation resorts and cultural achievements, or to advertise its industries.

(continued)

---

[11] As Boris Chertok – the brilliant engineer who designed most of Korolev's guidance systems – annotated in his book, the recurring early American humiliation by the USSR in space *"served to stimulate the beginning of competition on another plane, in a field that objectively led to the weakening of the positions of Cold War apologists. The historic paradox of cosmonautics was that the achievements of missile technology stimulated confrontation between the two superpowers, while the successes of the piloted space programs based on these achievements promoted rapprochement, cooperation, and a desire to exchange ideas and experience. The flights of our cosmonauts and American astronauts diverted a great deal of resources from weapons technology and did not contribute to meeting military challenges. Each new piloted flight around our shared planet objectively served as a call to unite and to reduce confrontation."* (Chertok [2009], p. 79).

14 Sputnik Triggers the USSR–USA Competition

(continued)

> Unquestionably, one of the key players in this regard was the Soviet Union. Once the Communists came to power, after the 1917 Revolution, they flooded the world with their stamps, almost invariably conveying the Soviet ideology and glorifying, in the most spectacular way, social and political milestones, such as the success of the Five Year Plans, or Soviet industrial achievements, the ideal citizens, workers, peasants and Red soldiers.
>
> Particularly impressive were the stamps issued during WWII, vaunting its military power – especially its air force, infantry and navy – and showing pictures of Soviet forces in action; soldiers throwing grenades, sharpshooters, and planes destroying tanks.
>
> Collectors and experts remark that old Soviet stamps quite often appear unused. Many of them have likely never been on sale in any Soviet post office but were distributed or sold by a special Soviet philatelic agency in Moscow to foreign buyers, as suggested by the high denomination of some of the most attractive stamps. There is no doubt that foreign markets were an important target. We know, for example, that the official commercial agency Mezhdunarodnaya Kniga, used duplicates of official Soviet postmarks to produce philatelic commemorative covers that had never been in an actual post office, or were never run through the mail service, for the foreign markets.
>
> During the era of totalitarianism in the USSR (under the rule of Stalin), stamp collectors were looked upon suspiciously because they had too many contacts and knew too much about foreign countries, while philatelic societies (as with any other unofficial community) were considered as potentially counter-revolutionary organizations and enemies of the people. Accordingly, active collectors were often prosecuted and either sent into forced labor or killed, with their collections confiscated and sold to finance the rising Soviet military industry.
>
> In the 1950s, things changed in USSR and a new generation of internal collectors appeared, but the main goal, especially at the beginning of the Cold War, was to influence its dependent Eastern European states, the Warsaw Pact satellites. Russia adopted the stance of having a superior space program – a 'We're the best' approach – that guided 1950s and 1960s propaganda. The USSR adopted the same approach with Western countries, sometimes more successfully as happened in France, always a strong supporter of the Soviet Union.

Soviet philately and liability

Soviet launches were decided and prepared in the strictest secrecy, and this was especially true for the launch of Sputnik 1, the first ever launch into space. Even the Soviet Post Organization were not prepared for the event and it was more than one month later – on November 5, 1957 – before they were able to issue the first set of two Sputnik commemorative stamps featuring the orbit of the satellite. The text, in Cyrillic, read "*4 October 1957 – World's first artificial satellite of Earth.*"

Figure 1.8: Fake commemorative cover for Sputnik 1, with a backdated cancellation.

One collector then doctored his commemorative cover (see Figure 1.8) by backdating the cancellation, a misuse not unusual in the USSR at that time, in order to have a cover seemingly issued on November 4, 1957, exactly one month after the launch of Sputnik. This of course did not take into account the fact that the stamp had not been officially issued until the following day. The cover was also cancelled in Moscow and not in Tyura-Tam, where the Sputnik was launched from. That was because the secretive site was absolutely unknown at the time and would be kept as a state secret until Gagarin's flight in 1961 when, for reasons we will see, it would be named 'Baikonur' (see Chapter 2, p. 102 'Baikonur and Soviet lies.')

(continued)

16   Sputnik Triggers the USSR–USA Competition

(continued)

Figure 1.9: Konstantin Tsiolkovsky, featured on a 1957 stamp.

Another commemorative stamp was put in circulation a few weeks later, on November 28, with the Cyrillic text "*4/10/57 – World Premiere: The First Soviet Artificial Satellite of Earth*" overprinted in black on the 40-Kopek stamp. The stamp had originally been issued back on October 7 (shortly after Sputnik's launch) in honor of the missile scientist Konstantin Tsiolkovsky. Unfortunately, many forged copies of this historic stamp exist on the market.

It is very hard to find covers "*cancelled at the exact site and on the exact date*" of the launch of a spacecraft – as prescribed by the rules of competitive astrophilately – from the USSR of the Cold War era. At that time, it was a rule to keep every piece of information related to the space program, including launches, absolutely secret until the authorities were certain about the success of the mission. This made it materially impossible to prepare envelopes or cancellations in advance, simply because there was no information available about a flight until it was all over.

The only information available, after the fact, was obtained from the stamps and from official postmarks issued after a considerable delay that gave – often in an emphatic tone – some vague idea of the spacecraft or the rockets and, for propaganda purposes, provided some data on missions, trajectories and so on.

Thus, while commemorative covers that celebrate anniversaries of space events are normally to be avoided in competitive astrophilately, for the early phases of the Soviet space program up to 1975, items that celebrate subsequent recurrences such as the 1000th or 3000th or 10,000th orbit of a satellite are often the only, somewhat belated, witnesses to these first space conquests.

The 'French' fakes. [13]

In retrospect, it has to be said that as far as the early Soviet space program is concerned, there are grounds to be suspicious of any items that fulfil astrophilately rules by bearing the exact date of a specific space event. The secrecy of the Soviet space program and the total unavailability of information made it virtually impossible for collectors to produce covers or cards in time to document the events on the same day that they happened.

Years later, however, that same secrecy and unavailability of information allowed unscrupulous individuals to invent plausible and attractive covers that were supposedly issued by the Soviets as 'witness' to the early phases of the Space Race. Thanks to the total lack of official data, nobody at that time would have been able to challenge the authenticity of those suspected forgeries, and therefore they sold successfully for years. Nobody really knows about the origin of those fakes that, in their own way, commemorated the Soviet space program between 1957 and the mid Seventies.

The 1985 catalogue *C.O.S.M.O.S. Catalogue des Oblitérations Spèciale et des Marques Officielles Spatiales* (6th Edition) – issued by Lollini, the French dealer of Space Philately – listed 300 of these 'vintage' Baikonur-Karaganda covers, as it called them. The number of forged covers continued to grow year after year, and they numbered 397 in the 7th Edition of the same catalogue (1994). The 8th Edition (1998) devoted 18 full-color pages to the "*old cancel covers*" (from page 299 to page 316), offering 400 quite expensive 'junk' covers to naive collectors. At that time, the 'commemorative' Sputnik 1 and Sputnik 2 covers were already sold out and the oldest items available were the covers for Sputnik 3, priced at $833 U.S. each.

It is possible to find a 'commemorative cover' for virtually every early Soviet space event, cancelled in the fictitious post office of Baikonur-Karaganda exactly on the day of that event. These covers were made out of an unusual semi-glossy paper, in an unconventional format that was smaller than the typical Soviet covers at the time (165 x 91 mm, which in the Lollini catalogue was named "*international format*" [14]). Everything about them – in particular the postmark – is forged.

Usually such covers carry tirage (printing), which is normally "*150*" for each cachet (sometimes only "*100*" or "*50*", as much as such figures could be meaningful or reliable), and an individual serial number. On this basis, it is easy to calculate that the family of Baikonur-Karaganda fakes should be quite significant globally and encompass no less than 58,000 items. There isn't a complete list, however, and it is entirely likely that the total number of such forgeries is actually considerably greater.

(continued)

18 Sputnik Triggers the USSR–USA Competition

(continued)

Alas, these fake Soviet covers are very prolific and so widely spread that it is hard to find a space collector who doesn't hold at least one of them in their collection. Paul Bulver noted in his book that he asked for clarification of these covers in 1972. [15] The French dealer replied that he was *"absolutely sure that the black cancel with date has been affixed by the local post-office of the town Baikonur."* He added, *"I wrote, already several years ago, to the Central Post-Office of Moscow, and they gave me confirmation of this fact,"* although when he was asked to provide proof of such a letter, the dealer stated that he had lost it. [16]

It was not until cooperation began between the USSR and the USA during the Apollo-Soyuz Test Project (ASTP) that verifiable data began to surface. In particular, it was established that no Baikonur-Karaganda post office had ever existed.

Figure 1.10: Fake Sputnik 1 commemorative cover with cancellation of Baikonur-Karaganda, on October 4, 1957.

How unlikely it is for a commemorative cover issued for the launch of Sputnik 1, like the one shown in Figure 1.10, to exist is immediately clear when one realizes the degree of absolute secrecy under which the early Soviet space program developed, especially during the first decade. As already mentioned, no collector could know in advance about the launches, nor would anyone be able to prepare commemorative covers in time. Most importantly, nobody could possibly have covers cancelled at the location of the launch since – for many years – the site was treated as a State secret (even though the American U-2 spy planes had pinpointed the R-7 launch pad in June 1957).

(continued)

The American collector Ray Cartier has referred to a casual encounter which happened during the *Pacific 97* meeting between his colleague Les Winick, himself a collector of space memorabilia, and Professor Oleg Vaisberg from Russian Science Academy. [17] Professor Vaisberg revealed that the fakes had been produced in Moscow by Boris Korichev (who passed away in the early 1980s), "*who had then sold them to a Frenchman.*" I discussed this topic with a renowned expert of Russian fakes, who told me that it was difficult to believe this story. As he explained to me, philately was used in the USSR as a strategic propaganda factor and it is unlikely that a 'business' of this size, if it originated in the USSR, would escape the attentions of the KGB. They KGB, who had a special Philatelic Commission and, as everyone knows, kept a very strict control over territory.

The notorious Baikonur-Karaganda postmark exists in two different versions: Type 'A' (the most popular one) and Type 'B'. The main peculiarity of the Type 'A' is the 'α' below the date, between the date bridge and the external crown. In Type 'B', the letter is replaced with a '*b*' in the same position. The shapes of the letters and digits in the two postmarks are significantly different.

Figure 1.11: Cover and cancellation for Baikonur-Karaganda fake Type 'A' (above) and 'B' (below).

(continued)

20 Sputnik Triggers the USSR–USA Competition

(continued)

> An annotation box on page 316 of the previously mentioned Lollini catalogue (1998 edition) noted that the "*old cancel*" was in use until June 14, 1975, when it was replaced by a "*new cancel*" (see Figure 1.12). June 14, 1975 was a fictitious date invented for this catalogue.

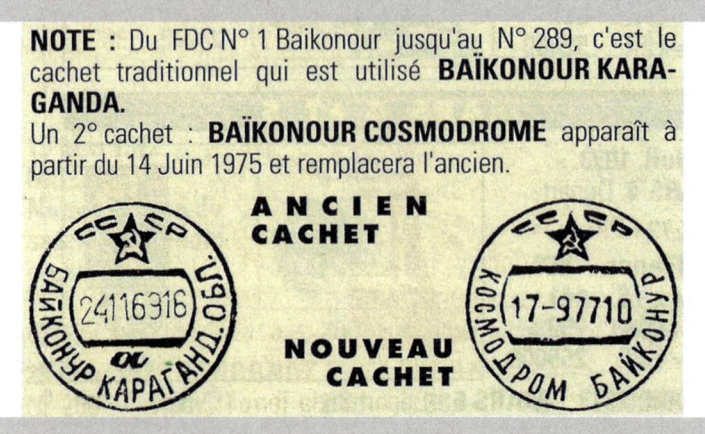

Figure 1.12: Annotation from the 1998 Lollini catalogue showing the 'old' and 'new' Baikonur cancellations.

Everybody is aware that this information is inaccurate, since a Baikonur Post Office didn't exist before 1975 (and consequently no 'old' cancel exists because a Baikonur-Karaganda Post Office has never existed). The Post Office was officially opened only on April 27, 1975, during preparations for the joint Soviet-American ASTP program. Baikonur-Karaganda fake postmarks continued to be produced even well after that date, however.

The fake cover shown in Figure 1.13 'commemorates' the launch of the Space Station Salyut-5 on June 22, 1976, when the official *Cosmodrome Baikonur* cancel had already been in service for more than a year. Since philately was very popular at that time in the USSR and there still remained little information available to collectors, forged cancelling devices could still be used with some impunity.

For a number of reasons, there is little doubt today that these fakes were issued in Moscow and the expert I spoke to regarding the Cartier story hinted that they could have been produced by somebody close to the KGB's Philatelic Commission that was strictly controlling this trade.

(continued)

(continued)

Figure 1.13: Similar fake covers with the forged postmark were also produced to 'commemorate' events even after the official Cosmodrome Baikonur cancel had been brought into service in 1975.

The French space memorabilia dealer kept these fakes listed in its catalogue for a long time and was still selling them until a few years ago, flooding the European market with fakes.

Both of the fake covers shown in the bottom half of Figures 1.14 are postmarked on March 27, 1968, to 'commemorate' the death of Gagarin.

The bottom-left cover in Figure 1.14 is cancelled with the fancy Baikonur-Karaganda postmark, while the bottom-right cover bears an even fancier "*advanced use*" of a forged duplicate of the postmark of Cosmodrome Baikonur, which would not be put into service until seven years later (as shown in the annotation from the Lollini Catalogue mentioned in Figure 1.12). One may spot, in the word 'БАЙКОНУР' on the second cover, the irregular letters ЙК, which are typical of another well-known fake that had already been reported in specialized books. [18]

The amazing similarity of the cachet of the two covers, their unusual format which is characteristic of the Baikonur-Karaganda fakes, the atypical semi-glossy paper, and the "*usual FIRST DAY logo*", strongly suggest that there is some connection between the producer of the two well-known families of forged items. Such similarities are also found in the two covers shown in Figure 1.15.

(continued)

(continued)

Figure 1.14: Baikonur-Karaganda fake covers for the flight of Yuri Gagarin (top pair; April 12, 1961 – printing 150) and for his death (lower pair; March 27, 1968 – printing 50).

Figure 1.15: Fake covers 'commemorating' the flights of Soyuz 11 in 1971 (above) and Soyuz 35 in 1980 (below).

(continued)

> Once again, the two covers have the same physical characteristics (glossy paper, 'international' format, etc.) and the same cachet to 'commemorate', respectively, two different events: the tragic flight of Soyuz-11 in 1971 (postmarked at Baikonur-Karaganda) and the flight of Soyuz-35 in 1980 (postmarked with a fake Baikonur Cosmodrome cancel). The same cachet is used in two different 'families' of fakes. One peculiarity of this particular Baikonur-Karaganda cover is the unusual – and unlikely – red postmark.
>
> Each of the two 'families' include several similarities. It is not yet clear who produced such covers and when. However, these two items suggest that the same hand is behind them. This still bears further investigation.

SPUTNIK 2: A ROCKET SIX TIMES MORE POWERFUL IN FOUR WEEKS!

The news about the launch of Sputnik 1 did not surprise the average Soviet citizen. For many years under Stalin, it was drummed into all of them that Soviet science was the most advanced in the world, and that all the major inventions worldwide had been made by Russians, including the lamp, the radio, the airplane, the locomotive, and the steamship. That the first artificial Earth satellite was a Soviet design was just another achievement that was taken for granted, but the average person also knew that the Sputnik would do nothing to improve their poor situation with regard to housing, clothes, food, wages, and other such everyday necessities.

Once Khrushchev realized the extent of the impact that Sputnik 1 had had on Western opinion, however, and that its disruptive, tremendous effect had far exceeded his expectations, he quickly summoned Korolev to his office in the Kremlin, barely three days after the launch of Sputnik 1[12]. Khrushchev ordered the Chief Designer to launch a new satellite a few weeks later, this time to mark the forthcoming 40th anniversary of the 1917 Bolshevik Revolution on November 7. The Premier wanted another spectacular space event. [19]

---

[12] Among other things, the Sputnik had the immediate effect of blotting out the memory of recent events in Hungary. The Soviet Army's brutal suppression of the Hungarian people's revolt had taken place less than a year before and had had a very serious effect on the Soviet Union's international prestige. But in some strange way, the Sputnik seemed to reconcile both Western statesmen and Western communist leaders with the Soviet Union. (see Vladimirov 1973], pp. 70-71).

24 Sputnik Triggers the USSR–USA Competition

Working round the clock, Korolev and his team built another spacecraft in less than a month. The result was Sputnik 2, which was launched on November 3, 1957, carrying the first living being in history to go into space: the mongrel dog named Laika. What concerned Western public opinion, however, especially in the United States, was not so much Laika (although animal-lovers the world over protested about the unfortunate dog which had been condemned to death in space), but the weight of Sputnik 2. The new satellite was reported to be six times heavier than the first, coming in at 1,118 pounds (508.3 kg,) compared with the 184 pounds (83.6 kg) of Sputnik 1. This implied that the Soviets had succeeded in constructing a rocket six times more powerful than the first one in the course of just one month.

Actually, as would only be revealed many years later, the second rocket was exactly the same as the launcher of the first Sputnik. The only difference was that on the second flight the Soviets used the *"innocent trick"* of including the whole of the rocket's second stage that went into orbit as part of the Sputnik 2 spacecraft. The second stage had also gone into space with the original Sputnik and circled the Earth in roughly the same orbit, but it was not regarded as part of the satellite. [20]

The first two Soviet stamps celebrating Sputnik 2 were issued on December 30, 1957 (followed by two more, issued in the same set in 1958). They featured an allegorical representation of progress, with no allusion to Laika, the dog that travelled with a one-way ticket as, at that time, there was no capability of recovering capsules from space.

Figure 1.16: Commemorative cover postmarked in Moscow to celebrate Sputnik 2 reaching 2000 orbits around the Earth.

Sputnik 2: A Rocket Six Times More Powerful in Four Weeks! 25

Laika would eventually be depicted in some stamps issued by Soviet satellite states, but they were very slow in marking the event philatelically (with the exception of Romania and Czechoslovakia). [21]

Figure 1.17: Romanian stamp commemorating the flight of the dog Laika.

During a space conference in Moscow on June 26, 1998, Oleg Gazenko – one of the leading scientists behind the Soviet animals in space programs and a former director of the Institute of Biomedical Problems in Moscow, who had selected and trained Laika – expressed regret for the manner of Laika's death: *"We shouldn't have done it... We did not learn enough from this mission to justify the death of the dog."*

In fact, this mission was more about the Cold War competition than it was about science. More so than with Sputnik 1, the goal of this new flight was propaganda. The launch of Sputnik 2 was a political decision, overruling the wishes of reluctant scientists who did not feel ready, and the science content was very much a secondary consideration. For example, Sputnik 2 crossed the Van Allen Belts and its simple onboard Geiger counter picked up the radiation, but no-one was assigned to find out what it was. The propaganda effect, however, was brilliant, and the Soviets would try to exploit its impact to highlight their perceived superiority.

Emblematic of this was the card issued by Radio Moscow, depicting the Earth smiling while looking at the two Sputnik satellites circling it (Figure 1.18). The meridian was tilted to show the southern part of the Earth, recalling in some ways the official logo of the IGY program. The accompanying text reminded everyone, in six languages, that *"the Soviet Sputnik is the first in the world."*

A similar design was featured on the postal cancellation which represented the first Soviet special cancel ever issued with a space-related subject. It was used in Moscow on March 21, 1958 to celebrate Sputnik 2 reaching 2000 orbits of the Earth. Below the satellite circling the globe were the three Cyrillic capital letters МГГ, which is the Russian acronym for International Geophysical Year or IGY, again directly recalling the official IGY logo. (see Figure 1.19)

26 Sputnik Triggers the USSR–USA Competition

Figure 1.18: *"The smiling Earth"*

Figure 1.19: (left) The Soviet cancellation commemorating the 2000 orbits of Sputnik 2. (right) The logo of the IGY

EXPLORER 1: ONE OF THE MAIN DISCOVERIES OF THE IGY

On July 29, 1955, American President Dwight D. Eisenhower half-heartedly announced – through his press secretary – that as part of the country's participation in the International Geophysical Year, the United States planned to launch a small, unmanned Earth-orbiting satellite called Vanguard, using the rocket designed by the U.S. Naval Research Laboratory (NRL). Nothing much had happened since then.

Suddenly, the USSR had provided an unexpected show of overwhelming technological capability and had opened a new battleground between the two superpowers. American reaction to the perceived Soviet lead was one of distress, and public opinion turned against the Eisenhower Administration. Attempts to respond to the Soviet successes were not helped by internal competition between the U.S. Army and U.S. Navy, which diverted resources and energies. Eisenhower's own attitude to the situation was highly criticized in Congress. The President barely supported the space program, snubbed Soviet achievements and seemed not to understand the reasons for such consternation. One of the loudest voices protesting at Capitol Hill against the Soviet lead in space was Senator Lyndon B. Johnson, the powerful moderate leader of the Democratic majority in the Congress, who proclaimed that *"Being first in space means to be first in everything"* and *"Governing space means governing the world."*

Part of President Eisenhower's reasoning was that he wanted to keep the military out of the IGY program, which was dedicated to scientific purposes. He wanted to keep the satellite effort separate and distinct from the country's military missile effort. This was viewed by many as an inadvisable *"division of the indivisible."* [22]

On September 20, 1956, the U.S. Army had launched an enhanced Redstone Jupiter-C test missile, known as Juno, from Patrick Air Force Base at Cape Canaveral. Juno could have put a satellite into orbit, if Medaris's[13] team had not been ordered by the Pentagon to use a dummy fourth stage loaded with sand instead of a live stage. [23] Eisenhower was reluctant to see the United States being represented by a satellite launched on a rocket built by a former Nazi[14].

---

[13] General John Bruce Medaris was the U.S. Army officer who was commander of the Army Ballistic Missile Agency (ABMA) during the 1950s. As such, he was the head and sponsor of Wernher Von Braun, who developed the Redstone, Jupiter-C, and Saturn I boosters there.

[14] According to some historians, there was another motivation behind this surface reason, ironically based on military strategy. Eisenhower had intelligence on Soviet intentions and strategic capabilities. In the early Cold War, he authorized the first top-secret high-altitude espionage use of balloons, involving overflights and photographic surveillance of the Soviet Union, as the balloons could reach altitudes unattainable at the time by airplanes. (The final report of Project Genetrix, also known as WS-119L (Weapon System 119L) has recently been declassified as CIA-RDP89B00708R000500040001-0.) The President was aware of the remarkable progress of U.S. rocket technology and was committed to averting nuclear war at a time when the threat was very real.

28 Sputnik Triggers the USSR–USA Competition

The launch of the Vanguard, originally scheduled for December 4, was shifted for technical reasons to December 6. It reached barely four feet off the ground before losing thrust and falling back in a tremendous explosion that destroyed the rocket and severely damaged the launch pad. The whole thing was watched by President Eisenhower and followed via live TV, thus showing the American fiasco to the entire world. 'Vanguard' had proven to be an embarrassing choice of name and public opinion once again expressed irritation and anger. The internal rivalries and tensions that had never been soothed flared up again into overheated debate. While the Soviets carefully kept their failures secret (and *TIME* magazine put Nikita Khrushchev on its cover as "*Man of the Year*"), America's troubles made worldwide headlines. Newspapers bitterly derided the U.S. efforts with names like "*Flopnik*", "*Puffnik*", "*Kaputnik*" and "*Stayputnik*" (see Figure 1.20).

The Vanguard program ground to a halt, and Eisenhower's attention and hopes unexpectedly turned towards Wernher Von Braun and the ABMA, the missile agency of the Army, who had continued to work on their Jupiter-C rocket. The missing ingredient was the satellite, for which Von Braun himself contacted William Pickering, director of the Jet Propulsion Laboratory (JPL) in Pasadena, California. Pickering was already designing a satellite together with James Van Allen, a scientist from Iowa State University. By assiduously working together, the ABMA and JPL completed the required modifications to the launcher and built Explorer 1 in only 84 days. The payload capacity that the launcher was able to lift into orbit was quite limited, necessitating drastic reductions to the size and weight of onboard equipment and intensive research into the miniaturization of components. This latter research would lead to the creation of microchips. Explorer 1 was a 13.5 kg cylindrical satellite, carrying scientific instrumentation for monitoring the satellite's temperature and for studying cosmic rays and micrometeorites.

After a weather delay caused by a sudden tornado, Explorer 1 (officially, 1958 Alpha 1, according to the Harvard designation) was launched from Cape Canaveral on January 31, 1958 at 22:48 Eastern Time, atop the first Juno booster derived – as in the case of Sputnik – from military technology. With both sides having launched satellites that passed over countries all around the globe, the Americans could hardly object to satellites orbiting over their homeland. Neither could the Soviets, who were in a difficult position to raise any such objections. Khrushchev's efforts to show off Soviet power had coincided with the interests of the American Administration. [24]

As his 'Open Skies' policy proposed at the 1955 Geneva summit meeting (which would allow both sides to conduct mutual inspection of military capabilities over each other's territory) had been resolutely rejected by Soviet Premier Khrushchev, Eisenhower became much more interested in launching surveillance satellites that could tell American Intelligence where every Soviet missile was located. From the beginning, the Central Intelligence Agency (CIA) had played a key role in early U.S. space policy (see NSC 5520, declassified as CIA-RDP86B01053R000100060039-7), including the development of the U-2 spy plane and spy satellites such as Corona and Samos for surveillance of Soviet military bases.

Figure 1.20: Newspaper headlines denouncing the failure of the first U.S. attempt to launch a satellite, Vanguard, in December 1957.

As had happened for the previous failed launch of Vanguard, the launch of Explorer 1 had been announced in advance, giving collectors enough time to prepare postal items to document the success. There was no post office at Cape Canaveral at the time, with the closest ones being the civil post office of the nearby Port Canaveral or the military post office at Patrick Air Force Base, which was about 40 miles away.

The Explorer 1 instrumentation package soon began to gather and transmit precious scientific data and continued to do so for four months, until May 23. The satellite discovered, among other things, the zone of energetic charged particles known today as the Van Allen Belts, which protects the Earth's atmosphere from destruction. This was one of the most important discoveries of the IGY and ironically could have been 'discovered' earlier by Sputnik 2 if the onboard science had been properly investigated.

30   Sputnik Triggers the USSR–USA Competition

Figure 1.21: Postal covers postmarked respectively at Port Canaveral and at the Patrick Air Force Base (with a manuscript annotation by Wernher Von Braun that recalls the exact time of the launch). (From the collection of Steve Durst, USA.)

VANGUARD 1: THE MOST ANCIENT SATELLITE IN ORBIT

On February 5, 1958, the pre-announced launch of another Vanguard was also unsuccessful, failing to reach orbit. One month later, on March 5, Explorer 2 was also aborted when the Jupiter-C's fourth stage failed to fire.

Finally, on March 17, 1958, the tiny 1.6 kg Vanguard 1 satellite was placed into orbit by the 'civilian' three-stage Vanguard launch vehicle. This was the first satellite powered by solar cells (which were directly mounted on its body), a pioneering technology that was far from being as efficient as it is today. The battery transmitter ceased operating in June 1958 when the batteries ran down but the solar-powered transmitter continued to operate until May 1964. The satellite is still in orbit today, making it the oldest man-made object in space, and has an expected lifetime of 240 years.

Maintaining the spirit of internal competition between his Army team and the Navy-backed Vanguard team, Von Braun successfully launched Explorer 3 only nine days later (March 26), the third successful U.S. launch in less than two months.

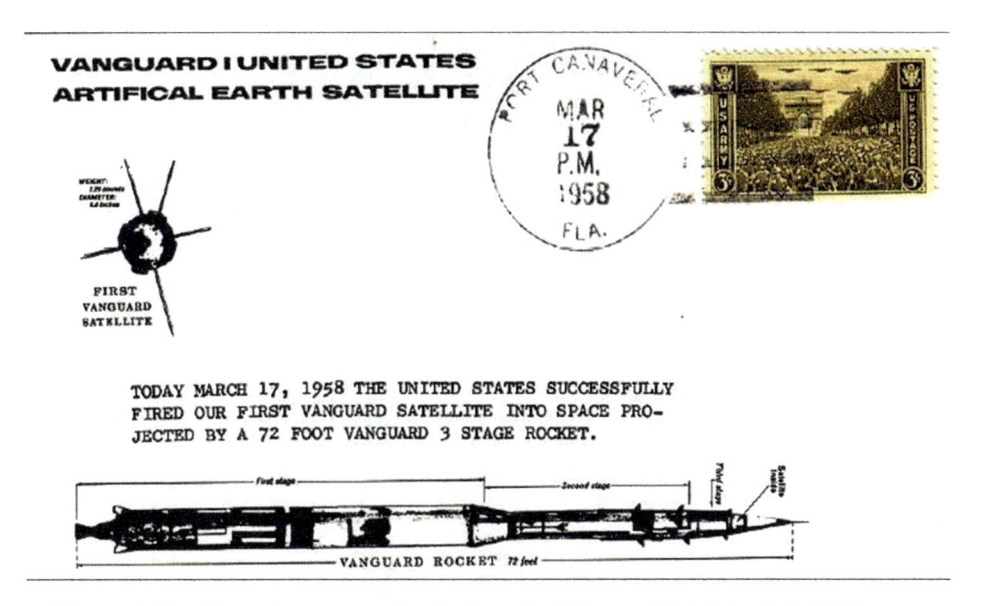

Figure 1.22: Cover commemorating the launch of Vanguard 1 in March 1958. The satellite is still in orbit to this day.

SPUTNIK 3: THE EMBLEM OF THE SOVIET SATELLITES

The American successes had angered Khrushchev and in one of his more irritable speeches, he had derided the U.S. satellites as *"oranges"*, underlining the fact that the Soviet Sputniks were much bigger.

Khrushchev ordered Korolev to put something more impressive into orbit, so the Chief Designer moved quickly to finalize the Tikhonravov satellite, Object D. This cone-shaped research laboratory included a large array of instruments for geophysical research that had originally been planned for the first Soviet satellite, Sputnik 1, until the rush to beat the Americans forced Korolev to put together an extemporary solution. Object D, housed within the upper stage of the launch rocket, weighed 1,327 kg and carried 12 scientific instruments. These included: a magnetometer and field-mill electrometer to measure fields in space; four space radiation detectors to study cosmic and solar particle radiation; a mass spectrometer and two pressure gauges to analyze the rarified outer atmosphere; an ion trap to measure plasma; and a piezoelectric microphone to count micrometeorite strikes. Also onboard was an experimental solar battery to power one of the transmitters. Sputnik 3 reached orbit on May 15, 1958. It was the only Soviet satellite launched that year, but it meant that the Soviets had finally succeeded in putting their first truly scientific satellite into orbit during the International Geophysical Year. Shortly after the launch of the satellite, the famous early space poster "Motherland! You Were the First to Spark a Star of Peace and Progress Above the Earth… Glory to Science, Glory to Labor, Glory to the Soviet System!" was produced by Valentin Viktorov (see Figure 1.23).

32   Sputnik Triggers the USSR–USA Competition

Figure 1.23: Valentin Viktorov's poster was published in Moscow by the Izogiz photo studio in 1958.

As usual, knowledge of the launch was disclosed only when it was all over, and once again genuine postage covers could not be produced on the day of the launch. Two months later (July 16), a commemorative stamp and corresponding 'First Day Cover' were released (Figure 1.24), postmarked in Moscow as the launch site remained a State secret.

Over the following months, commemorative covers were prepared with special postmarks to celebrate the 3000th, the 6000th and then the 10,000th orbit (Figure 1.25).

The silhouette of Sputnik 3 would be widely used for many years as a metaphor to represent Soviet satellites. [25] As Soviet censorship would not allow realistic representations of rockets or satellites in order to maintain secrecy about the space program, the stamps of the early Sixties – as well as the annotations and vignettes of the commemorative envelopes – often featured stereotypical propagandistic drawings that referenced the silhouette of Sputnik 3. Sometimes, these were modified to show a more cylindrical body, a more rounded nose, an additional cylindrical collar to the base, or additional radial antennas (Figures 1.27, 1.28 and 1.29).

Figure 1.24: 'First Day Cover' for the issue of the Soviet stamp commemorating the launch of the Sputnik 3 satellite. The stamp can be seen in more detail on the right.

Figure 1.25: Commemorative covers celebrating, respectively, the 3000th orbit (on the left; cover cancelled in Moscow on December 19, 1958, with a special postmark) and the 10,000th orbit (on the right; cover cancelled with a special postmark in Leningrad on April 4, 1960).

Figure 1.26: For several years afterwards, the silhouette of Sputnik 3 would remain one of the favorite subjects of greetings cards for the New Year. These images show cards issued in 1958, 1959 and 1975 respectively.

Figures 1.27 and 1.28: The silhouette of Sputnik 3 would be widely used for many years as a metaphor to represent Soviet satellites, as realistic representations were not allowed in order to maintain secrecy.

Figure 1.29: A selection of stamps and covers showing stylized drawings meant to represent the silhouette of Sputnik 3, as realistic representations of the actual spacecraft were not allowed in order to maintain secrecy.

Sputnik 3 'French' fake

There is a 'French' fake available for Sputnik 3, which carries a Baikonur-Karaganda postmark bearing the exact launch date of May 15, 1958 but was produced many years later. The cover has the anomalous format of 16.5 x 9.5 cm, which was unusual for Soviets covers of that time, as was the atypically glossy paper from which it was made. A print run limited to 150 copies was declared, as usual.

(continued)

(continued)

Figure 1.30: A 'French' fake cover produced for the mission of Sputnik 3.

RIVALRY AND INEFFICIENCY IN THE SOVIET SPACE PROGRAM

Contrary to popular perception, the Soviet space program was not managed by a goal-oriented, centralized organization with a systemic management and long-term plan. Sergei Korolev developed the world's first intercontinental ballistic missile, the R-7 or 'Semyorka' (Little Seven), which was never deployed for military operation but was launched, in different configurations, to carry Sputnik, Luna, Molnya, Vostok, Voskhod, and, later, the first Soyuz. While Korolev was referred to as the mysterious, top-secret 'Chief Designer', he was not the only designer of missiles in the Soviet Union. When the confusion of the post-Stalin months prompted a major restructuring of advanced technology industries such as nuclear and rocket weapons in 1953, the position of Korolev remained uncertain and he was still trying to overcome the deleterious effects of his 'rehabilitation' in prison[15]. He had been

---

[15] Korolev was one of the last of the major rocketry designers to join the Communist Party of the Soviet Union. A convicted *"enemy of the State"*, he first had to be formally rehabilitated for his 'crimes' of the 1930s. He attended classes on Marxism-Leninism at the Mitishtinskiy Evening University in 1950, finishing his coursework with distinctions. Unable to forget the toils of his past, however, Korolev remained unsure whether to join the ranks of card-carrying

unjustly convicted at the height of Stalin's purges in 1938 and had been sent to a gulag camp in the far east of Siberia, in the Kolyma River region. Even after joining the Party, he continued to hold ill feelings towards many of the leaders of the old Soviet government. [26]

Korolev's situation only changed under Khrushchev, who considered him to be at the heart of the successes of the early Soviet ballistic missile program and the one who brought together the abilities and talents of thousands. As former cosmonaut Alexei Leonov recalled in his book, *"There was no authority higher. Korolev had the reputation of being a man of the highest integrity but also of being extremely demanding. Everyone around him was on tenterhooks, afraid of making a wrong move and invoking his wrath. He was treated like a god."* [27] Korolev's biographer described him as *"the concretization of the history of our land in one man."* [28] But he was not the only one.

Korolev's main rival under Khrushchev was the highly credited Vladimir N. Chelomey who, as an 'external specialist', was 'supervisor' of the projects that Korolev had developed during his detention in Siberia. Following his release, Korolev was decorated with the badge of honor for his *"contribution to the design and implementation of rocket engines for military aeronautics."* Surprisingly, Chelomey was also decorated, with a superior honor.

Pleased with the liberation of Korolev, Chelomey obtained sanction from Stalin to open his own research institute – named NII 88 – and offered Korolev the position of chief engineer. The newly-freed Korolev declined the offer, unwilling to work under the direction of his former overseer, and his place was instead assigned to Valentin Glushko. Chelomey eventually outmaneuvered his rival during Khruschev's leadership by hiring the Premier's son, Sergei. That family link offered a great advantage in a political system in which personal connections were often all-important. With Khruschev's blessing, Chelomey soon had the biggest project budget of all the bureaus in the USSR and gained access to any secret document concerning German rocketry. He began to expand his research institute, encroaching into what had been Korolev's domain.

Chelomey soon secured a valuable ally in Glushko, the primary designer of Soviet rocket engines. He had been sent with Korolev to Germany in the aftermath of WWII to study the German V-2 rocket and had subsequently worked with Korolev to develop the strategic R-2 missile and then the R-7 (the first ICBM and

---

communists. In 1952, prompted by several local Party officials at Kaliningrad, Korolev finally decided to begin the application process, formally applying for full Party membership in early June 1953. Despite his worries, as it was by no means certain that a former prisoner would automatically be accepted as a Party member, he was accepted as a candidate member the following month (Siddiqi [2000], p. 160).

satellite launcher), with Glushko's bureau designing the engines and Korolev's bureau designing the rockets. But there were long-standing resentments between the two, dating back to the 1930s when the testimony of Glushko – himself a prisoner – helped to have Korolev sentenced to six years imprisonment in a Siberian correctional labor camp.

Glushko also opposed Korolev's technical choices, favoring the storable but highly toxic hypergolic propellants over the highly energetic new cryogenic fuels such as liquid hydrogen preferred and chosen by Korolev. The hypergolic chemicals were based on nitric acid and hydrazine and while they were easier to manage, Korolev declined to use them due to their toxicity. Glushko, who would never be able to count on important political support by himself, found a sympathetic ear in Chelomey and took his new, highly efficient RD-253 rocket engine to NII-88, where it was adopted into Chelomey's series of Universal Rockets (Universalskaya Raketa, or UR.)

Up to this point, Chelomey had designed and developed missiles for military purposes and had no experience with space launchers. On April 29, 1962, however, he was given the go-ahead for this program, with the initial goal being a three-stage space launcher called UR-500K, later to be known as the Proton rocket. This was created simply by taking the UR-500 ICBM first stage and adding a small two-stage UR-200 rocket to the top of it. In 1964, Chelomey was also entrusted by Khrushchev with the preparation of the LK-1, the spacecraft designed for circumnavigation of the Moon.

In early 1954, Khrushchev had already instructed Minister Ustinov to draw up a plan to dilute Korolev's power and absolute monopoly in the rocket-building business. [29] Ustinov created a new independent group in the Ukraine (OKB-586 Design Bureau), led by Mikhail Yangel, a strict Party man who was another recognized pre-eminent designer of strategic ballistic missiles in the Soviet Union. At the start of the 1950s, Yangel had also been a supervisor of Korolev (at that time, still officially an *"enemy of the State"*). He would have liked to work with Korolev again, but Chelomey prevented this. Yangel dealt with the development and mass production of intercontinental ballistic missiles and was a pioneer in the use of hypergolic propellants. The office he was to direct would design the R-12 (SS-4 Sandal), R-16 (SS-7 Saddler) and R-36 (SS-9 Scarp) missiles. During the development of the R-16 missile, Yangel barely escaped death in the Nedelin catastrophe[16].

In 1962, Yangel designed the R-56 rocket to perform a manned mission to the Moon. The design was to use a cluster of at least four long, pencil-like first and second stages to create a heavy-lift lunar booster. As with Chelomey's proposal,

---

[16] See Chapter 2, page 79.

the R-56 was intended to use the engines produced by Glushko, including the giant 7,000kN thrust RD-270, which was as powerful as the F-1 rocket used on the American Saturn V first stage. Little is known of Yangel's proposal, but it was abandoned in 1964 in favor of Chelomey's R-500 rocket, which was supposed to service an orbital mission around the Moon, and the N-1 (or *Nositel-1*) rocket of Korolev to be used for the actual landing. Yangel turned his attention to the design of the lunar module LK, the landing craft that was supposed to detach from the Soyuz spacecraft to take cosmonauts to the lunar surface. Following the failure of the N-1 rocket, however, the project to land Soviet cosmonauts on the Moon was at first suspended and then definitively cancelled.

Personal rivalries and the dispersal of resources characterized the management of Soviet space projects and as the years went by, the situation did not improve. Parallel projects were prepared and developed by rival design offices, but that was welcomed, as it allowed the politicians to keep the situation under control according to the old principle "Dīvide et imperā" (divide and rule). Any decisions about what would fly were taken by the Central Committee of the Party, according to the recommendations of the Academy of Sciences.

In an interview released at the beginning of the 1990s, Vasily Mishin (Korolev's deputy, who replaced him after his death) blamed underinvestment in the space program (only U.S.$4.5 billion compared to Apollo's U.S.$24 billion) and the lack of cooperation between design bureaus for the Soviet failure to beat the Americans to the Moon. He recalled: "*Five hundred organizations were involved in the space industry, referring to 26 ministries and government agencies. Only nine of them were under the direct control of the Military-Industrial Commission. All the others had to be constantly convinced and prodded and there was no government resolution to keep. The assignments given to the organizations were not their specialty and they often failed to fulfill them in time.*" [30]

The resulting duplication of effort was something that the Soviets could afford even less than the Americans. In the end, failure to control its competing schools of rocket and spacecraft designers, and dispersing efforts and resources, were significant contributing reasons why the USSR – which misjudged American intentions and resources and mobilized its own resources far too late – lost the race to the Moon.

NASA STARTS ITS ADVENTURE

In parallel to the Soviet situation, there were also internal rivalries and competition in the USA, between the U.S. Army, U.S. Navy and U.S. Air Force. Senator Lyndon Johnson, the chairman of the Senate Armed Services Preparedness

Subcommittee (who eventually founded NASA's largest research center that would ultimately bear his name) was alarmed by the possible loss of technology leadership and the danger to national security. It was Johnson who called upon the space-sceptic President Dwight D. Eisenhower and the Democrat-controlled Congress to put aside their differences in order to claw back the Soviet advantage.

In a bipartisan cooperation, Johnson succeeded in having the Space Act (NASA's founding law) approved by Congress and signed off by Eisenhower on July 29, 1958. The National Aeronautics and Space Administration (NASA) officially began operating on October 1, 1958, as a civil government administration which directly reported to Congress. It was derived from the staff and research facilities of NACA, the civilian National Advisory Committee for Aeronautics.

Figure 1.31: The historic decision of November 4, 1959 is immortalized on an envelope by Clyde J. Sarzin, one of the most active producers of astrophilatelic covers. The cover is signed by Wernher Von Braun, the father of the Saturn project, who became the first NASA Technical Director.

With the establishment of NASA, American space policy was finally consolidated and, for the first time, explicitly affirmed its non-military stance, although military research facilities such as the Army's Ballistic Missile Agency in Huntsville, Alabama, were integrated into the new space agency.

The Redstone and Jupiter missiles remained under the control of the Army, as did all the missile research activities which were strongly linked to military and strategic interests. Since 1957, under the Army, Wernher Von Braun had been working on a new concept of space propulsion using a cluster of relatively small rocket engines to develop a booster with a thrust of 6.7 million Newtons. The project was approved as Juno-5 by the Department of Defense in August 1958, and in February 1959 it was decreed that the project would not have military purposes. The rocket was renamed Saturn. That decree ultimately led to the historic decision taken in November 1959 to transfer the project to the control of NASA, and to transfer Von Braun (who became the NASA Technical Director) and his team, including the ninety scientists who built the V-2 in Germany, to the new agency. Somewhat reluctantly, President Eisenhower approved the funding for the ambitious program.

While NASA stressed the peaceful uses and scientific purposes for such rockets, it was also true that their use in space projects would advance research towards the obvious secondary objective of developing and testing new, more powerful missiles and implementing relevant infrastructures. This was precisely what the Soviets were also doing with their R-7 program.

Figure 1.32: Commemorative stamp issued by Monaco to celebrate the 50th anniversary of the foundation of NASA (signed by Claude Andreotto, the line-engraver and designer of the stamp).

PIONEER AND LUNIK IN A RACE TO THE MOON

The establishment of NASA created new expectations in the States and its main players were now invigorated with a new spirit. The United States Air Force (USAF), whose projects were not favored by Eisenhower, entered into the International Geophysical Year arena – in competition with the Army and the Navy – and launched a probe called Thor-Able 1. The probe, which would retroactively be renamed Pioneer-0, carried a TV camera and other scientific instruments. It was the first launch ever attempted to go beyond Earth orbit, its ambitious goal being a lunar mission to study the surface of the Moon. The mission was also intended to study the lunar far-side, with an innovative TV system using infrared scanning, as well as studying magnetic fields and micrometeorites in both Earth and lunar orbit.

The probe was launched from Cape Canaveral on August 17, 1958, but unfortunately its first stage blew up after just 77 seconds, while it was merely 16 km above the Atlantic Ocean. A month later, on September 23, 1958, the USSR secretly launched Ye-1, its own first 'lunar rocket'. The missile exploded 93 seconds after lift-off. According to some reports, if Pioneer-0 had launched successfully, then the Soviets would have made their first lunar attempt the following day. They would have launched Ye-1 on a shorter lunar trajectory mapped out by Korolev and Tikhonravov. If all had gone well, the Soviet probe would have reached the Moon before the American one. The Space Race was already on!

Once Korolev, who had been closely following the early preparations in the United States, realized that the American mission had failed, he brought his rocket back to the shed for more careful testing, returning it to the launch pad once again on September 23. Ye-1 was part of a modest Soviet lunar exploration plan – endorsed by the Soviet government in March 1958 – that included four probes. Ye-1 was intended to impact the lunar surface; Ye-2 and Ye-3 were designed to photograph the Moon's far side; and Ye-4 was intended to be armed with a nuclear bomb to blast the lunar surface. [31]

A specially-modified launcher, Semyorka 8K72, was designed for mission Ye-4, and a full-scale mock-up was built. But when all the first three launches failed and exploded after a few seconds, there was a fear that the probe with the nuclear warhead could fall back to Earth. In addition, nuclear experts had warned that a nuclear explosion on the lunar surface, with its lack of atmosphere, would be difficult to observe. The mission was quietly dropped. Needless to say, all of these missions remained top secret until the Fall of the Wall.

The Americans had a similar plan at about this time. In parallel with the preparations for the Pioneer project, the USAF developed a top-secret plan to detonate a nuclear warhead on the Moon as a display of military might. The documents of the U.S. project remained secret for nearly 45 years, and the existence of Project

Pioneer and Lunik in a Race to the Moon 43

A119 – euphemistically named "*A Study of Lunar Research Flights*" – first emerged in 1999 in a biography of U.S. scientist and astronomer Carl Sagan. It was confirmed the following year by Leonard Reiffel, the former NASA executive who had led the project in 1958, with the support of Sagan[17]. "*It was clear,*" according to Reiffel, "*that the main aim of the proposed detonation was a PR exercise and a show of one-upmanship. The Air Force wanted a mushroom cloud so large that it had to be visible on Earth... I made it clear at the time there would be a huge cost to science of destroying a pristine lunar environment, but the U.S. Air Force were mainly concerned about how the nuclear explosion would play on Earth.*"

Figure 1.33: The launch of Pioneer-1 was openly announced, and Collectors had time to prepare commemorative covers for the mission.

---

[17] The project was first revealed in a British newspaper (see: "*U.S. Planned One Big Nuclear Blast for Mankind,*" Anthony Barnett, *The Guardian*, May 14, 2000). The news was then spread by American newspapers (see "*U.S. Planned Nuclear Blast on the Moon, Physicist Says,*" William J. Broad, *New York Times*, May 16, 2000; and "*U.S. Weighed A-Blast on Moon in 1950s,*" Associated Press, *Los Angeles Times*, May 18, 2000).

44   Sputnik Triggers the USSR–USA Competition

On October 11, 1958, the USAF and the newly-formed NASA attempted another Pioneer mission, under the supervision of Charles F. Hall of the NASA Ames Research Center. Due to a software bug in the upper stage, there was a slight error in the burnout velocity and angle, and the probe – intended to achieve lunar orbit – flew only a ballistic trajectory with a peak altitude of 113,800 km (70,712 miles) before returning to Earth. It did, however, confirm the existence of the Van Allen Belts. This was NASA's first space mission, just a few days after its formation.

A few hours later on the same day, the Soviets also attempted a repeat mission, launching their Ye-1 No. 2 from Baikonur, intending, as with Ye-1 No.1, to impact on the lunar surface. Korolev was confident that he would beat the Americans to the Moon by a few hours, but the Luna booster exploded 104 seconds into the flight, scattering debris over the steppes of Kazakhstan. The investigation into the crash revealed that vibration had set up oscillations in the boosters. Once again, documents commemorating the failed Soviet mission do not exist due to the policy of secrecy. In contrast, commemorative covers would be produced for the next U.S. missions – Pioneer-2 launched on November 8 and Pioneer-3 launched on December 6, 1958 – even though both failed due to different malfunctions.

Figure 1.34: Commemorative covers celebrating the launch of Pioneer-2 and Pioneer-3. The elegant Goldcraft Cachet covers were designed by George Goldey, a pioneer space cover maker who used an innovative thermography technique.

Pioneer and Lunik in a Race to the Moon 45

On December 18, 1958, America successfully orbited the Signal Communications by Orbiting Relay Equipment (SCORE) satellite using an Atlas rocket. This was the world's first communications satellite as well as the first successful use of the Atlas as a launch vehicle.

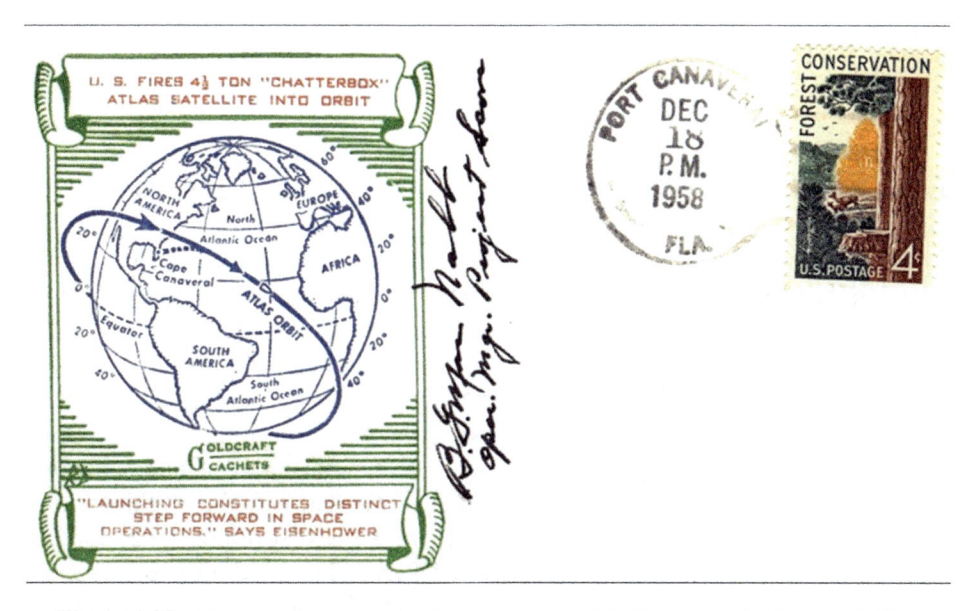

Figure 1.35: Cover commemorating the launch of SCORE, cancelled at the Post Office of Port Canaveral and signed by Brig. Gen. McNabb, who was responsible for operation of the project. (From the collection of Dennis Dillman, USA.)

The *"Talking Atlas"* as it became known, captured the world's attention by broadcasting a Christmas message from U.S. President Eisenhower via short wave radio and an onboard tape recorder. While Sputnik had communicated with the world by transmitting simple radio beeps, the SCORE satellite transmitted a human voice from orbit for the first time: *"This is the President of the United States speaking. Through the marvels of scientific advance, my voice is coming to you from a satellite circling in outer space. My message is a simple one: Through this unique means, I convey to you and to all mankind America's wish for peace on Earth and goodwill toward men everywhere."*

The payload weighed 68 kg (150 pounds) and was built into the fairing pods of the last stage of the Atlas missile. The combined weight of the on-orbit package was 3,969 kg (8,750 pounds). Now, the Americans could claim that they had put four tons into orbit.

By the end of 1958, there had been five Soviet launches – four of which had failed – and 17 American launches, of which only seven had been successful.

Despite the fact that the Americans had more resources available, they were still having great difficulty in terms of transport capability to space. Ironically, much of their difficulty had derived from their technological superiority. All the launchers were evolutions of missiles designed to carry nuclear warheads. The USA, with the most advanced nuclear technology and lighter bombs, did not therefore need particularly powerful launchers. In contrast, the USSR had bulkier and heavier nuclear devices and thus had to develop more powerful rockets. They would retain this power superiority over the next few years.

LUNA 1: THE YEAR BEGINS WITH A NEW SOVIET RECORD

The beginning of 1959 would be a triumph for the Soviet Union, as it accumulated a number of new space records over the United States. On January 2, 1959, the *"Cosmic Rocket"* (as the Soviet press dubbed what would retroactively be named Luna 1, or Lunik 1 in imitation of Sputnik, after 1963) became the first man-made object to reach escape velocity and overcome Earth's gravitational force. The following day, some 119,500 km away from Earth, Luna 1 released a large cloud of sodium gas, thus creating the first artificial comet in human history. The cloud, visible for a while over the Indian Ocean and with the brightness of a sixth-magnitude star, allowed astronomers to track the spacecraft. It was another significant propaganda coup for the Soviet Union. The cloud also served as an experiment to investigate the behavior of gas in outer space. During its trip, Luna 1 detected the existence of high energy particles just beyond the Van Allen Belt, which suggested the existence of a solar wind. This would later be confirmed by Luna 2.

Luna 1 missed its main objective of crashing into the Moon and passed within 5,995 km of the lunar surface on January 4, after 34 hours of flight. It went into a heliocentric orbit, between the orbits of Earth and Mars, and thus became the first man-made object to reach heliocentric orbit and the first artificial satellite in the Sun's orbit. It was dubbed *"the Tenth Planet,"* and propaganda art promptly exploited this unplanned 'success' (see Figure 1.37).

On April 13, after 100 days, the Soviet Post Administration celebrated Lunik 1, or 'Solnik' as it was later nicknamed, with the release of two stamps depicting Earth and the orbits of Lunik 1 (Figure 1.36). Thus, the first late commemorative covers of this mission appeared more than three months after the launch, with the cachet bearing the usual emphatic allegorical representations. This began the Lunik program, the forerunner of Soviet space exploration that would allow the USSR to accumulate several new records.

Figure 1.36: Cover and stamps commemorating the mission of Luna 1, the first man-made object to reach escape velocity and leave Earth orbit.

Figure 1.37: The success of "*the Tenth Planet*" (Luna 1) was widely celebrated by Soviet propaganda art. (left) The well-known poster "The Tenth Planet Symbolizes the Victory of Communism!" (1959) by Viktor Ivanov, an important Soviet artist who worked as a political poster designer and won numerous State prizes and international awards. (right) The poster "We are Born to Make a Fairy Tale Come True!" (1960) by Valentin Viktorov (1909-1981).

LUNA 2: THE FIRST MAN-MADE OBJECT ON THE MOON

The Soviets set another new record with the September 12 launch of Luna 2. Two days later, it became the first spacecraft to reach the surface of the Moon, landing east of Mare Imbrium near the craters Aristides, Archimedes and Autolycus on September 14, 1959. Before its impact with the lunar surface, Luna 2 released its own vapor cloud of sodium, which served both as a way of studying the behavior of gas in a vacuum and in zero gravity, and to check the accuracy of the vehicle's trajectory. The sudden cessation of the signal sent by the probe on September 14 marked the impact of the spacecraft on the Moon. For the first time, an artificial human-made artifact had landed on another celestial body. Thirty minutes later, the third stage of the carrier also arrived on the Moon.

As would happen with subsequent Soviet interplanetary missions, the probe carried two small spheres (see Figure 1.38) whose surfaces were made of identical titanium alloy pentagonal elements, featuring the State Seal of the USSR and the Cyrillic letters СССР СЕНТЯБРЬ 1959 (USSR September 1959). [32]

Figure 1.38: An example of the titanium alloy spheres carried aboard Luna 2.

It is likely that the spheres were fitted with an explosive charge, designed to be fired from the spacecraft to disperse pentagonal pennants across the lunar landscape as a symbol of Soviet scientific might. However, they most likely vaporized on impact. It was later calculated that they had struck the Moon at a relative velocity of 3.3 km/sec and that the kinetic energy was converted into heat, generating a temperature of almost 11,000 degrees C.

On September 15, 1959, Soviet Premier Nikita Khrushchev presented American President Dwight D. Eisenhower with a replica of the spherical pennant as a gift. That sphere is kept at the Eisenhower Presidential Library and Museum in Abilene, Kansas. For the next few years, the pennant became the preferred subject of propaganda posters, postcards and commemorative philatelic covers created to celebrate the triumphant success of Luna 2.

Two Soviet stamps were issued on November 1, 1959, featuring the trajectory of the probe (Figure 1.42).

The "Pennant on the Moon", the pentagonal USSR Seal, Khrushchev's glory, became very popular on the commemorative covers and was also used in postal postmarks (Figures 1.39, 1.40 and 1.41). Luna 2's arrival on the Moon made it the first item of lunar debris. It has been estimated that humankind has so far sent and abandoned approximately 170 tons worth of objects since then, most of which are no longer useful and can be considered lunar garbage.

LUNA 3: A NEW SOVIET TRIUMPH

A month after Luna 2, on October 4, 1959, the second anniversary of the original Sputnik, the Soviets launched the *"Automatic Interplanetary Station"*, which later became known as Luna 3 (Figure 1.43). This was one of the first real triumphs of space exploration. On October 7, Luna 3 became the first mission ever to photograph the far side of the Moon, sending back to Earth pictures of something no human had ever seen before: the hidden face of the Moon. The 29 shots, taken in about 40 minutes, albeit in low definition, covered approximately 70 percent of the far side, at distances ranging from 63,500 km to 66,700 km above the lunar surface. At the time, the surface was perfectly illuminated by sunlight[18].

---

[18] The imaging system was developed by P.F. Bratslavets and I.A. Rosselevich at the Leningrad Scientific Research Institute for Television, using the temperature-resistant and radiation-hardened photographic film invented by Kodak for the top-secret American Genetrix Program (previously mentioned – see footnote 14, p. 27). Genetrix was the mid-1950s precursor to spy planes and satellites, and used stratospheric balloons launched from sites in Scotland, Norway, Germany and Turkey for overflights and photographic surveillance of the Soviet Union, under the cover of "Meteorological Survey" missions. The Soviets recovered a number of such American photographic films and copied the technology, thus solving the problem they were having with producing a film resistant to cosmic radiation.

50   Sputnik Triggers the USSR–USA Competition

Figure 1.39: Propaganda postcards paying tribute to Luna 2. (left) The poster "42nd October. Glory to the Soviet System of Governance and to the Heroic Soviet People!" by Anatoly Antonchenko (published by Izogiz, Moscow) was created for the 1959 celebration the "Pennant on the Moon" and was chosen as the perfect expression of the success of the October Revolution. (center) The "Pennant on the Moon" is also the main message conveyed by another popular propaganda poster, "In the Name of Peace" (1959) by Iraklii Toidze. This poster was inspired by another well-known poster by the same artist, "The Motherland is Calling!" (July 1941). It features the same female figure, in the same red outfit and headscarf, which was already familiar and recognizable to the Russian citizen from the earlier artwork. (right) The "Pennant on the Moon" was also featured in the "Glory to the Soviet Scientists, Engineers, Technicians and Workers" poster by Mikhail Soloviev. This poster hails the two major Soviet achievements of 1959: the successful Moon probe Luna 2 and the launch of *Lenin*, the world's first nuclear-powered icebreaker ship.

Figure 1.40: (above) The "Pennant on the Moon" is also featured on the postcard "Glory to Soviet Science" by Viktor Semenovich Klimashin 300,000 copies of which were distributed. Klimashin was a famous and talented Russian watercolor artist who created posters, postcards and stamps. For many years, he designed the covers of the popular Soviet magazine *Ogonyok*. (below) The first anniversary of the launch of Luna 2 was celebrated with a commemorative philatelic cover and a special postmark featuring, once again, the pentagonal "Pennant on the Moon" as well as the orbital track of the Luna 2 probe. The text along the orbit reads: "First anniversary of the launch of the Soviet cosmic rocket to the Moon." Below the word 'Moon' is the identifying mark of the post office that used the postmark, in this case Leningrad. Four different versions of this special cancellation were produced, for the postal facilities of Leningrad, Kiev, Minsk and Moscow.

Figure 1.41: (left) While simultaneously hailing the latest Soviet achievements in space exploration (the first Moon probe) and in the peaceful use of nuclear energy (the launch of the nuclear-powered icebreaker *Lenin*), the poster "Let There be Peace" by Nikolai Litvinov features a common motif of the Soviet Union's propaganda, depicting them as peace loving and proclaiming Soviet support for the Peace movement. (right) In another popular poster, the Soviet space achievement of Luna 2 is used to suggest that "Science and Communism are Inseparable" (Anatoly Antonchenko, 1959).

Figure 1.42: Two Soviet stamps depicting the lunar trajectory of the Luna 2 probe.

On October 12, a stamp was issued in Moscow showing the satellite's trajectory around the Moon. It was accompanied by a 'First Day Envelope' which adopted the emphatic standard cachet that had already been used to celebrate Luna 1.

Figure 1.43: First day cover and stamp (enlarged on the right) commemorating the Luna 3 mission, the first to photograph the far side of the Moon.

News of the photographs taken by Luna 3 first began to circulate in the press on October 19, but the first images of the lunar far side were not revealed until October 26. An *"automatic photographic laboratory"* aboard the satellite enabled the film to be developed, fixed, dried and finally scanned with a low-resolution camera for transmission of the images back to Earth. It was some time later before it was revealed that the photos received in the early days were very disappointing due to the low signal strength. In fact, it was not until October 18, after repeated attempts, that the satellite had been able to convey 17 (some say 12) readable images successfully back to Earth – via the tracking stations in the Crimea and Kamchatka. All contact with the probe was lost on October 22, 1959.

The pictures showed a hemisphere completely different to the familiar one, consisting mainly of valleys and mountains, with much smaller and denser craters separated by two 'seas'. Up to the very end, Eisenhower continued to snub this Russian success, calling it a *"stunt."*

A Soviet Academy of Sciences Commission studied the Luna 3 pictures and created the Atlas of the Far Side of the Moon. This Atlas included names for 19 lunar features, seven of which were identified in the design of the stamp issued by

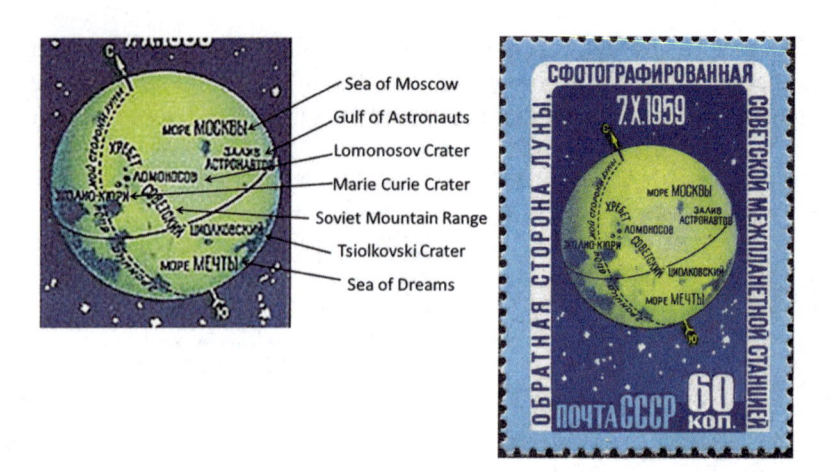

Figure 1.44: Soviet stamp issued on April 30, 1960, detailing seven of the 19 features named by the Soviet Academy of Sciences from the images provided by Luna 3.

the USSR on April 30, 1960. Of the 19 identified lunar features, 15 were craters, which the Soviets named after scientists and other international personalities. [33]

Six of those individuals were Russian: I.V. Kurchatov (atomic scientist); N.I. Lobachevski (mathematician); M.V. Lomonosov (scientist, astronomer); D.I. Mendeleyev (chemist); A.S. Popov (radio pioneer); and K.E. Tsiolkovsky (spaceflight theorist). When the IAU convened its General Assembly in 1961, 18 of the 19 lunar features and most of the names proposed by the Soviet Union were adopted. The main modification made to the Soviet names was the translation of the feature names from Russian into Latin (as seen in the stamp). For example, "*Sea of Moscow*" became "*Mare Moscoviense*". The one feature whose name was amended was the "*Sea of Dreams*", which became the "*Mare Ingenii*", with "*Ingenii*" translating into English as "*Cleverness*".

The Luna or Lunik Soviet program began in 1959 and ended in 1976, with 17 successful launches (out of the 24 Luna missions) and 38 failures. The probes that failed to launch or remained in low Earth orbit however were never officially named Luna, but instead were grouped under the 'Cosmos' label[19].

---

[19] The 'Cosmos' label was a clever umbrella title to use with any 'inconvenient' Soviet satellites. Like their American counterparts, the Soviets soon learned that creating scientific cover stories for programs, when they were hardly announcing any results of the missions in public, generated expectations within the scientific community that had the potential to become problematic for the military. To hide their intentions, the Soviets devised the 'Cosmos' label for all of their satellites, ranging from spy systems to oceanographic mapping to failed deep-space probes. By the end of the program, the USSR had launched more than 2,400 satellites under the Cosmos designation (see *Lies, damned lies, and cover stories*, Asif Siddiqi and Dwayne A. Day, in thespacereview.com, accessed in February 2018).

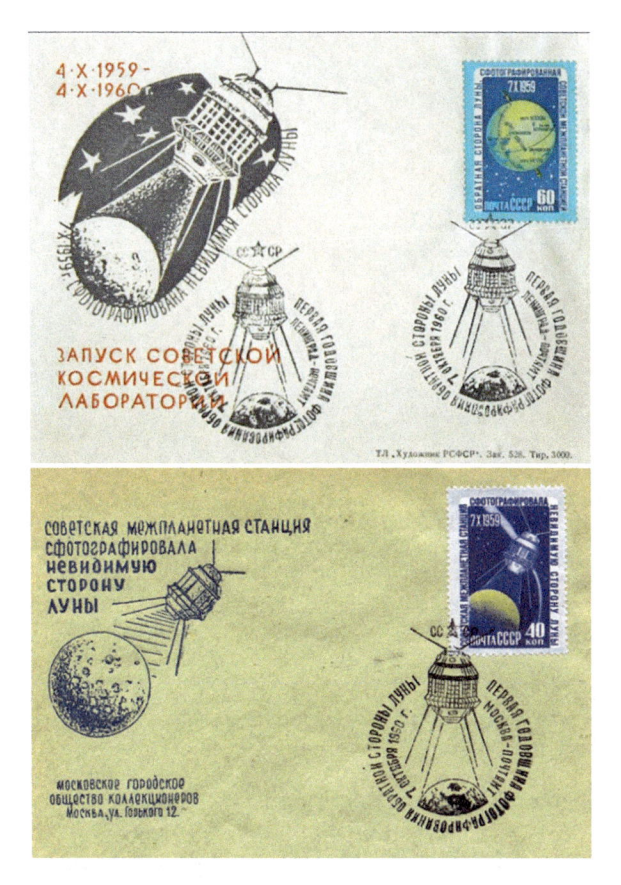

Figure 1.45: Numerous commemorative covers were prepared on October 7, 1960, to celebrate the first anniversary of the Luna 3 mission. A special postmark was made available for the event at the Main Post Office in Moscow, as well as Kaliningrad, Kiev, Leningrad and Minsk.

The Luna program provided the Soviets with many world records, including the first probe on the Moon's surface, the first orbit with lunar images, the first soft landing, and the first probe to circumnavigate the Moon and return to Earth. A few lunar landing missions (Luna 9, 13 and 22) would eventually be designed to capture images of the lunar surface that could be used to determine the possibilities of landing with human teams.

CORONA: EYE IN THE SKY

With fear of the threat of an imminent surprise nuclear attack increasing following Soviet propaganda successes in space, the need to uncover what the Soviets were actually doing behind the impenetrable Iron Curtain became more urgent in the

56 Sputnik Triggers the USSR–USA Competition

Luna 3 'French' fake

Figure 1.46: A 'French' fake cover produced to 'commemorate' Luna 3.

As with all the main Soviet space missions of the pioneering era, a 'French' fake appeared for the Luna 3 mission many years later. In its usual limited edition run of 150 items, it bore the cancellation of the launch site at Baikonur-Karaganda and the exact date of the launch, October 4, 1959. As we know, the launch site remained a secret at that time – and would continue to do so for many years – and, as usual for the Soviets, news about the mission only began to be released after the government authorities were certain of its success. Once again, it would have been impossible for a collector or cachet maker to prepare a cover in time.

United States. After an American U-2 aircraft was shot down in May 1960 and its CIA pilot, Francis Gary Powers, had been captured and forced to confess to spying on the Soviet Union, the Paris Summit of the Big Four collapsed as Khrushchev, the first speaker, demanded an apology from the U.S. and Eisenhower refused to do so. The U-2 program was discontinued.

However, in the meantime, in cooperation with the Air Force and private industry, the CIA had developed a better, more secure and more effective space-based reconnaissance system: The Corona program. Corona would incorporate 144 U.S. reconnaissance satellites, equipped with sophisticated imaging systems. They would be launched between 1959 and 1972 to monitor the Soviet bloc countries and China and, above all, to identify missile launch sites and production facilities. Corona camera systems were integrated into an Agena upper stage and launched into polar orbit aboard a Thor booster. Corona satellites, which were launched under the cover name 'Discoverer', used an innovative, constant-rotation panoramic camera system, which provided a stable platform that was constantly pointed toward the Earth. The basic camera technology was a breakthrough developed as part of the Genetrix Project. Film was loaded into a recovery capsule and returned to Earth, for air recovery by a USAF C-119 aircraft while the capsule was floating back to Earth under a parachute. [34]

Figure 1.47: Cover commemorating the test flight of Discoverer 1, launched from the Vandenberg Air Force Base on February 28, 1959. 'Discoverer' was the cover name of the Corona reconnaissance satellite, officially described as a scientific research program.

Unfortunately, the first Discoverer/Corona test mission was a failure. Discoverer 2 (April 14, 1959) carried a recovery capsule for the first time and was the first satellite to be placed into polar orbit. The main bus performed well, but the recovery capsule was lost. It apparently came down near Spitsbergen Island in the Arctic (Norway), but was never found. Rumors circulated that it had been recovered by the Soviets, but specialists were quite skeptical about this after examining many incorrect details in Russian reports that appeared to describe objects that did not match Discoverer 2. Another reason was that no images were ever made available and experts wondered why the Soviets, who had used and displayed the U-2 wreckage, camera and film to embarrass the Americans publicly, did not do the same with this satellite if they had recovered it.

58 Sputnik Triggers the USSR–USA Competition

Between 1959 and 1960, Corona experienced 13 failed mission attempts in succession. The CIA and its partners endured these setbacks and kept persevering through the endless frustration, until their persistence finally paid off. The first successful recovery of Corona film from space (Discoverer 14) occurred on August 18, 1960. The mission yielded 3,000 feet of film and stereoscopic space imagery, covering 1,650,000 square miles of Soviet territory and including 64 Soviet airfields and 26 new surface-to-air (SAM) sites. Discoverer 14 provided more overhead photographic coverage than all of the U-2 flights over the USSR combined. Reassuringly for the Americans, the new intelligence revealed that the Soviets had greatly exaggerated their military capabilities and that the 'missile gap' in fact favored the United States.

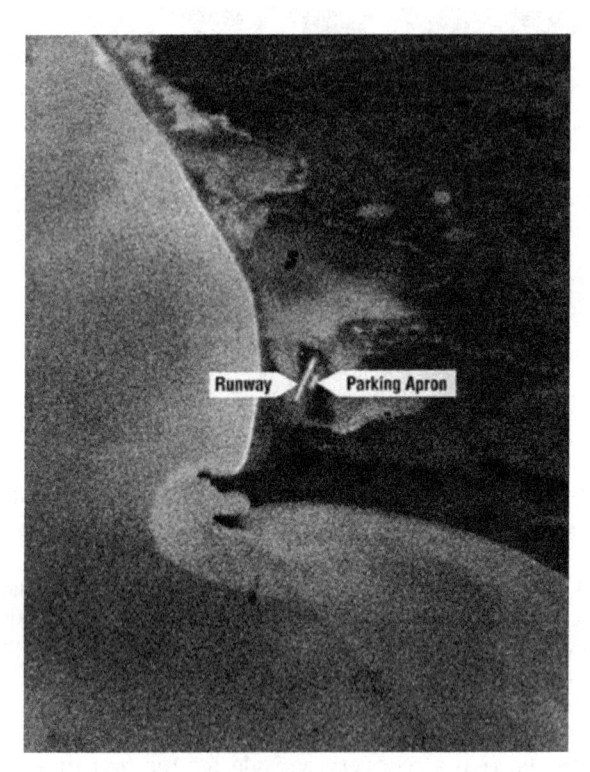

Figure 1.48: The first picture recovered from Discoverer 14 (Corona) on August 18, 1960, showed the Soviet Mys Shmidta Airport (Мыс Шмидта, also known as Cape Shmidt) with a military airport for bombers in the far north-east of Siberia (north is roughly toward the bottom of the picture). (Credit: SRO, KH-1 Corona).

Over the next 12 years, more than 100 Corona satellites collected 800,000 (unacknowledged) pictures over areas of eastern Europe and Asia (individual images on average covering approximately 100 x 120 miles of the Earth's surface), enabling the U.S. to monitor all deployed missiles, bombers and fighter forces, sorting them by type and location and providing mapping for Strategic Air Command targeting and bomber routes.

The imagery allowed the Americans to determine the precise locations of Soviet air defense missile batteries, identify those missile batteries located to protect the Suez Canal, and prove the Soviet contribution to the Chinese nuclear program. They also identified the Plesetsk missile test range, north of Moscow, and provided information about which missiles were being developed, tested and deployed. In addition, the missile launching sites of the People's Republic of China were identified and the Soviet surface and submarine fleets were mapped[20].

The entire Corona program was carried out with the utmost secrecy, disguised as a function of the Discoverer research and engineering spacecraft. Even the Discoverer Project staff did not know about Corona. They were unaware that the scientific instruments placed on Discoverer satellites were secretly removed beforehand and replaced by reconnaissance cameras. It was not until President Bill Clinton's decision to declassify the Corona archives in 1995 that details of the operation became widely known, some 35 years after the event.

The Corona program allowed the United States to accumulate a number of records – undisclosed at the time, of course – including: the first photoreconnaissance satellite; the first satellite in polar orbit (Discoverer 2); the first recovery of an object from space (Discoverer 13); the first mid-air recovery of a vehicle returning from space (Discoverer 14); the first mapping of Earth from space (Discoverer 14); the first stereo-optical data from space; the first multiple reentry vehicles from space; and several others.

> **The first document in the history of 'space mail'**
>
> The Discoverer/Corona program is also remembered because it represents the first milestone in the history of 'space mail'. On November 12, 1960, Discoverer 17 carried 28 letters into space, under the auspices of the U.S. Air Force. The 28 envelopes, shipped from Vandenberg Air Force Base in California, contained letters addressed to President Eisenhower, Vice President Richard Nixon, and 26 other high-ranking dignitaries.
>
> The letter specified that to reach the recipient, the envelope had traveled over 800 million kilometers, making 17 orbits around the Earth at a speed of 29,000 kilometers per hour. It also emphasized the this was *"the first time that letters have been sent by a satellite and is in the tradition of airmen who less than thirty years ago pioneered in the first use of airmail."*

---

[20] In the end, it turned out that America had an indisputable nuclear superiority and the Soviets had a mere 25 missiles capable of reaching American soil (Russian experts today put the number as low as four), which would have taken so long to be fueled that they could be caught on the ground if America struck first.

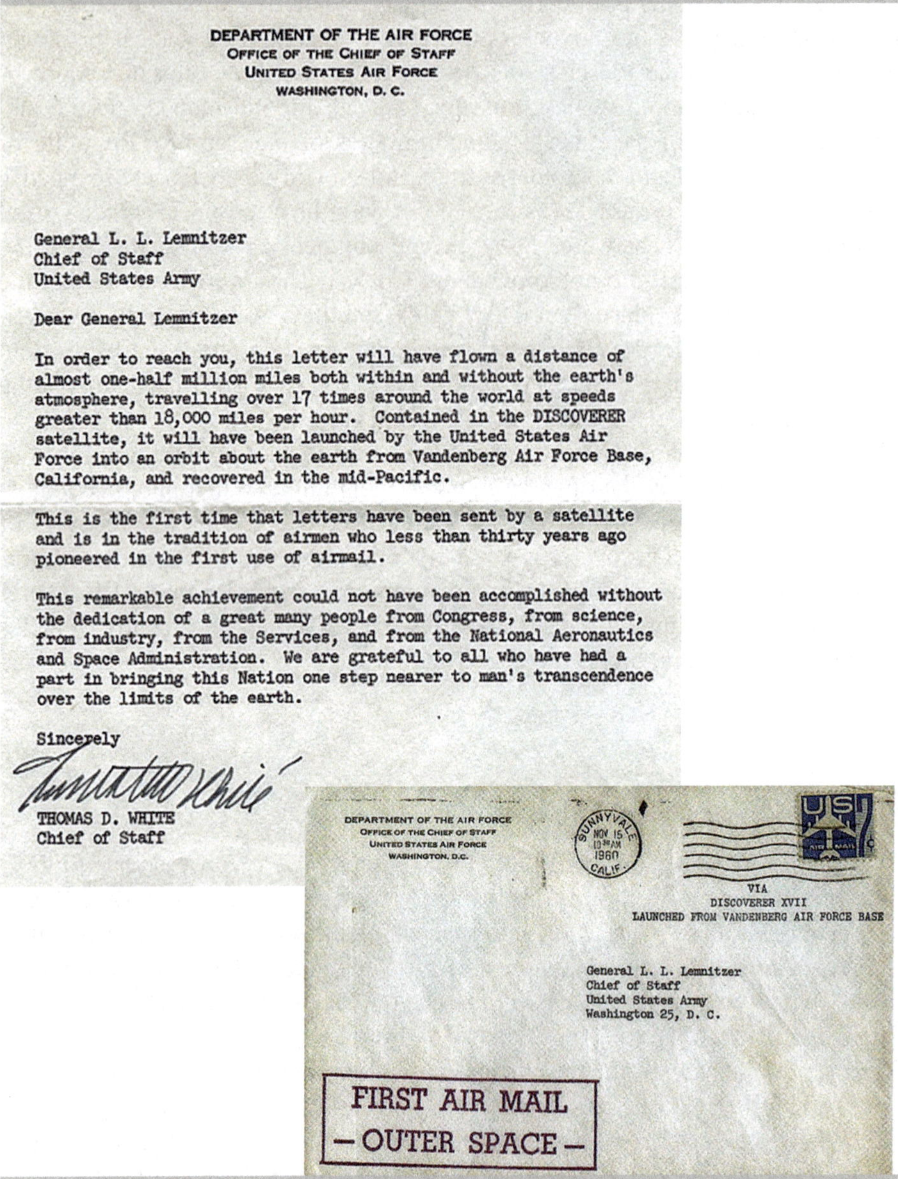

60 Sputnik Triggers the USSR–USA Competition

Figure 1.49: One of the 28 letters transported to space aboard Discoverer 17. (From the collection of Walter Hopferwieser, Austria.)

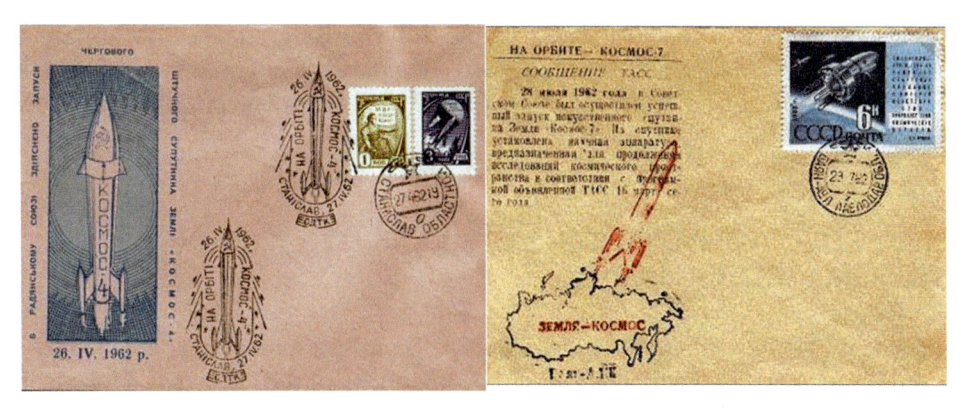

Figure 1.50: Covers commemorating the launches of Cosmos 4 (left) and Cosmos 7, the first two Soviet spy satellites of the Zenit program.

The initial reaction by the USSR to the Corona overflights was with formal protests. That attitude soon changed, however, when they began their own Zenit program.

ZENIT: THE SOVIET CORONA

Having incorporated over 500 spy satellites in a 33-year period, the Soviet Zenit program (Зенйт) flew the largest number of satellites in the history of spaceflight, most of them under the 'Cosmos' designation.

Despite some initial tensions and protests, both superpowers soon realized that spy satellites would become an important stabilizing factor: each of the contenders would be fully aware of the military capabilities of the other. In contrast, anti-satellite systems (ASATs) tested by the USA – including the 12 Bold Orion tests in 1958–59 yielded poor results. In 1961 and 1962, both the USA and the USSR performed several nuclear tests outside the atmosphere, and there were some bizarre proposals such as using nuclear warheads to intercept and destroy satellites. President Eisenhower put a stop to such operations by the U.S., by declaring that the principal uses of space should be peaceful.

Even though they would never be used effectively, the USSR conceived sophisticated programs to intercept and destroy enemy satellites. In March 1961, the OKB-52, headed by Chelomey, developed the fearsome co-orbital 'Istrebitel Sputnik' (lit. 'fighter satellite'), a missile interceptor guided by an onboard radar.

This would take between 90 and 200 minutes (one or two orbits) to reach its target and could then explode a warhead in close enough proximity to kill it. The interceptor, which weighed 1,400 kg, could be effective up to one kilometer from its target.

KENNEDY: THE SPACE PROGRAM LEADS TO A WINNING PRESIDENTIAL CAMPAIGN

President Eisenhower's term of office drew to a close at a time of heightened tension in America, living with the perceived superiority of the Russians and the fear of an imminent nuclear attack. In the context of growing international fears, John F. Kennedy, the Democrat competitor to the Republican candidate Richard Nixon, cleverly used the first electoral campaign ever dominated by TV to leverage the apparent technological gap in space, and the perception that America was trailing behind, to support the need for space technological innovation in the U.S. He evoked a vision of a *"new frontier"*, beyond which were *"uncharted areas of science and space"* to revitalize the American people.

This was a new approach, with space taking on a leading role. Many people began to feel emotionally involved with space programs and the potential threats coming from space, and began to demand the recovery of national prestige. They started to ask questions, avidly began reading books and magazines, and watched the TV reports about space more closely. This quickly evolved into a 'movement' to which the politicians had to respond carefully. The space program became an election campaign that successfully brought John Fitzgerald Kennedy to the White House. However, once he had achieved office, President Kennedy realized that the gap between the two superpowers was less significant that he believed and began to push concerns about space to the background.

Within a very short time, space soon pushed to the forefront once again, for two reasons: the need to divert attention away from the failed military attempt to invade the Bay of Pigs and overthrow the increasingly communist government in Cuba led by Fidel Castro; and the need to restore confidence to the nation once more after a new humiliation by the Soviet Union when, on April 12, 1961, they sent the first man into Earth orbit.

References

1. *We Shocked the World - Nikita Khrushchev's son recalls the night Sputnik made history,* Sergei Khrushchev, in *Air & Space Magazine,* October 2007 www.airspacemag.com (accessed in February 2018).
2. *How Russia lost the moon,* Sergei Khrushchev, in *The Guardian* (October 2, 2007) www.theguardian.com (accessed in February 2018).

References

3. **The Russian Space Bluff – The inside story of the Soviet drive to the moon**, Leonid Vladimirov, Dial Press, London 1973, p. 7.
4. *Sputnik's Secret History Finally Revealed*, Associated Press, (October 2, 2007) in foxnews.com (accessed in February 2018).
5. **Rockets and People – Vol. 3: Hot days of the Cold War**, Boris Chertok, NASA SP-4110, Washington D.C., 2009, p. 2.
6. Reference 3, p. 56.
7. This description of the Sputnik signal comes from **The All-American Boys**, Walt Cunningham, iBooks, New York, 2003, p. 21.
8. **Vanguard: A History,** Constance McLaughlin Green and Milton Lomask, NASA SP-4202, Washington D.C., 1970, p. 186.
9. *Le Figaro*, Paris, October 7, 1957, pp. 4–5 (quoted by James Harford, *Korolev's Triple Play: Sputniks 1, 2 and 3*, p. 10 in Launius (ed) [1997]).
10. Reference 8, p. 188.
11. Neil Armstrong in **Two Sides of the Moon. Our Story of the Cold War Space Race**, David Scott and Alexei Leonov, St Martin's Press, New York, 2004, p. IX.
12. *Postage Stamps as Propaganda*, Carlos Stoetzer, NASA Public Affairs Press, Washington D.C., 1953, p. 1.
13. *The Baikonur-Karaganda fakes*, Umberto Cavallero, in *Orbit* (quarterly Journal of the ASSS - UK), #95, (October 2012), pp. 25–27.
14. *C.O.S.M.O.S. Catalogne des Oblitèrations Spèciale set des Marques Officielles Spatiales*, Lollini, 1998, p. 316.
15. **Study of Suspect Space Covers**, Paul C. Bulver, Reuben A. Ramkissoon and Lester E. Winick, ATA Space Unit, Dallas, TX, 2nd Edition 2001 (CD version).
16. Reference 15, p. 8.1.
17. *Fake Baikonur Cancel Story Uncovered*, Ray E. Cartier, Astrophile, July 1997, Vol. 42, No. 4, pp. 8–9.
18. **Typy a Padělky Ruskỳch Razítek Tématu Kosmos** (*Russian space postmarks and fakes* [in Czech]), Julius Cacka, Prague 2006, pp. 3-8; and **Outer Space Mail of the USSR and Russia**, Viacheslav Klochko, Zvezdnyi Gorodok, Moscow, 2009, p. 99.
19. *Designer Mishin speaks on early space Soviet programmes and the manned lunar project*, Vasily Mishin interview in *Spaceflight*, vol. 32, March 1990, p. 104.
20. Reference 3, p. 73.
21. **Philatelic Study Report 2013-1: Space-related Soviet Special Postmarks 1958-1991**, James G. Reichmann, American Astrophilately, Framlingham, MA, 2013, p. 121.
22. Reference 8, p. 196.
23. **Challenge to Apollo: The Soviet Union and the Space Race, 1945-1974**, Asif Siddiqi, NASA SP-4408, Washington D.C., 2000, pp. 153–4.
24. *Race to the Moon 1957-1975*, Chapter 5.6 *American Response: Vanguard & Explorer*, Greg Goebel, vc.airvectors.net
25. *Soviet propaganda-design satellites*, Don Hillger and Garry Toth, rammb.cira.colostate.edu, Colorado State University, 2001-2017.
26. Reference 23, pp. 109-118.
27. Reference 11, p. 53.
28. **Sergei Korolev: The Apprenticeship of a Space Pioneer**, Yaroslav Golovanov, Mir Publishers, Moscow, 1975, p. 293.
29. Reference 23, p. 114.

30. Reference 19, p. 105.
31. *The E-4 Project – Exploding a nuclear bomb on the Moon*, Aleksandr Zheleznyakov, in *Space History Notes*, by Sven Grahn (www.svengrahn.pp.se – accessed in February 2018).
32. *Soviet Spacecraft Pennants*, Don P. Mitchell, in www.MentalLandscape.com.
33. *Russians on the Moon*, Jim Rechman, in Ad* Astra (quarterly journal of ASITAF), #7, July 2010, pp. 5–10.
34. *Corona: America's First Satellite Program*, Kevin C. Ruffner (Ed.), History Staff Center for the Study of Intelligence, Central Intelligence Agency, Washington D.C., 1995, pp. XIII–XIV.

2

Man in Space

USA IN THE RUNNING TO PUT A MAN IN SPACE BEFORE THE SOVIETS

The Space Race had well and truly begun. It was still unclear what the ultimate goal would be, but the USA was in the running. The USAF began the secret Man-In-Space-Soonest program (MISS) which – in competition with the Adam project of Von Braun's Army team and the Navy's Manned Earth Reconnaissance program – was intended to put a man into space before the Soviet Union, using a rocket-boosted, winged manned space vehicle that would follow on from the X-15 rocket plane. [1]

On June 25, 1958, eight months after Sputnik, nine test pilots were chosen for the project as the first astronaut selection group in history:

- Neil A. Armstrong, NACA (age 27)
- William B. Bridgeman, Douglas Aircraft Company (42)
- Scott Crossfield, North American Aviation (36)
- Iven C. Kincheloe, USAF (29)
- John B. McKay, NACA (35)
- Robert A. Rushworth, USAF (33)
- Joseph A. Walker, NACA (37)
- Alvin S. White, North American Aviation (39)
- Robert 'Bob' M. White, USAF (33)

© Springer Nature Switzerland AG 2018
U. Cavallaro, *The Race to the Moon Chronicled in Stamps, Postcards, and Postmarks*, Springer Praxis Books,
https://doi.org/10.1007/978-3-319-92153-2_2

The MISS program was planned with the goal of launching a man into orbit, as well as conducting investigations into the human factors of spaceflight, man's ability to function in a weightless environment, and the capability of recovering both man and machine after the flight. But MISS would never fly. The program was cancelled on August 1, 1958 and two months later, NASA was set up to take responsibility for all manned space flight. Within seven days of its creation, NASA initiated Project Mercury. Only two men from the MISS program would actually reach space. The first, Joseph A. Walker, would do so twice during X-15 rocket plane tests in 1963. The other was of course Neil Armstrong, who would be selected as a NASA astronaut in 1962 and would become the first person to walk on the Moon in 1969.

Although tests had demonstrated that both the USAF Atlas and the Army Redstone launchers were reliable[1], the Americans proceeded with great caution. The rockets had been constructed for essentially military missile purposes and would have to be meticulously adapted if they were going to be used in the new Mercury program. Not only would the missile technology have to be fine-tuned, but there would also be many technical and scientific issues to address and answer before venturing into a completely unknown environment. Issues such as whether a human body could tolerate the gravitational acceleration required to escape the atmosphere, whether it was possible to survive the absence of gravity, and how long a human could survive in space. The initial plans for Project Mercury envisaged 25 missions.

Of particular importance to the project was the Launch Escape System (LES), the crew safety system connected to the spacecraft, which would be used to separate the capsule quickly from its launch rocket in case of emergency. The first test of the LES was carried out on May 9, 1960, but the mechanism did not perform well and the test had to be repeated several times.

The Mercury spacecraft was designed by Max Faget[2], one of the original 35 members selected to form the NASA Space Task Group (STG) on November 5, 1958. The STG would later develop into the Manned Space Center (MSC), now the Johnson Space Center (JSC) in Houston, Texas. For almost 20 years, Faget would be the group's Director of Engineering and Development, and he would solve one of the thorniest problems for late 1950s engineers eager to put a man

---

[1] The team of German former V-2 scientists led by Wernher Von Braun worked on the Redstone project.

[2] With the advent of Apollo, Maxime 'Max' A. Faget would be appointed chief engineer at MSC, with responsibilities for the design, development and proof of performance of manned spacecraft and their systems. Faget's numerous accomplishments include patents on the Aerial Capture Emergency Separation Device (escape tower), the Survival Couch, the Mercury Capsule and the Mach Number Indicator. (See Swanson [1999], Chapter 14, pp. 347–49.)

Figure 2.1: Covers commemorating the tests of the Mercury Launch Escape System (LES).

into space: how to protect the spacecraft and its occupant from the severe hazards of reentry into Earth's atmosphere.

Twelve companies bid to build the Mercury spacecraft and in January 1959, McDonnell Aircraft Corporation, based in St. Louis, Missouri, was chosen to be the prime contractor, supported by some 600 subcontractors. With barely 100 cubic feet (2.8 m^3) of habitable volume, the Mercury capsule was only just large enough for a single crewmember, but it would be a tight fit. A standing joke among the astronaut candidates was that they did not so much ride in the Mercury capsule as put it on. Inside the capsule were 120 controls: 55 electrical switches, 30 fuses and 35 mechanical levers to allow the spacecraft to be controlled across the three axes.

The Mercury spacecraft did not have an onboard computer. Amid concerns that weightlessness could potentially cause the pilot to become disorientated, the spacecraft was designed to be fully controlled from the ground, relying on computations calculated back on Earth by computers housed in NASA's ground facilities to provide accurate reentry trajectories.

68 Man in Space

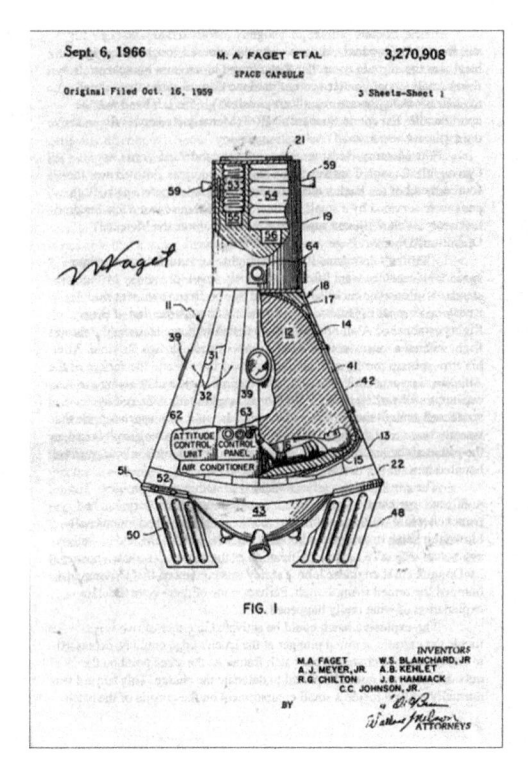

Figure 2.2: Front page of the Patent Application filed by Max Faget for the Mercury spacecraft on October 16, 1959.

The first Mercury astronauts were selected from a group of 110 military pilots. The criteria established at the beginning of 1959 stated that the job required candidates with a high level of intelligence and exceptional stamina, together with advanced training in science or engineering and the psychological ability to be able to perform effectively in situations of high stress. In addition, the engineering constraints dictated the maximum height and weight of potential applicants. Consultants recommended the following basic requirements: Maximum age of 40; maximum height of 5 feet 11 inches (180 cm); maximum weight of 180 pounds (82 kg); excellent physical fitness; college education in engineering or a physical science; graduation from test pilot school; and a minimum of 1,500 hours flying time as a qualified jet pilot. [2]

NASA originally planned to select its first astronauts in open competition, but President Eisenhower's decision to limit the search to test pilots within the military services greatly simplified the selection procedure. From a total of 508 service records screened in January 1959, 110 men were found to meet the minimum standard requirements. The list of names included five U.S. Marines, 47 from the U.S. Navy and 58 USAF pilots. Testing and training for the candidates began in

February 1959 and on April 9, seven of the candidates were officially chosen to become Astronaut Group 1, or *"The Original Seven"* as they became known. Those seven were:

- M. Scott Carpenter, USN (age 34)
- L. Gordon Cooper Jr., USAF (32)
- John H. Glenn Jr., U.S. Marines (38)
- Virgil I. 'Gus' Grissom, USAF (33)
- Walter M. 'Wally' Schirra Jr., USN (36)
- Alan B. Shepard Jr., USN (36)
- Donald K. 'Deke' Slayton, USAF (35)

Almost immediately, they and their families become worldwide celebrities. Their fame was further enhanced with an exclusive contract with *Life* magazine worth $500,000 collectively (or more than $4 million today). The stories painted the astronauts as American heroes, fighting communism with their space missions.

Figure 2.3: The Original Seven Mercury astronauts with a U.S. Air Force F-106B jet aircraft. From left to right: M. Scott Carpenter, L. Gordon Cooper Jr., John H. Glenn Jr., Virgil I. 'Gus' Grissom, Walter M. 'Wally' Schirra Jr., Alan B. Shepard Jr., Donald K. 'Deke' Slayton. Courtesy NASA.

While Project Mercury was designed with the generic goal of sending a man into space and placing him in orbit around the Earth, the program did not initially proceed according to a well-defined scientific plan. It was only after tests using both monkeys and dummies yielded satisfactory results that the first U.S. manned suborbital flight was announced.

USSR: DETERMINED TO KEEP PRE-EMINENCE IN SPACE AT ANY COST

If anything, the Soviet manned space program was even less focused. The USSR had also begun preparations for human space flight in January 1959 and the piloted component of the Soviet space program had developed hand-in-hand with that of the first Soviet reconnaissance satellite. When it came to selecting candidates for the first voyages into space, the Soviets initially considered individuals from a variety of professional backgrounds, including aviation, the Soviet navy, rocketry and even motor racing. [3] However, physicians from the Soviet Air Force insisted that the potential cosmonaut candidates should be qualified air force pilots, arguing that they would have the relevant skills required, such as experience with higher g-forces and ejection seats. Korolev decided that the initial cosmonauts should be male and that the ideal candidate would be no taller than 174 cm, weigh between 70 and 75 kilograms (necessary in order to fit the small 3KA capsule) and be aged between 25 and 30 years. The final candidate criteria were approved in June 1959.

As with the Americans, it would be the military services, in this case the Soviet Air Force, that would provide the pool of potential candidates[3], even if the Soviet capsule would merely have "*a man onboard*" as opposed to a pilot flying it. The selection process began in August by inspecting the records of more than 3000 fighter pilots. Most were eliminated at an early stage due to height, weight and medical history. A special Air Force commission led by military physician Yevgeny Karpov eventually selected 200 candidates. This group was summoned to the

---

[3] See Chertok [2009], p. 61. Actually, the decision had more to do with politics and concepts of heroism as, following WWII, the fighter pilots were regarded as the most heroic of the Soviet forces. Among the factors that may have affected the decision to choose pilots was the news that NASA had selected its first astronauts from those with aviation backgrounds in the American armed forces. In addition, drawing the first cosmonauts from among the ranks of the military was essential to Korolev's ongoing campaign to win over reluctant military leaders. The space capsule would be totally automated and no piloting skill would be required to fly it. Nothing would depend upon the decisions of the cosmonaut within, partly because the doctors were concerned about the psycho-physical integrity of the 'pilot' in weightlessness, but also because the KGB wanted to prevent the cosmonaut from being tempted to land the capsule outside Soviet territory.

mysterious *"Commission for the Theme No. 6"* at the Central Research Military Hospital in Moscow in October 1959.

The cosmonaut candidates underwent a series of rigorous medical tests. Apart from general health and physical condition, the commission also tested the candidates' professional suitability, moral and ethical characteristics, memory, mental agility, resourcefulness in stressful situations, and powers of observation. [4] This was the first Soviet selection, so no one was entirely sure what the cosmonauts would have to be trained for. Although the pilots were not told they might be flying into space, one of the physicians in charge of the selection process perceived that some of the candidates had figured this out. They were all forbidden to reveal details of this top-secret project, however.

By February 25, 1960, the list of candidates had been reduced to 20, who formed the first group of Soviet cosmonauts: [5]

- Senior Lieutenant Ivan N. Anikeyev (age 27)
- Major Pavel I. Belyayev (34)
- Senior Lieutenant Valentin V. Bondarenko (23) *
- Senior Lieutenant Valery F. Bykovsky (25)
- Senior Lieutenant Valentin I. Filatev (30) *
- Senior Lieutenant Yuri A. Gagarin (25)
- Senior Lieutenant Viktor V. Gorbatko (25)
- Captain Anatoly Y. Kartashov (27) *
- Senior Lieutenant Yevgeny V. Khrunov (26)
- Captain Engineer Vladimir M. Komarov (32)
- Lieutenant Alexei A. Leonov (25)
- Senior Lieutenant Grigory G. Nelyubov (25) *
- Senior Lieutenant Andrian G. Nikolayev (30)
- Captain Pavel R. Popovich (29)
- Senior Lieutenant Mars Z. Rafikov (26) *
- Senior Lieutenant Georgy S. Shonin (24)
- Senior Lieutenant Gherman S. Titov (24)
- Senior Lieutenant Valentin S. Varlamov (25) *
- Senior Lieutenant Boris V. Volynov (25)
- Senior Lieutenant Dmitry G. Zaykin (27) *

* did not fly in space

Five of the group did not meet the age criteria of between 25 and 30 but, as Chief Designer Korolev had insisted on having a pool of candidates three times larger than NASA's seven-strong team, this condition was waived because of their performance in the selection procedures. Two in particular, Belyayev and Komarov, were the most educated and experienced members of the team.

72 Man in Space

Unlike NASA's first astronaut group, the Soviet cosmonaut group did not particularly consist of experienced pilots. As mentioned, piloting skill would hardly be needed in the early spacecraft as the capsules were more automated than their American counterparts. Only one of the final 20 candidates had flown on the newest Soviet jet at that time, the MIG-19; the rest had flown the older MIG-15 or MIG-17. The most experienced pilot was the 34-year-old Belyayev, who had logged 900 hours of flight time in both piston and jet aircraft. In contrast, Gagarin had a log of only 230 hours. The group would be placed under the command of Nikolai Kamanin, a legendary Soviet pilot and polar explorer[4].

The Soviet trainers seemed to want to prepare for any contingency and adopted a quite grueling training process. Discipline and surveillance were good starting points and they transformed the cosmonauts into lab rats. Among the most dreaded tests were the centrifuge to simulate the effects of extreme gravitational pull and the 'rotor', a spherical cage that could spin wildly around three axes, into which the cosmonaut would be positioned with arms and legs splayed. This was an extreme training device and would be eliminated from the process following Gagarin's flight.

THE FIRST ACCIDENTS AND CASUALTIES

While the centrifuge and rotor were designed to probe the boundaries of human physical endurance, the outer limits of the human psyche were tested by the isolation chamber. Each cosmonaut spent between 10 and 15 days in a *"public loneliness"*, where they could neither see nor speak to anybody, but were constantly under observation via television cameras. During one such exercise on March 23,

---

[4] In February 1960, General Nikolai Petrovich Kamanin was appointed overall head of cosmonaut selection and training. In November 1960, he became head of space exploration for the armed forces. Kamanin was not just any general, but a KGB general with a proven track record in both military aviation and intelligence work. A pilot himself, Kamanin was the first to receive the honor 'Hero of the Soviet Union' in 1934, the year of Gagarin's birth. He received the honor for his dramatic flight to rescue the crew of the Cheliuskin ice breaker that had been crushed in Arctic Sea ice. He rose through the ranks after flying numerous sorties during WWII, including daring reconnaissance missions into enemy territory. Like his protégé, future hero Gagarin, Kamanin knew what it was like to live simultaneously in the shadowy world of secret police, behind the gates and checkpoints of the Soviet military-industrial complex, and in the public eye as a Soviet hero. During training, he was a constant presence with the cosmonauts, observing them, compiling reports on their character and maintaining close connections with engineers, KGB informants and political authorities. He also kept a detailed diary which would be published posthumously after the collapse of the Soviet Union. It would become one of the most important sources for understanding the daily lives of the cosmonauts.

1961, three weeks before Gagarin's flight on Vostok 1, the youngest member of the first cosmonaut group, 24-year-old Valentin Bondarenko, was killed. He perished in a fire he accidentally started on the tenth day of his routine, 15-day isolation and endurance exercise in the low-pressure altitude chamber.

Because of the pressure difference, it took the doctor on duty several minutes to open the chamber door (some reports suggest half an hour). Bondarenko died eight hours later from the shock of the burns he had sustained. It was the very first death of a space trainee in the history of the Soviet space program, a tragedy for the Soviet cosmonaut group – and in particular for his friend Gagarin. But the experiment that claimed his life effectively had nothing to do with the preparations for the first crewed Vostok launch and the program continued as scheduled, although all experiments with heightened partial oxygen pressure were stopped. Some heads would roll as a result of the tragedy, but Bondarenko's death remained a secret for more than two decades. [6]

As he had already appeared in group films and images of the first cosmonaut selection, Bondarenko's unexplained disappearance sparked rumors of cosmonauts dying in failed launches. Rumors reached the West about the existence of secret graves of anonymous dead cosmonauts, killed on unannounced missions. But when the International Astronomical Union (IAU) updated the map of the Moon in 1970, the Soviets, even though they could assign three lunar craters to their fallen cosmonauts, named only two of them: Komarov and Gagarin.

Details of the Bondarenko incident first appeared in April 1986 in an article in Izvestia – then celebrating the 25th anniversary of the first piloted Vostok mission – by science writer Yaroslav Golovanov, himself a former cosmonaut candidate (and biographer of Korolev)[5].

Colonel Yevgeny Anatoliy Karpov, the newly appointed chief of the future Cosmonaut Training Center, selected what in his view were the six most promising candidates for top priority training on January 25, 1961. Captains Pavel Popovich and Andrian Nikolayev, and Senior Lieutenants Yuri Gagarin, Gherman Titov, Valentin Varlamov and Anatoliy Kartashev, were given priority during

---

[5] See Oberg [1988], pp. 156-176 and Siddiqi [2000], p. 266. Some have speculated that, had the Soviets been open about the circumstances of Bondarenko's death, then NASA might have been alerted to the hazardous design of the early Apollo Command Module and would have made changes that could have prevented the deaths of the three Apollo 1 crewmembers in January 1967. As Leonov reported in his book: *"The Soviet Union did not alert those in charge of the U.S. space program about our tragedy. In those days, there was only a limited exchange of information between our two programs via international forums and congresses, and no bilateral mechanism for exchanging information of this sort. What had happened to Bondarenko was considered an internal matter, in any case, not a matter we wanted openly discussed. Like nearly every aspect in our space program, it became a closely guarded secret."* (Scott-Leonov [2004], p. 57.)

training sessions and access to the first Vostok simulator. The remaining candidates followed a less intensive training program. During the training process, Kartashev and Varlamov were injured and dropped from the first group of six. They were replaced by Senior Lieutenants Grigory Nelyubov and Valery Bykovsky. [7] The six were awarded the title of 'Pilot-Cosmonaut' but were prohibited from using that title in public.

Figure 2.4: The first group of Soviet cosmonauts in May 1961 at the Black Sea resort of Sochi, following Gagarin's flight. Sitting in front from left to right: Pavel Popovich, Viktor Gorbatko, Yevgeny Khrunov, Yuri Gagarin, Chief Designer Sergei Korolev, Korolev's (second) wife Nina Ivanova with Popovich's daughter Natasha, Yevgeny Karpov (Director of the Cosmonaut Training Center), Nikolai Nikitin (parachute jumping instructor) and Yevgeny Fyodorov (physician). Standing in the second row, from left to right: Alexei Leonov, Andriyan Nikolayev, Mars Rafikov, Dmitry Zaikin, Boris Volynov, Gherman Titov, Grigory Nelyubov, Valery Bykovsky and Gyorgy Shonin. Third row, from left to right: Valentin Filatyev, Ivan Anikeyev and Pavel Belyayev. (This image features only 16 of the original selection of 20. The missing four are: Valentin Bondarenko, who had died in a training accident three weeks before the flight of Gagarin; Anatoliy Kartashov and Valentin Varlamov, who had both been dropped from training on medical grounds; and Vladimir Komarov, who was on medical leave. Eleven of the remaining 16 would fly in space. This photo – likely taken by I. Snegirev – couldn't be published for many years because of the "*secret people*" it depicted, including the top-secret Chief Designer, Sergei Korolev and the training officials to his left). (From the archive of Boris Chertok. See Chertok [2009], p. 59. Image courtesy of Roscosmos.)

The First Accidents and Casualties

At his research lab in Moscow, directed by Grigory I. Voronin, Sergei Korolev had a group of engineers and assistants whose job was to monitor the American press in real time and study all the American literature on space. As soon as the Americans announced a launch, it was mandatory for the Soviets to precede them and maintain their perceived dominance in space flight, even if that meant taking enormous risks. The announcement of the first human flight by NASA once again induced the Soviets to speed up their own plans, by eliminating all the experiments not deemed indispensable from their own human flight and squeezing those left in order to maintain the Russian lead. As always happens with hasty decisions, the accidents and casualties began to rise.

One of the critical problems was the return of the cosmonaut to Earth at the end of the mission. Korolev knew from American reports that the Mercury spacecraft would come down in the sea under parachutes. For this reason, the Mercury spacecraft was being made from light alloys of sufficient strength to withstand the splashdown. Originally, Korolev considered following the same path, but that plan was immediately scrapped by Khrushchev, who ordered that *"A Soviet spacecraft must land on Soviet territory."*

It is not difficult to understand the reasons why Khrushchev did not want a Soviet cosmonaut to land in international waters. Access to the area where the spacecraft would splash down would be open to everybody and Western experts and the world's press would rush to the spot. At the same time, it would be impossible to prevent Korolev and his close colleagues from going aboard to welcome the cosmonaut back. This would make it inevitable that they would come into contact with foreigners, and the identities of the designers who built the spacecraft would be revealed.

The main problem for Korolev with the decision to return to Soviet soil was the tremendous weight that would be required to build the spacecraft. A capsule destined to parachute down to the ground would have to be far stronger than one which returned to water. That alone would necessitate increased weight, but on top of that was the fact that the speed of the touchdown would have to be reduced to a minimum. That would mean equipping the capsule with a very powerful parachute system, thus increasing the weight still further. Korolev partially solved this dilemma when he decided that the pilot of the spacecraft would have to eject out of the capsule before reaching the ground and complete his descent under his own parachute. This would allow the empty craft to descend at a much greater speed and consequently the dimensions of its parachute system could be reduced to become much smaller and lighter. [8] However, the cosmonaut would still have to be ejected from a spacecraft moving at tremendous speed and completely uncontrolled at the moment of ejection. Korolev wanted to test this procedure thoroughly, using dogs and monkeys, but due to time constraints, the tests had to be minimal.

Once the Americans had announced that their first suborbital manned flight would take place in the spring of 1961, orders arrived from Moscow to review all research programs and to speed everything up. All experimental work not directly connected with manned flights had to be dropped and every effort had to be made to ensure that the first man into space was a Soviet one. The ejector seat and catapult system were tested using two dummy figures. In both tests, everything worked normally, but pressure to speed up the trials increased. Under these time constraints, when the test was repeated by an expert parachutist, his helmet hit the edge of the hatch as he was ejected, with fatal consequences.[6] The secrecy surrounding this fatality provided further material for the persistent rumors that some Soviet *"phantom cosmonauts"* had lost their lives before Gagarin's successful flight.

HUMANS IN SPACE: TWO DIFFERENT PATHS

In their attempts to achieve the same purpose of dominating outer space, the USA and USSR chose different paths, in competition surrounded by the strictest secrecy, seeking technical solutions that were sometimes very different. In the USSR, Korolev had been working on the Vostok project since 1958, incorporating military concepts that had previously been developed for unmanned probes, such as the Zenit camera platform.

While the Mercury spacecraft was bell-shaped (see Figure 2.5), the Vostok capsule consisted of two main parts: the spherical module that housed the cosmonaut and a conical section that housed the equipment, such as the braking retro-rockets and the fuel tank (see Figure 2.6). Both the Mercury and Vostok capsules had parachute systems for recovery, but Mercury was designed to splashdown in the ocean, while Vostok would 'crash down' on the land.

The Vostok spacecraft weighed over 10,000 pounds. The Mercury spacecraft – with its escape tower that was ejected during the launch sequence when it was no longer required – weighed only about 3,500 pounds, scarcely more than a third of

---

[6] According to Vladimirov, the task of carrying out the test in October 1960 was entrusted to the most experienced Soviet parachutist, Air Force Colonel Piotr Dolgov (Vladimirov [1973], pp. 89–90). Dolgov was not a member of the Soviet cosmonaut corps but had approximately 500 test jumps to his credit, including a significant number carried out at speed with the aid of an ejector seat. He had been ejected several times from new types of aircraft to test the efficiency of their escape systems. Vladimirov inferred that he knew Dolgov very well while conducting his own parachute practice over several years.

However, other sources have stated that Dolgov was killed in an accident in February 1961 (others state November 1, 1962), during a high-altitude parachute jump from a Volga balloon gondola. This version was apparently confirmed by the obituary that appeared in German in *Der Spiegel* on November 21, 1962).

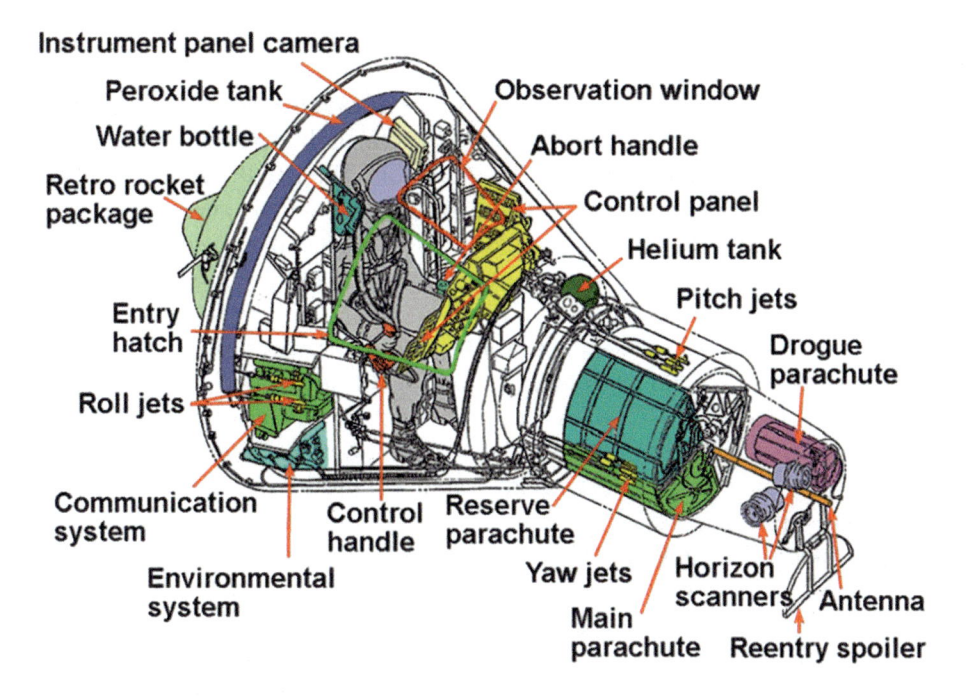

Figure 2.5: Mercury diagram – NASA Mercury Spacecraft Familiarization Manual, December 1962. Courtesy NASA.

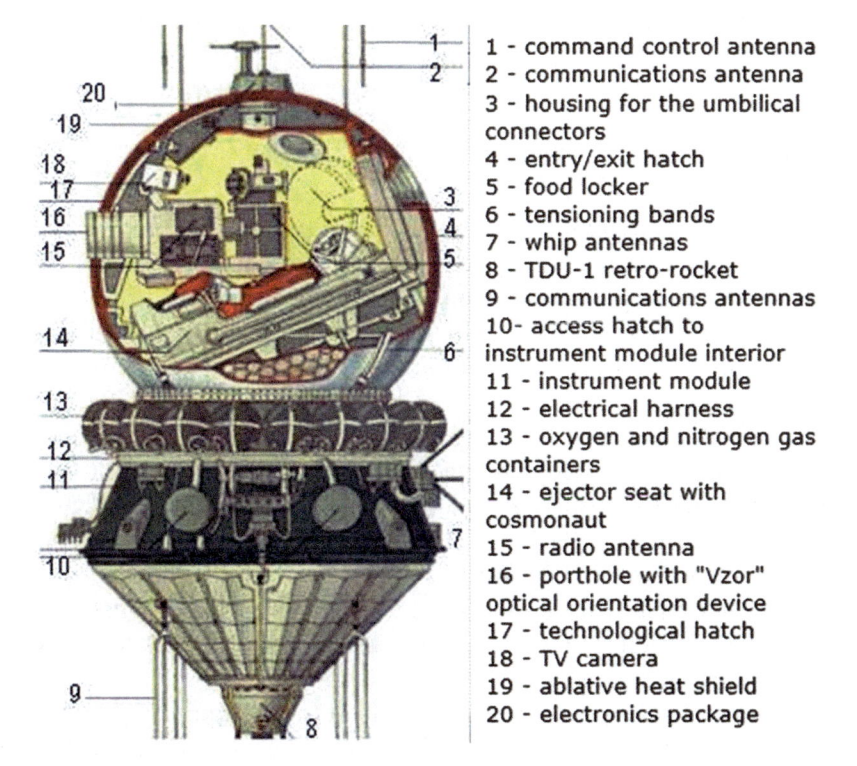

1 - command control antenna
2 - communications antenna
3 - housing for the umbilical connectors
4 - entry/exit hatch
5 - food locker
6 - tensioning bands
7 - whip antennas
8 - TDU-1 retro-rocket
9 - communications antennas
10 - access hatch to instrument module interior
11 - instrument module
12 - electrical harness
13 - oxygen and nitrogen gas containers
14 - ejector seat with cosmonaut
15 - radio antenna
16 - porthole with "Vzor" optical orientation device
17 - technological hatch
18 - TV camera
19 - ablative heat shield
20 - electronics package

Figure 2.6: Vostok Diagram – Courtesy of *The Soviet Manned Space Program*, Phillip Clark (Orion Books, New York, 1988, p.15).

the weight of the Vostok. Another reason for the divergence, apart from the different landing options, was the variation in booster capability that existed at the time between the United States and the Soviet Union. The small Mercury-Redstone used for the initial Mercury flights was 25 m tall, weighed 30 tonnes and produced just 350 kN of thrust. In contrast, the Soviet Vostok rocket that would take Gagarin into orbit for the first time was five times as heavy and generated six times as much thrust.

Inside the Mercury capsule, the astronauts would breathe a pure oxygen atmosphere, which had several advantages but one big disadvantage, namely that a fire in that atmosphere could be a disaster (as it was in the Bondarenko accident). The cosmonauts in Vostok capsules breathed a normal atmosphere of nitrogen and oxygen. They would not have to worry about the fire hazard but would need to wear a pressurized spacesuit to prevent the possibility of embolism, caused by nitrogen bubbles forming in the bloodstream when the capsule depressurized during reentry.

Vostok capsules included four switches and 35 indicators, while the Mercury capsule, as previously mentioned, had 120 controls. But neither spacecraft would rely on input from the occupant. Neither Vostok nor Mercury could perform orbital maneuvers; they could only be translated around their axes. Nor could the main engines of either capsule be restarted. They were used only at the end of the mission for the reentry braking maneuver, which was handled automatically by radio commands.

1960: BAD LUCK FOR THE SOVIET PROGRAM

In 1960, the Soviet program went through a dark period. The first two test craft of the Vostok program did not return to Earth. The first of these, dubbed 'Sputnik 4' in the West, was the Sputnik-Korabl-1, launched on May 15, 1960 with a dummy onboard. After four days of flight, the reentry cabin was separated from its service module and the retro-rockets fired, but something went wrong with the rocket's braking system and instead of reentering the atmosphere, the craft was wrongly orientated as the retro-rockets fired it into a higher orbit. Perhaps unsurprisingly, the Soviets claimed that they had never planned to bring it back anyway. The retro-rockets silently orbited the Earth for two years before burning up on reentry on September 6, 1962. A piece of space junk fell to Earth in Manitowoc, Wisconsin, in front of the Rahr-West Art Museum (Figure 2.7). The spacecraft carrying the dummy remained in orbit as an Earth satellite until October 1965, when atmospheric friction slowed it down enough to bring it back to a fiery reentry and a landing in the sea.

The problem was easily resolved and the Soviets tried again on July 28, 1960. But this second trial, carrying the two dogs Chaika (Seagull) and Lisichka (Little

Figure 2.7: Plaque commemorating the Sputnik crash in the street in front of the Rahr-West Art Museum in Manitowoc, WI. (Courtesy Wikimedia commons).

Fox, Korolev's favorite dog) was also unsuccessful, with the spacecraft disintegrating 19 seconds after launch and crashing into the steppe. [9] Flight controllers sent a command to jettison the descent module, but the parachute only partially deployed due to the low altitude and the two dogs were killed on impact with the ground. This event caused the whole launch escape philosophy to be reassessed. [10]

Korolev was very disturbed by what had happened and he immediately set a group of designers to work on developing an independent system aboard the capsule as an Emergency Rescue System (in Russian: Sistema Avariynogo Spaseniya, or SAS). Future cosmonauts would therefore be provided with the possibility of engaging the braking system themselves, if necessary. [11]

But the worst setback happened at Baikonur on October 24, 1960, with the pad failure that would become known as the Nedelin Catastrophe. This was by far the worst ever disaster in the history of rocketry.

THE NEDELIN CATASTROPHE

Numerous important officials – including Marshal Mitrofan Ivanovich Nedelin, the Commander-in-Chief of the Strategic Missile Forces – were in attendance at Tyura-Tam on October 24, 1960, to witness the first launch of the prototype of the R-16, the new strategic missile that was designed to replace Korolev's 'old' R-7. The new missile was intended to mark an historic turning point, as the first truly operational intercontinental ballistic missile in the Soviet Union and an effective and large-scale strategic deterrent against the United States.

As with the earliest ballistic missiles, the R-7 'Semyorka' (Little Seven) used kerosene and a cryogen such as liquid oxygen as fuel and, while powerful, was vulnerable to enemy attack because of its large launch pad and slow refueling procedure. It was "*a great booster but a poor weapon.*" [12] The propellant it utilized was not storable and was unsuited to missiles that had to be kept launch ready in firing position for any length of time.

The new R-16 intercontinental ballistic missile was intended to counter the Atlas missiles that the U.S. were deploying. [13] Designed by Mikhail Yangel, the R-16 used the "*more practical*" hypergolic fuel, a mix of unsymmetrical dimethylhydrazine as a bipropellant in conjunction with red-fuming nitric acid. Although commonly used, as they can be stored as liquids at room temperature, hypergolic propellants are difficult to handle because of their extreme toxicity and corrosiveness, and because they are very volatile and capable of exploding if even only slightly neglected. The hypergolic rocket motor has the advantage of being simple and reliable because it requires no ignition system, since the components of the fuel spontaneously ignite when they come into contact with each other.

Looking to score political points, Nedelin[7] pressured Yangel and his R-16 team to accelerate their timetable in order to have the rocket prototype ready for the forthcoming anniversary of the Bolshevik Revolution. The maiden launch was originally set for October 23, but a major propellant leak that evening forced it to be postponed until the next day, Monday, October 24. The management of the launch team proposed draining the highly toxic and corrosive propellants from the rocket, then flushing the tanks with inert nitrogen as per the procedure to 'safe' the rocket.

On the orders of Marshal Nedelin, however, who was apparently under pressure from Moscow, all the repairs to the missile were carried out in a fully-fueled state and without stopping the prelaunch operations. This violated all the basic safety rules and created a remarkably dangerous situation at the pad. The repairs were almost completed successfully overnight and everything was proceeding as planned, until several technical difficulties apparently arose as the time of launch approached. Nedelin ordered a driver to take him to the launch pad, where he wanted to supervise matters personally and "*figure out what's going on.*" Against all safety procedures, which prescribed that all non-essential personnel should leave the area during prelaunch operations, approximately 200 officers, engineers and soldiers were around the launch pad, including Chief Designer Yangel and Marshal Nedelin himself, who scoffed at suggestions that he should leave the pad.

As the commission members arrived around the launch pad, supervisor Konstantin Gerchik ordered a chair for Nedelin – ignoring safety rules again – so that he could sit within 15–20 meters of the rocket! The presence of Nedelin and his entourage created a sense of tension among the engineers and military personnel involved, while multiple final tests were being conducted at the same time.

---

[7] Nedelin was described as a respected commander, a very thorough and careful individual and extremely cautious in judgements and actions. (Siddiqi [1994], p. 39.)

Thirty minutes prior to the set launch time, shortly after 6:45 pm, the inevitable occurred, probably when a technician plugged the umbilical cable from the first stage into the receptacle for the second stage, triggering the ignition followed by an enormous explosion of the fully-fueled rocket and its 500 tons of propellant.

A giant fireball, up to 120 meters in diameter, engulfed launch pad 41, with the fire and heat increasing in intensity as all the propellant ignited. Within seconds, the rocket broke in half and fell on the pad, crushing anyone who might have still been alive. Eyewitnesses described horrifying scenes of people burning alive or being vaporized altogether and reduced to ashes while running from the rocket across asphalt that was melting under their feet. Those who did not burn were suffocated by the poisonous propellant fumes released by the inferno. Nedelin himself would eventually be identified only by a pin that was attached to his uniform. Powerful explosions continued for about 20 seconds and the resultant fire raged for two hours. The flashes of light could be seen from as far away as 50 kilometers from the pad.

Yangel survived the catastrophe by sheer chance. A few moments before, he had been invited into a safe bunker for a cigarette break. Dozens of soldiers, specialists and technical personnel were not so fortunate. [14] Many of the USSR's spaceflight pioneers perished in the accident, including Aleksandr Nosov, who had pushed the launch button for Sputnik three years earlier, and Evgeny Ostashev, who had been instrumental in developing the Sputnik booster. In Leninsk, streets named after Nosov and Ostashev can be found among the usual 'Marx', 'October' and 'Red Army' street names.

Figure 2.8: The R-16 rocket explodes on the launch pad. (Courtesy Russianspaceweb.com)

Figure 2.9: Wreckage of the rocket and the launch site structures in the aftermath of the R-16 explosion. (Courtesy Russianspaceweb.com)

According to the official report of Artillery Major General Grigory Yerofeyevich Yefimenko, 74 people were killed on the launch pad (57 military and 17 civilians) and 49 injured. With 16 more people later succumbing to their injuries, the official death toll rose to 90. The bodies of two soldiers were also found outside the perimeter of Site 41 after the official list of victims was submitted, bringing the number of dead to 92 (74 military and 18 civilians). In his book, Boris Chertok reported that 126 people had died, while other sources have cited the total number of deaths as high as 165 or 180. [15] Upon hearing the news of the accident, Premier Khrushchev imposed total secrecy over the entire incident and directed Leonid Brezhnev to head to Tyura-Tam immediately with a group of experts to investigate.

The investigative commission, led by Brezhnev, concluded that no one was to be punished for the incident, because *"All the guilty had been punished already."* Nobody learnt anything about the terrible catastrophe and the *Pravda* newspaper, the official mouthpiece of the Communist Party, reported that Marshal Nedelin had been killed in an aircraft crash. He was given a hero's burial in the Kremlin Wall.

European journalists in Moscow soon picked up rumors of a *"gigantic rocket explosion in Siberia"*, killing hundreds, but those stories were soon dismissed alongside other oft-embellished legends of dead cosmonauts, super weapons and similar folklore. [16] The truth about the accident would emerge in bits and pieces over the years. The tragedy was first heard about in the West in the 1970s, and as the story surfaced during a period favorable for the launch of Mars probes, analysts were at first confused about the identity of the rocket that blew up, considering it to be a failed Soviet attempt to send a space probe to Mars. It was only in the 1990s that the full account was revealed, and it became clear that the tragedy had nothing to do with Mars exploration.

Figure 2.10: Marshal Nedelin's burial in the Kremlin Wall.

The first published account of the disaster appeared in April 1989 in the popular Soviet pro-glasnost weekly magazine *Ogonek*, under the title "*Sorok Pervaya Ploshadka*" (or "*Site 41*" in English), authored by Aleksandr Bolotin, who was one of the young officers at the time who witnessed the accident and miraculously survived. "*At the moment of the explosion,*" Bolotin reported, "*I was about 30 meters from the base of the rocket. A thick stream of fire unexpectedly burst forth, covering everyone around. Part of the military contingent and testers instinctively tried to flee from the danger zone. People ran to the side of the other pad, towards the bunker … but on this route was a strip of new-laid tar, which immediately melted. Many got stuck in the hot, sticky mass and became victims of the fire … The most terrible fate befell those located on the upper levels of the gantry: the people were wrapped in fire and burst into flame like candles blazing in mid-air. The temperature at the center of the fire was about 3,000 degrees. Those who had run away tried while moving to tear off their burning clothing, their coats and overalls. Alas, many did not succeed in doing this.*"

By coincidence, on the same day three years later, a fire at a launch pad killed another seven testers. In the wake of these two accidents, October 24 has become known as "*a black day*" for space exploration and Russian officials commemorate the memory of all those who dedicated their lives to the space program. Even though the Nedelin Catastrophe had nothing to do with the space program in itself, space officials do not schedule any launches for that October day.

84 Man in Space

FURTHER YEAR-END MISADVENTURES

On December 1, 1960, the Sputnik-Korabl-3 mission (also known as Sputnik 6) also failed. This was a test flight of the Vostok spacecraft, carrying the two dogs Pchelka and Mushka aboard. The flight was intended to last one day and it was planned to recover the spacecraft after 17 orbits around the Earth. Due to a malfunction in the retro-rockets, the spacecraft did not reenter until one-and-a-half orbits later than planned, resulting in the spacecraft reentering the atmosphere on a trajectory which could allow non-Soviet parties to recover and inspect it. To prevent this, the reentry sphere containing the dogs was deliberately destroyed by activating an automatic self-destruct mechanism. [17] Soviet official media announced that because *"the descent went along an unplanned trajectory, the satellite spaceship ceased to exist when entering the dense layers of the atmosphere,"* suggesting that the spacecraft had been destroyed because of overheating.

Figure 2.11: Unusually, on the first anniversary of the ill-fated Sputnik 6 mission, a commemorative cover was produced. It was canceled with a special postmark in the Main Office in Moscow.

The following day, Korolev was taken to hospital having suffered a heart attack. Once he was in hospital, doctors realized that he was also suffering from a serious kidney disorder – an illness which often resulted from detention in Soviet prisons and camps. Korolev was warned that if he continued to work at the same pace as before after his release from hospital, it would be the equivalent of a death sentence.

But taking a long convalescence at this point would probably have meant failing in his attempt to beat the Americans in the race to put a man into orbit. The prospect of being second in that race was not in itself of great concern to Korolev, because he knew that it would be difficult, if not impossible, to maintain the lead in space indefinitely. However, he also knew how Khrushchev would likely react to the loss of leadership in space. The Premier would have stopped releasing precious military materials to him and would have cut his financial backing, so that Korolev would have been able to do very little with what was likely to be the short time he had left to live. Three weeks later, Korolev was out of hospital and working more furiously than ever before, trying to make up for lost time. [18]

Unfortunately, testing problems were not yet over, because a follow-up launch to Korabl-Sputnik-3 also failed, on December 22. The dogs Damka (Little Lady) and Krasavka (Beauty) were launched aboard a modified booster (the 8K72K) that was to be used for manned Vostok launches, but the spacecraft failed to reach orbital velocity due to a malfunction of the third stage. [19] The mission was aborted, as planned in the event of an unscheduled return to Earth. In fact, the craft was programmed to eject the dogs and self-destruct, with the self-destruct mechanism being set to a 60-hour timer. However, the ejection seat failed and the two dogs hit the ground in the Siberian frost after a rough ballistic reentry. The temperature on the ground was minus 40 degrees C and it seemed unlikely that the dogs would survive.

A team was immediately sent out to locate and recover the capsule and they reached it in deep snow at the end of the first day, too late to disarm the self-destruct mechanism and open the capsule in the remaining daylight. The team reported that the window was frosted over and that they had detected no signs of life. The following day, the rescuers opened the hatch and could hear quiet barking. They found the animals alive, despite the shock of crashing to the ground and the cold they had endured since the landing. Due to the failure of the ejection system, the two female dogs had been trapped inside the capsule, where they were protected from the freezing Siberian winter weather[8]. The two frozen, exhausted little dogs were wrapped in sheepskin coats and flown to Moscow alive. All the mice aboard the capsule were found dead due to the cold.

---

[8] Krasavka, also nicknamed Kometka (Little Comet) would be a space hero on her third mission, when she tested the forerunner of the spaceship that would be used for Yuri Gagarin's flight. Krasavka/Kometka was then adopted by Academician Oleg Gazenko – one of the leading scientists behind the Soviet animals in space programs – and remained with him for 14 years. Despite her Siberian adventure, the dog would still have puppies. The incredible story remained a secret for more than 50 years and was only revealed in May 2013 by Kate Baklitskaya as *"The remarkable (and censored) Siberian adventure of stray dog cosmonauts Comet and Shutka"* in *The Siberian Times* (www. siberiantimes.com). Korolev wanted to make the story public but was prevented from doing so by state censorship. As a result, the heroic adventure of Damka and Krasavka, surviving a space failure, is not properly recorded in many histories of space animals.

86 Man in Space

THE AMERICANS REGAIN CONFIDENCE

Following a failed attempt on November 21, 1960, NASA successfully launched the first mission of the Mercury program – Mercury 1A – into space on December 19. The mission tested some of the flight controls and the operations involved with launching, tracking and recovering a spacecraft for the first time.

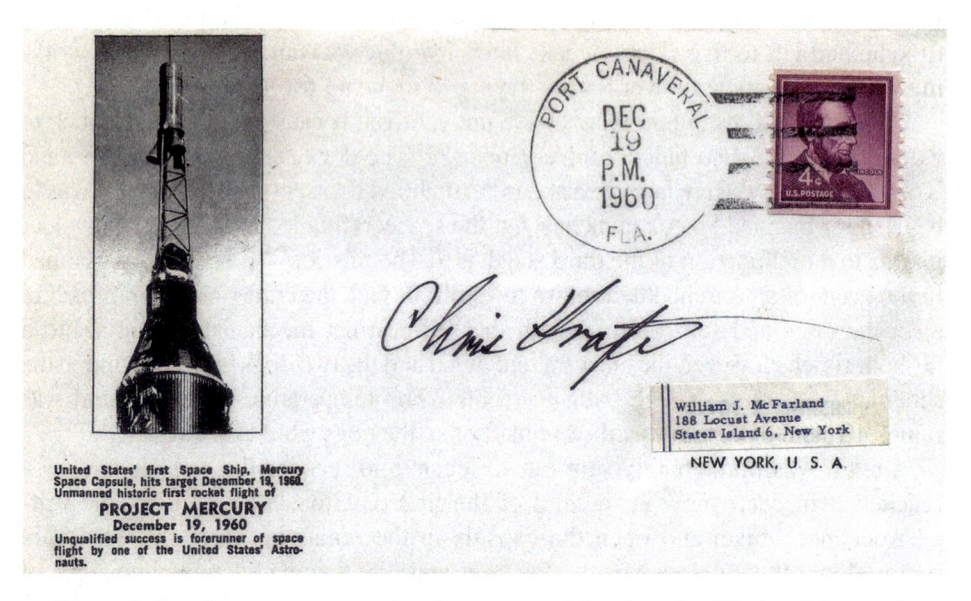

Figure 2.12: Cover commemorating the unmanned first launch of Project Mercury in December 1960.

The spacecraft achieved a maximum altitude of 210 km and a maximum velocity of just under 8,000 km/hour, in a flight that lasted 15 minutes and 45 seconds. Fifteen minutes after landing in the Atlantic Ocean, the recovery helicopter picked up the spacecraft. The mission was a complete success, one which the Americans were confident a live occupant would survive.

Following the success of the mission, the next step was to test out the spacecraft with a living organism inside. Forty days later, on January 31, 1961, in preparation for the first American astronaut's journey into space, Mercury-Redstone 2 (MR-2) was launched carrying a chimpanzee named Ham (in honor of the Holloman Aerospace Medical Center), chosen from a colony of six 'astrochimps'. The mission was intended to test several new designs in the Mercury spacecraft, including the Environmental Control Systems (ECS) and a pneumatic landing bag intended to absorb much of the impact shock when the returning capsule hit the water. The tests would help to confirm that humans could safely make the trip. Researchers decided to send chimpanzees into space because their organs and skeletal structures are similar to those of humans, and chimps can be trained. [20] During the flight, Ham experienced 6.5 minutes of weightlessness and performed well. He moved levers in response to flashing lights, just as he had done in the laboratory, and his response times in space were as good as on Earth.

Figure 2.13: In the medical laboratory, Ham was identified as 'Subject 65' until after his safe recovery following the mission. Courtesy NASA.

However, a faulty valve feeding too much fuel into the Redstone's engine caused the flight to overperform, and MR-2 achieved a velocity of about 9,400 km/hour (as opposed to the intended 7,000 km/hour) and an altitude of 157 miles instead of the planned 117 miles (approximately 250 km rather than 185 km).

Ham's trip took two minutes and 24 seconds longer than intended (16 minutes and 39 seconds total flight time) and reentered the atmosphere at an incorrect angle and higher than intended speed, with a peak g-load during reentry of 14.7, almost 3G more than planned. The spacecraft came down with such a force that the heat shield punctured the capsule and water began to enter the cabin. When MR-2 splashed down, there were no rescue ships in the vicinity because the craft landed some 60 miles (100 kilometers) from the nearest recovery

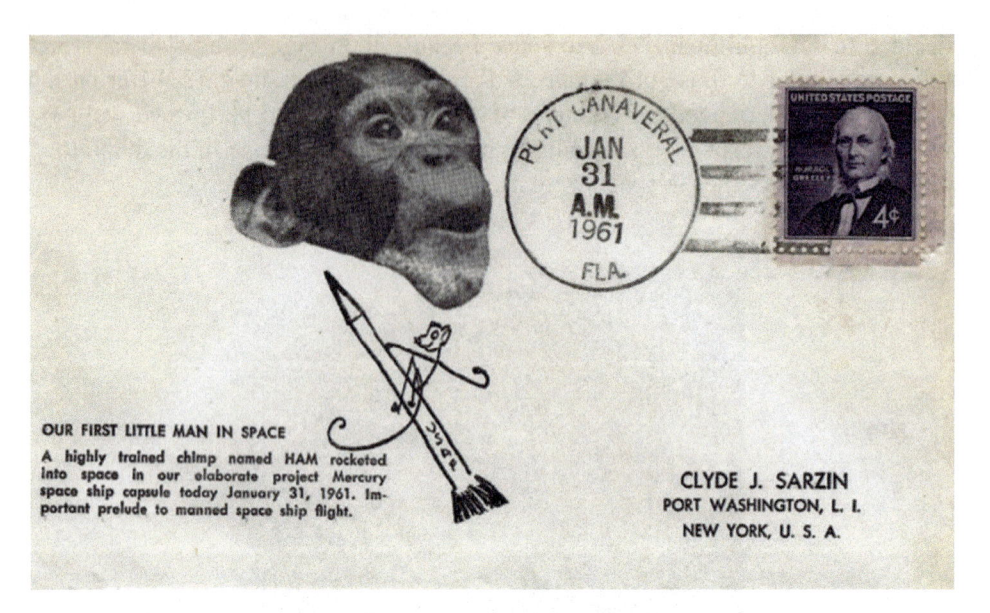

Figure 2.14: Cover commemorating the launch of the Mercury-Redstone 2 mission, carrying the chimpanzee Ham.

vessel, the destroyer USS Ellison. A P2V search plane located the capsule about 27 minutes after splashdown. Helicopters were dispatched from the USS Donner but it would be a further two hours before the USS Ellison arrived. When the spacecraft was opened, Ham appeared to be in good condition and readily accepted an apple and half an orange. Ham eventually passed away on January 19, 1983, at the age of 26.

The heat shield and landing bag mechanism were quickly redesigned to be able to withstand stronger impacts. A few weeks later, on February 21, 1961, the Mercury-Atlas 2 (MA-2) mission – using a more powerful, modified Atlas rocket that would eventually be used to take American astronauts into orbit – was launched from Cape Canaveral, Florida. The Atlas was designed to launch payloads into low Earth orbit, and NASA had already used it as a space launch vehicle in 1958 for Project SCORE, the first communications satellite that had transmitted President Eisenhower's pre-recorded Christmas speech around the world. Atlas had also been used for all three robotic lunar exploration programs and for the Pioneer planetary probes.

The goal of the MA-2 mission was to check maximum heating and its effects during the worst-case reentry design conditions. The trajectory was designed to provide the severe reentry heating conditions that might be encountered during an emergency abort of an orbital flight attempt. All the objectives of the mission were fully completed.

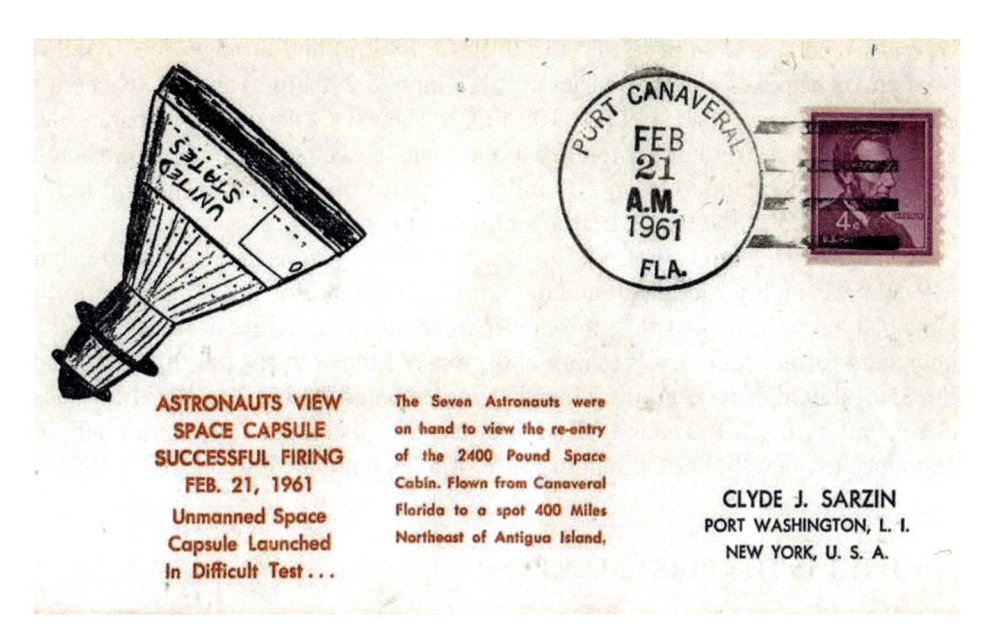

Figure 2.15: Cover commemorating the launch of the Mercury-Atlas 2 mission.

HOW THE AMERICANS LOST THE RACE TO PUT A MAN IN SPACE FIRST

Despite the successful MA-2 mission, numerous technical malfunctions on previous flights, especially Ham's flight, led to a reassessment of the reliability of the overall system. The scrupulous attention to detail of the NASA Space Task Group (STG) regarding reliability led to requests for significant modifications. During the first two weeks of March 1961, seven technical changes were implemented in the Mercury-Redstone capsule-booster combination, after which one final test was requested before the vehicle would be trusted to carry a human pilot.

Although Alan Shepard, as an experienced test pilot, was convinced that he could handle and overcome any problems that might arise similar to Ham's mission, Wernher Von Braun himself opted for inserting another unpiloted flight into the launch schedule between the MR-2 and MR-3 missions. Shepard was privately furious, seeing the decision as a costly mistake that would lose America the opportunity to beat the Soviet Union into space. "*We were furious,*" recalled Chris Kraft, who was serving as a flight director within Project Mercury at that time. "*We had timid doctors harping at us from the outside world and now we had a timid German fouling our plans from the inside.*" [21]

Shepard's launch was postponed until the end of the month and on March 24, 1961, the MR-BD, or Mercury-Redstone Booster Development mission, lifted off from Launch Complex 5 at Cape Canaveral. The mission lasted 8 minutes and 23

90 Man in Space

seconds, reaching an apogee of 113.5 miles (183 km) and a range of 307 miles (494 km). The peak velocity was about 5,123 mph (8,245 km/h) and the spacecraft experienced a peak load of 11G (108 m/s^2). There was no plan to separate the Redstone rocket from the Mercury boilerplate spacecraft and they impacted together in the Atlantic Ocean 307 miles (494 km) downrange, some five miles (eight km) short of the target, before sinking to the bottom.

In the Soviet Union, the mission was erroneously considered a failure, but MR-BD was highly successful and demonstrated that all the major booster problems had been eliminated. The mission marked the completion of the tests of the launchers for the Mercury Program and prepared the way for the flight of Alan Shepard aboard MR-3, as the changes implemented had solved all the issues uncovered by the MR-1A and MR-2 flights. Now, the Americans were ready to announce the first flight of a man in space officially, planned for April 28, 1961.

A SOVIET IS THE FIRST MAN IN SPACE!

The response from the Soviets did not take long to arrive. On April 12, 1961, the news made front page headlines in newspapers around the world: Yuri Alexeievich Gagarin, the Soviet cosmonaut, was the first to make an orbital voyage around the Earth, aboard the Vostok spacecraft. Once again, the Soviets had beaten the Americans.

Figure 2.16: Mission emblem for Yuri Gagarin's historic Vostok flight.

Unlike what had happened with Sputnik, while the first human flight in space by Gagarin was indeed a deep humiliation for the United States, it was not a big surprise. As early as May 1960, the USSR had been testing prototypes of Vostok capsules in orbit. In February 1961, Premier Khrushchev had proclaimed that "*the time to launch the first man in space is close.*"

On the day of the mission, Gagarin and his back-up, Gherman Titov, had been awakened at 05:30. They had 'space food' for breakfast, which was followed by routine medical checkups. They were then assisted into their spacesuits. One of the onlookers in the dressing room semi-jokingly suggested that upon landing in his futuristic outfit, Gagarin could be mistaken for the pilot of an American spy plane like the one that had been shot down over the USSR the previous year. The idea was taken seriously and officials made the urgent decision to paint CCCP (USSR) on the front of Gagarin's helmet in big red letters, in order to establish his identity. A number of photographs showing Gagarin in his helmet before and after the letters were painted confirm the authenticity of this story. [22]

Gagarin and Titov were then transported to the launch pad. Folklore has it that mid-way to their destination, the bus made a stop in the middle of the steppe, letting Gagarin out to relieve himself onto a rear tire of the transport van through the suit's urine tube, thus establishing a 'good luck' ritual for all those who would follow. Gagarin entered the Vostok 1 spacecraft at 07:10 local time and the hatch of the spacecraft was closed. It was soon discovered that the seal was not complete, however, so the technicians spent nearly an hour fixing the problem. During this time, Gagarin requested some music to be played over the radio. Chief Designer Sergei Korolev was very nervous in the control center, but Gagarin was described as "*calm.*" About 30 minutes prior to launch, his pulse was recorded at 64 beats per minute.

In 108 minutes, Vostok 1 completed a single orbit around the Earth, reaching a speed of 27,000 km/h – a speed that no human had ever experienced before – and a maximum distance of 327 km from the surface of the planet. For the first time, Gagarin would enjoy the experience of weightlessness and the spectacular view through the spacecraft's small viewport. He would also be the first to learn that in microgravity, any unsecured item swiftly and inevitably migrates to the most inaccessible and inconvenient place. TASS released its first bulletin while he was still in flight.

Public congratulations promptly arrived from President John F. Kennedy and Wernher Von Braun, conceding defeat. Alan Shepard complained that he had missed the opportunity to beat the Soviets on March 24, muttering over and over again: "*We had them by the short hairs, and we gave it away.*"

Gagarin instantly 'skyrocketed' to fame and became both an international celebrity and the emblem of the Soviet space program. A bronze bust of him was

92 Man in Space

Figure 2.17: The world's newspapers react to Gagarin's history-making flight.

soon unveiled in Moscow and posters showing his portrait were distributed to the cheering crowds. A new Soviet postage stamp was issued with his likeness[9] and the Soviet propaganda machine began to make him into a demi-god. Soon, Soviet daily TV programs began a tradition of starting off with the announcement "*Poyekahli!*" ("*Let's go!*" the words spoken by Yuri Gagarin at launch). In honor of his extraordinary enterprise, the date of April 12 became a Soviet public holiday, to celebrate Cosmonautics Day. But it had been a hard road for Gagarin to achieve all this.

For propaganda purposes, it was important to the Soviet leadership that the first cosmonaut candidates had a "*clear*" relationship with the Party and an unblemished past. Gagarin fulfilled all the criteria and soon rose to the top of the pile. Doctors and psychologists were impressed by his mental toughness, fantastic

---

[9] Yuri Gagarin is one of the historical figures featured most often on postage stamps from different countries. According to collector expert Peter Hoffman, as of April 2018, he has appeared on more than 605 stamps worldwide.

Figure 2.18: (left) The first stamp issued in Russia to celebrate Gagarin's flight in April 1961. (right) One of the most recent Gagarin stamps was issued in Italy in 2011, at the request of the author, for the 50th Anniversary of the historical flight.

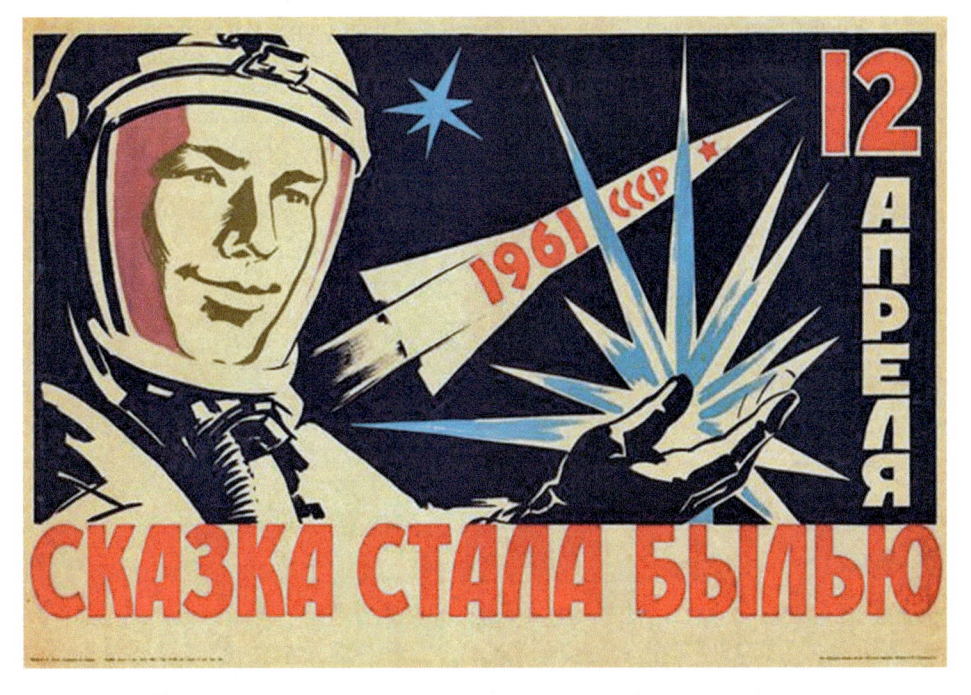

Figure 2.19: A few days after Gagarin's flight, dozens of posters and postcards were put up in cities and villages, and distributed to the cheering crowds. This image shows the popular poster "The Fairy Tale Became Truth", designed by Boris Staris and published by Molodaya Gvardia (The Young Guard). The poster depicts Gagarin as a modern-day Prometheus – the Greek god who gave fire to man.

Figure 2.20: (left) Educational poster "Glory to the Party's Son!" a photomontage poster by the artist Boris Berezovsky and photographer S. Raskin, produced by Izogiz, Moscow. (center) "A Soviet Man is in Space!" designed by the Lesegri Collective in 1961. (The Lesegri Collective (in Russian Лесегри), was formed in 1957 by a team of three Soviet graphic artists: Boris Lebedev, Leonard Sergeev and Mark Grinberg. They worked for the Znanie (Knowledge) Publishing House and its popular science magazine. The artists designed posters, stamps, illustrated postal envelopes and postcards for the Communications Ministry. One of the most well known is the stamp designed to celebrate the spacewalk of Alexi Leonov in 1965 (see Chapter 4, page 172, Figure 4.17). (right) "Glory to the Communist Party! Glory to Our People, the Conqueror of Space!", a poster by Viktor Dobrovolsky, issued by Sovetskaya Rossiya, Moscow.

Figure 2.21: (left) "Glory to the Soviet People, the Conquerors of Space!", a poster by A. and E. Kruchina, printed by Voyenizdat Moscow just ten days after the flight. (center) "Glory to the First Cosmonaut Yu. A. Gagarin!" designed by the artist Valentin Viktorov. (right) "The Road to the Stars is Being Built by Communists!" a photomontage poster based on the iconic Gagarin-Khrushchev photograph taken by journalist V. Smetanin. Issued by Izogiz, Moscow.

Figure 2.22: There is much symbolism in the postcard "Peace for the World!" of which 380,000 copies were distributed. Artist Irakli Moissejewitsch Toidse features a worker holding world peace in his hands and who seems to be releasing the rocket as if he had built it himself.

memory, wide-ranging attention to his surroundings, well-developed imagination and quick reactions: *"He prepares himself painstakingly for his activities and training exercises, handles celestial mechanics and mathematical formulae with ease [and] excels in higher mathematics; does not feel constrained when he has to defend his point of view if he considers himself right; [and] appears that he understands life better than a lot of his friends."* [23] The final selection for Vostok 1 would be tortuous, however.

The nation's first spaceman would become a national icon and represent the Soviet Union with honor. As he would play an important political role, the candidate chosen would be required to show that he was not simply a *"Russian by passport"*, but that both his parents and all his grandparents were Russian as well. That was why the candidatures of, for example, Popovich who was Ukranian, Nikolayev who was a Chuvash, and Bykovsky who was half Ukranian, were immediately rejected for the first flight. They would be used for the subsequent flights, to demonstrate the *"friendship of peoples"* in the USSR, but the first man in space had to be one hundred percent Russian.

96 Man in Space

Figure 2.23: The 'Vanguard Six' or the 'Sochi Six' in the picture, taken at the Black Sea resort of Sochi in May 1961, a few weeks after Yuri Gagarin's history-making flight. The six cosmonauts officially involved in the Vostok program were the top picks of the first class of 20 space pioneers; the best and boldest of their nation, the ones destined to ride the first manned missions. In the original picture (above), sitting in front from left to right: Andriyan Nikolayev, Yuri Gagarin, chief designer Sergei Korolev, Yevgeny Karpov (Director of the Cosmonaut Training Center) and Nikolai Nikitin (parachute jumping instructor). Standing in the second row, from left to right: Pavel Popovich, Grigory Nelyubov, Gherman Titov and Valery Bykovsky. In the picture below, the image of Nelyubov has been artfully airbrushed out after he had been discharged from the cosmonaut corps for disciplinary reasons. Many versions of this doctored picture remain in circulation. (Courtesy Wired.com)

Among the first group of twenty to go through the training, there were several candidates that would have satisfied this requirement. It was originally supposed that Cosmonaut No. 1 would be Alexei Leonov, a very skillful and capable man, a brilliant pilot and a professional parachutist. But Leonov was more heavily built than Gagarin, and after the disaster with Dolgov (or whoever it was), Korolev's options were narrowed. In practice, there remained only two possible candidates: Gagarin and Titov, although another possibility was Vladimir Komarov, who was only slightly larger than these two. Unlike Titov, Gagarin's name was indisputably

Russian, whereas Gherman Titov suggested a 'Germanic' heritage (although that was not the case).

A second major factor that came into effect for the selection was that the future cosmonaut had to be of genuine "*proletarian*" origins. Khrushchev planned to organize an international propaganda tour for the cosmonaut after the flight and he wanted to stress the point that only the Soviet system could provide the son of a worker or a peasant with a ticket to outer space. According to Kamanin, Titov in some respects was the stronger candidate and was certainly better educated, but Gagarin had the advantage over Titov, and even more so over Komarov: He was born in a village and had "*impeccable*" peasant parents. Titov was also one hundred percent Russian, had also been born in a village and was even shorter than Gagarin. But he was also the son of a village teacher, an "*intellectual*", and that did not fit in as well with Khrushchev's ideal. For Komarov, the prospects were not good at all on these criteria. He was also the son of an intellectual, but he lived in a town and was an engineer himself. So Gagarin got the nod for the first flight, with Titov as his backup.

As Kamanin would reveal in his posthumously-published diaries, consideration had to be given to the possibility that selection to the flight could be a death sentence, since nearly everyone believed that the first cosmonaut would have a roughly 50-50 chance of survival[10]. After all, two of the three spacecraft launched so far, prototypes of the future Vostok, had failed to return to Earth: one had remained in orbit and the other had burnt up in the atmosphere. On top of that, there was the ejection procedure to consider, for which one test pilot with the highest qualifications had already lost his life. [24]

Gagarin Kiev postmark

Because of the secrecy surrounding every aspect of the Soviet space program, as previously mentioned, collectors were unable to prepare their commemorative covers or cancels until those prepared for the ASTP program in 1975. But there was one exception: the so-called 'Kiev postmark' dedicated to Gagarin's space flight on April 12, 1961. This is *the first – and probably only – Soviet space postmark which was actually used on the exact day of the flight*. This came about because it had been known for some time that a Soviet cosmonaut would fly into space (it was actually expected for February), so the cancellation was produced as early as January 1961, without defining a date, pending finding out the exact day of the launch. [25]

(continued)

---

[10] Chertok summarized the annotation that Kamanin put in his diary on April 5: "*So, who will it be – Gagarin or Titov? It is difficult to decide who to send to certain death…*" (see Chertok [2009], p. 71).

98 Man in Space

(continued)

> This special postmark – designed and produced in the Kiev area by an artist named A. Levin on the initiative of the local Philatelic Club – features a meridian globe with a Soviet worker in overalls standing behind it. The worker is holding up a large space rocket with a Soviet star on its side. The text encircling the globe and the worker reads, in Ukranian: "Triumph of Soviet science – man in space!" Two lines of text in front of the globe identify the postmark date of April 12, 1961, and the postal facility that used the postmark, i.e., the "Kiev Post Office".
>
> When the news that Gagarin was flying was broadcast, the club managed to have the date April 12 engraved in the cancel, and obtain authorization for it from Moscow, in just a few hours.
>
> The special postmark could therefore only be used officially at the post office in Kiev for a relatively short time, in the afternoon of April 12. Only black ink was used and covers that have been genuinely cancelled are quite scarce. The canceling device was improperly used in the following days, however, with different colors including red. All of these have to be considered as favor and back-dated covers. Favor cancels, created after the date in the postmark, were not an uncommon practice in the USSR. Some covers

Figure 2.24: The well-known 'Gagarin Kiev postmark' was the first Soviet space postmark which was actually used on the exact day of the flight.

A Soviet is the First Man in Space!

(continued)

also exist postmarked in gold, but they are probably memorabilia made by the official Soviet foreign trade organization Mezhdunarodnaya Kniga (also known as 'Kniga' for short). They were responsible, among other things, for creating and distributing Soviet philatelic products outside the USSR. Real gold dust was used for this postmark. Since the cancellation made with this material was probably not successful, only eight covers of this type are known. The covers bear the 3-kopek Gagarin stamp (which was not issued until April 17) and they are therefore clearly backdated.

The fact that this special postmark was quite scarce and not available on the philatelic market is confirmed by the several attempts made to counterfeit it. Cacka lists three different types of fakes, with details on how to identify them. [26]

The Moscow Gagarin Special Postmark

Even though the Moscow post office probably had their own postmark designed and ready, and had huge resources and political clout, they could not finalize their commemorative mark for the Gagarin flight until the following days. In fact, their special postmark was used in Moscow only on April 13 and 14, 1961, even though the cancel had been engraved with the fixed date of April 12. [27] Two versions of this postmark are known (see Figure 2.25), which differ from each other with small detail variations in their design, the main differences being the graphics and the shape of the fonts in the word "CCCP" (Russian for "USSR").

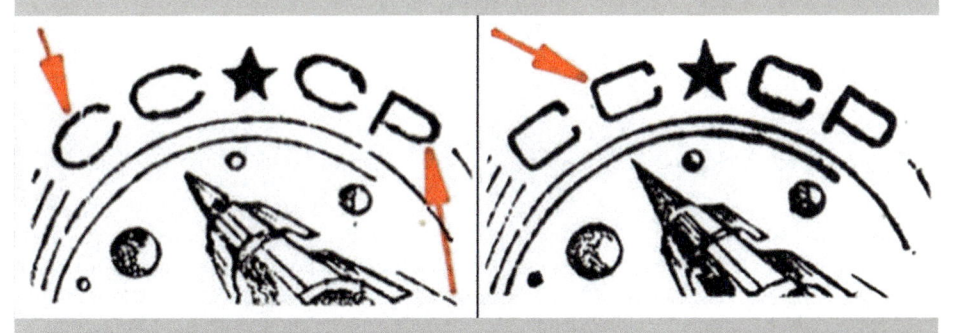

Figure 2.25: The two versions of the Gagarin special postmark; Type I (left) and Type II (right).

(continued)

> The Type I cancel (Figure 2.25 left), used in the main Moscow post office has rounded letters. The Type II cancel (Figure 2.25 right) used at the Moscow 'K-9' post office branch[11] has edged letters and was used on April 13 and 14 in 'really run' correspondence.
>
> After being used in the two post offices, these two cancelling devices were transferred to the 'Kniga' trading company for the subsequent production of commercial postal items. The Type I cancellation was used for the production of first day covers carrying the 3-kopek Yuri Gagarin stamp, although this, as mentioned previously, was only issued on April 17, three days after the official usage of the cancel. The Type II cancel was also used on this stamp, against postal regulations, and this has caused some confusion, as some of those envelopes may have been subsequently offered to collectors as 'really run'.

Figure 2.26: Backdated Kniga cover, with the Gagarin postmark used on April 29, 1961.

---

[11] The K-9 branch was the Moscow post office that provided postal support to the offices of the Moscow City Society of Collectors (abbreviated MGOK) and the national All-Union Philatelic Society (abbreviated VOF), making it the most convenient place for those organizations to get their commemorative covers postmarked (see Reichman [2013], p. 62.)

Postmark duplicates and covert meaning of the design variations

For almost every postal canceling device for space topics (and others), a duplicate exists. Normally, the two postmarks are not identical but, as seen here in the Gagarin Moscow cancel, they sometimes only differ in small details. Certainly, the Soviet engravers had the skills to make two canceling devices that were so similar in design that the cancellation images from both were virtually indistinguishable. So why are they often different? This is another dark side of the usage of Soviet special postmarks.

The different canceling devices were used in different places, for example in the Main Post Office in Moscow (or later at the Baikonur Cosmodrome) for normal postal services, and at the Kniga offices for producing philatelic items for the foreign market. Under special circumstances, three or more duplicates could exist, used in different places including temporary postal facilities. The most likely hypotheses are that the design changes were required so that someone would be able to determine specifically where an item was cancelled after the fact. This was probably driven by the security organizations or administrative reasons of the former Soviet Union. [28]

This was certainly not the only time that the Soviet security organizations had covertly created postal markings so that they could keep track of the mail processing. Specifically, in the 1950s and 1960s, an international mail hand-stamp was applied to almost all envelopes entering or leaving the USSR. This was called the MEZhDUNARODNOE mark. A typical use of this hand-stamp can be seen in Figure 2.27, featuring a close-up scan of the hand-stamp image. At the time, everyone thought these were just a mail marking to help speed the international mail through the Soviet postal system.

It was not until after the breakup of the Soviet Union that these imprints were identified as clandestine marks used to indicate that the envelope and its contents had been checked by a mail censor. Many of these hand-stamps were made, enough so that there would be one for each censor. Each censor's hand-stamp design was similar but had deliberate design modifications to make each one unique (using similar techniques to those described for the postmark mentioned previously). These design variations included small changes in the font style and size, the overall size of the hand-stamp image, and carefully-crafted breaks or gaps in the outline, in order to identify precisely which censor had examined a particular envelope. Supervisors would then randomly double check to see if the proper mail censoring had been performed. If not, the supervisor (or anyone else who needed to know) could readily determine which censor had failed to perform his/her job by looking at the international mail hand-stamp design. Such preemptive precautions were somewhat analogous to the Security Agency's collection of so many telephone records just in case they needed to go back, at some later time, to determine when and where events had occurred.

(continued)

(continued)

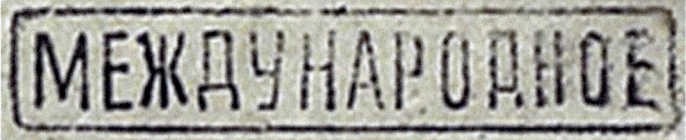

Figure 2.27: Cover bearing the International mail hand-stamp (seen in close-up at the bottom of the image).

"BAIKONUR" AND SOVIET LIES

Even after the USSR had launched Sputnik in 1957, there had been no need to make any reference to the location where the satellite had launched from, to maintain secrecy. But when Gagarin was launched into space in April 1961, the Soviets wanted to homologate the flight with the International Aviation Federation as a world record for height and distance. In order to qualify – according to the international standards for registration of a record – they would have to specify the starting point, flight itinerary and landing location of the mission. However, the Kremlin leaders wanted to maintain maximum secrecy about the location of their launches, near the village of Tyuratam in the Kazakh steppe (despite its discovery by American U-2 spy planes in June 1957, with the CIA naming it Tyura-Tam, based on a German Nazi map of 1939)[12]. The Soviets therefore did not mention this village, nor the unknown remote village of Kzyl Orda, which was the actual postal district to which the secret Tyuratam facility belonged (and whose address, we now know, was designated with the codename 'Company Kzyl-Orda-50').

Attempting to maintain the secrecy of the launch facility, the Soviets fictitiously used the name of Baikonur, a village known for its copper and coal mines, which was located some 370 kilometers to the north-east in the province of Karaganda, along the trajectory followed by most of the missiles on their way into orbit[13].

The true location of the future cosmodrome was chosen following a reconnaissance mission north of the Syr-Darya river in 1954. It was a desolated area in which, should a rocket go awry on launch, there was a reasonable chance that it would land in Soviet territory and thus avoid revealing the secrets of Soviet technology to foreign powers. Its coordinates (46°N and 63°E) were kept as a military secret. It was an area in the middle of a vast emptiness in the Kazakh steppe known in the Kazakh language as "Töre-tam" ("Töre's grave") after Töre-Baba, who was the local nobleman and a descendent of Genghis Khan. The name was spelled "Tiura-Tam" on the Nazi maps prepared around 1939. When the Soviets arrived in 1955, it was a forgotten place with an isolated railway stop on the important route linking Moscow and Tashkent that was simply called 'the water pumping station' after the old water pump that served to refill steam locomotives. Within a few years, the railroad stop had grown to house mechanics that helped with the railroad, and a small settlement appeared with a couple of two-story houses for the railwaymen, a couple of dozen small, mud-plastered houses, and the tents of geologists prospecting for oil.

The tsars had also used the location as a place of exile for undesirable citizens. One of those who happened to be banished here in the late nineteenth century was Nikifor Nikitin. Ironically, he had been sent to this gulag for *"his seditious plans for a flight to the Moon!"* [29]

In his book, Alexei Leonov recalled: *"Conditions at the Baikonur Cosmodrome, in the barren steppes of Kazakhstan in Central Asia, were enough to test the endurance of any human being in the early days of spaceflight. The desert, where the 2000 square-mile complex was located, swarmed with scorpions, snakes and poisoning*

---

[12] See *A Tale of Two Baikonurs*, Jim Reichman, AD*ASTRA #31, December 2016, pp. 15–21. Based on the findings of the CIA, President Eisenhower directed that Western leaders should be briefed privately on what the U.S. knew about the Tyura-Tam launch facility. The general public, however, was only informed in December 1957, through an announcement attributed to Tadao Takenouchi, a Japanese professor who had identified the location mathematically by using orbital dynamics and applying the equations to the orbits of Sputnik 1 and 2.

[13] The Soviets had chosen the location of the future cosmodrome in 1954, when they realized that the location and facilities of Kapustin Yar would be inadequate for the R-7 ICBM. A major concern was the proximity of the site to radar stations operated by U.S. intelligence services in Turkey. In the early days, the site was informally dubbed by locals with unofficial names like Zarya ('Dawn' or 'Beginning' in English), Zvezdagrad ('Star City'), or Kalingrad (named after a certain Major Kalin who ran a wooden shack that served as an improvised club. The town being built next to the Tyura-Tam rail stop was then officially given the name Leninsk.

spiders. I once witnessed a technician, a young captain, being bitten on the neck by a spider. He collapsed and died within minutes. There was nothing we could do.

"Few allowances were made for the brutal extremes of weather when accommodation was built for cosmonauts in the lead-up to their missions and for the many engineers and designers permanently stationed there. The brick apartment blocks and small houses were constructed according to Moscow specifications. Years later, the chief developer of the complex was awarded with a high honor by the state. I would have had him severely punished." [30]

Figure 2.28: A seldom-seen postal cover mailed from the Cosmodrome, postmarked at the Kzyl-Orda-50 Post Office. (From the collection of Jaromir Matejka, Austria).

The Soviets also lied on a further point. As we know today, Gagarin ejected from his capsule at an altitude of about 7,000 meters (23,000 feet) and landed separately. But international aeronautical rules stipulate that to claim a record, a pilot must remain with their craft from takeoff to landing. The Soviet media misleadingly announced that *"Gagarin and his capsule landed at 10:55 near the village of Smelovka, not far from Saratov,"* without making any mention of the fact that they had landed separately.

During the busy press conference that followed Gagarin's flight, one of the correspondents asked the cosmonaut directly how he had landed. Gagarin, repeating what censor 'Comrade Kroshkin' was whispering behind his back, began to offer an ambiguous, confused explanation about the fact that the construction of the spacecraft allowed the landing to take place in a variety of ways. It was only much later, after the Voskhod flight of October 1, 1964, that the Soviets would proudly announce that cosmonauts had landed inside their capsule *for the first time*, without making use of their personal parachutes. [31]

References

1. **This New Ocean: A History of Project Mercury**, Loyd S. Swenson (Ed.), NASA SP-4201, Washington D.C., 1966, pp. 91–93; and *First Up?* Tony Reichhardt, *Air & Space Magazine*, September 2000; also see *Man-In-Space-Soonest* and *Project Mer*, Mark Wade, *Encyclopedia Astronautica*, www.astronautix.com.
2. **The Human Factor: Biomedicine in the Manned Space Program to 1980**, Chapter 2, *The Human Factors of Project Mercury*, John A. Pitts, NASA SP-4213, Washington D.C., 1985, pp. 13–20.
3. **Challenge to Apollo: The Soviet Union and the Space Race 1945–1974**, Asif Siddiqi, NASA SP-4408, Washington D.C., 2000, p. 243.
4. **Rockets and People – Vol. 3: Hot Days of the Cold War**, Boris Chertok, NASA SP-4110, Washington D.C., 2009, pp. 56–57.
5. Reference 3, p. 246.
6. **Two Sides of the Moon: Our Story of the Cold War Space Race**, David Scott and Alexei Leonov, St. Martin's Press, New York, 2004, p. 56.
7. Reference 4, p. 60.
8. **The Russian Space Bluff – The inside story of the Soviet drive to the Moon**, Leonid Vladimirov, Dial Press, London, 1973, pp. 81–84.
9. Reference 4, p. 41.
10. Reference 4, p. 42.
11. Reference 8, p. 84.
12. *The Nedelin Catastrophe*, Mark Wade, Encyclopedia Astronautica (www.astronautix.com); see also **Rockets and People – Vol. 2: Creating a Rocket Industry**, Boris Chertok, NASA SP-4110, 2006, pp. 600–601.
13. *Disaster at the Cosmodrome*, Jim Oberg, in *Air & Space Magazine*, December 1990, p. 77.
14. From the Memoirs of Nikita Khrushchev, published in 1974 in the USA and quoted in *Mourning Star* by Asif Siddiqi in *Quest Vol. 2*, Winter 1994. See also Reference 3, pp. 256–7.
15. The figure '165' is taken from **Two Sides of the Moon: Our Story of the Cold War Space Race**, David Scott and Alexei Leonov, St. Martin's Press, New York, 2004, pp. 55–57 and p. 97. The figure '180' is quoted in **Rockets and People – Vol. 2: Creating a Rocket Industry**, Boris Chertok, NASA SP-4110, 2006, pp. 598–599.
16. Reference 13, p. 74.
17. Reference 3, p. 259.
18. Reference 8, p. 87.
19. *Roads to Space: An Oral History of the Soviet Space Program*, John Rhea, Aviation Week Group, 1995, pp. 197–199 and 415–417.
20. **Freedom 7: The Historic Flight of Alan B. Shepard Jr.**, Colin Burgess, Springer-Praxis books, 2013, Chapter 2, *The Mercury flight of chimpanzee Ham*, pp. 29–64.
21. *Timid Doctors and Timid Germans: How the United States Lost the Race for a Man in Space*, Ben Evans, www.americaspace.com (March 19, 2016).

22. *Vostok Lifts Off!* Anatoly Zak, www.russianspaceweb.com (April 12, 2016)
23. Reference 3, p. 262.
24. Reference 8, pp. 91–94.
25. *Typy a Padělky Ruskỳch Razítek Tématu Kosmos* (*Russian space postmarks and fakes* [in Czech]), Julius Cacka, Prague 2006, pp. 48–55. An English translation can be found in AD*ASTRA #26, September 2015, pp. 17–20
26. Reference 25, pp. 52–53.
27. Reference 25, pp. 55.58. In Czech. An English translation can be found in AD*ASTRA #30, September 2016, pp. 23–24.
28. *Why do Soviet Special Postmarks have design variations?* J. Reichman, AD*ASTRA #20, March 2014, pp. 5–8.
29. Reference 3, pp. 134–5.
30. Reference 6, p. 96.
31. Reference 8, pp. 102–105.

3

The Space Race changes direction

KENNEDY AND AMERICA'S PRESTIGE

Once he had arrived in the White House, America's new President John F. Kennedy had to address the problem of restoring lost pride and prestige back to the nation, after the double setback and humiliation of being beaten to put the first man into space by Yuri Gagarin's mission on April 12, 1961, and then, five days later, the failed Bay of Pigs military invasion of Cuba, the CIA's biggest ever fiasco.

A few days later, Kennedy held a brainstorming session with a few aides and confidantes, to identify a way that the U.S. could demonstrate strength without raising tensions. Many projects were discussed and it was space that provided the potential *"new frontier"* in which the U.S. could play a successful major role. Kennedy addressed a question to his Vice-President, Lyndon B. Johnson: *"Is there any chance of beating the Soviets? Is there any space program that promises spectacular results in which we could win?"*

Johnson immediately organized a two-week assessment of the options, with Wernher Von Braun an important contributor to the debate. In a memorandum dated April 29, 1961, Von Braun advised Johnson that: *"We do not have a good chance of beating the Soviets to a manned laboratory in space… [but] we have a sporting chance of sending a three-man crew around the Moon ahead of the Soviets… [and] we have an excellent chance of beating the Soviets to the first landing of a crew on the Moon."* [1]

A landing on the Moon seemed to be the right challenge: difficult, open, and a test to see who really had the best technology. In his book, Boris Chertok annotated: *"Yuri Gagarin's flight was the strongest stimulus for the development of the*

108   The Space Race changes direction

American piloted programs, which were crowned by the lunar landing expeditions. I contend that if Gagarin's flight on April 12, 1961 had ended in failure, U.S. astronaut Neil A. Armstrong would not have set foot on the Moon on July 20, 1969." [2]

ALAN SHEPARD: THE HIGH-TECH COLD WAR GLADIATOR IN A SILVER SPACE SUIT

Finally, on May 5, 1961, a new Mercury spacecraft stood on the pad, ready to take the first of the 'Original Seven' American astronauts into space. Alan B. Shepard climbed aboard the Mercury Production Model #7 capsule atop the Redstone rocket, which he had patriotically named *Freedom 7*. While the '7' represented the seventh Mercury capsule built, it also represented the 'Original Seven' astronauts selected by NASA as 'Group One' and would be subsequently incorporated into the naming of each Mercury mission, all of which were flown by that group. Although Shepard's flight was originally scheduled for October 1960, it had been postponed several times. It was delayed one final time from its planned launch of April 28, 1961 due to adverse weather conditions.

Figure 3.1: Two stamps issued in 2011 to commemorate the 50th anniversary of Alan Shepard's suborbital flight aboard *Freedom 7*.

The countdown on May 5 progressed very slowly, with several stoppages to resolve minor problems because no one wanted to risk the safety of the astronaut. Minutes stretched into hours and Shepard began to get nervous. He needed to pee and nobody seemed to have taken such a 'human factor' into consideration. As his flight was going to be very short, no space diapers or urine collection device had been included in the required equipment. Shepard asked to be let out of the

Alan Shepard: The High-Tech Cold War Gladiator in a Silver Space Suit 109

capsule to relieve himself but Von Braun, who was supervising the launch, flatly refused permission. That left only one possible solution, but the medical team feared that it could short-circuit his body-monitoring cables. The scientists at NASA did not think that this would happen, but there was a quick exchange of opinions, at the end of which the decision was taken to shut off the power and the monitoring system for a few minutes so that Shepard could relieve himself.

The countdown was restarted, very slowly. *"Why don't you guys fix your little problems and light this candle?"* an increasingly impatient Shepard growled at the ground crew[1]. Finally, the countdown reached zero and *Freedom 7* rose into the sky. The 15-minute suborbital flight carried Shepard to an altitude of 116 statute miles (187 km) and he became the second astronaut in history, surpassing the 100 km barrier that marks the border between sky and space, according to IAF rules. After reaching the top of its arc, *Freedom 7* carried Shepard back into the atmosphere, experiencing a peak of 11.6G.

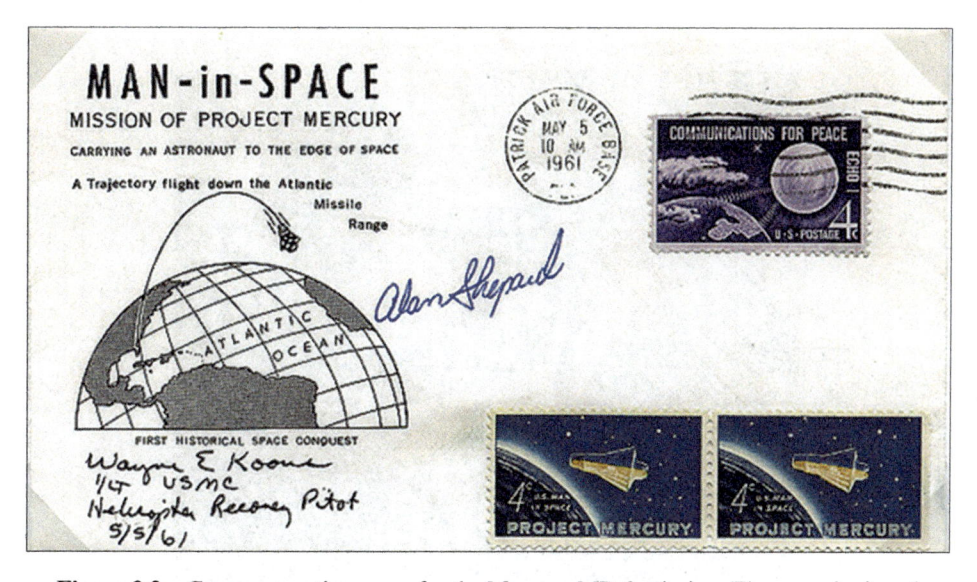

Figure 3.2: Commemorative cover for the Mercury MR-3 mission. The cover is signed by astronaut Alan Shepard.

The 15-minute ballistic flight into the skies above Florida was quite a modest flight compared to the epic odyssey of Gagarin. He had flown almost a complete orbit, while Shepard's mission had been just a short lob from Cape Canaveral in

---

[1] Taken from Cunningham [2004] p.21. When reporters asked Shepard what he had thought about as he sat atop the Redstone rocket awaiting liftoff, he replied *"The fact that every part of this ship was built by the low bidder."* Kranz [2000], pp. 200–01.

110 The Space Race changes direction

the general direction of Bermuda. Shepard's Redstone rocket – actually a redesigned V-2 – was incapable of delivering a capsule into Earth orbit. But while the launch of Gagarin was done completely in secret – and was announced only when it was known the mission was going to be a success so that people only heard about it on the radio or read about it later in the newspapers – Shepard's mission was preannounced and seen live on television by millions. It made a very favorable impression on both Americans and international opinion. The fact that the mission was conducted in full public view had the effect of galvanizing the American people and giving them back some trust. Many future astronauts would later admit that they had been inspired by this success.

After Shepard had been recovered from the ocean, President Kennedy called him aboard the carrier USS Lake Champlain to congratulate him on behalf of the nation. On May 8, at a ceremony in the Rose Garden at the White House, Kennedy personally pinned a Distinguished Service Medal on the astronaut, NASA's highest award.

Figure 3.3: President John F. Kennedy pins the NASA Distinguished Service Medal onto astronaut Alan B. Shepard. (Courtesy NASA.)

On the other side of the Space Race, however, opinions were less than complimentary. Soviet Premier Nikita Khrushchev teasingly commented that the Americans were getting much more publicity out of a short hop than the Soviets had done out of a full orbital flight[2]. The Cuban press, meanwhile, suggested that as his flight was only a short hop into space, Shepard should be labeled a *"cloudnaut"* rather than an astronaut. Many people around the world, not just in the Soviet Union, were still of the opinion that the Americans were lagging behind and that, for all their colossal effort, they had only launched a man into space 23 days after the Russians and had not even achieved orbit. Strangely, it seemed not to occur to most that the American flight had been announced in advance and had taken place exactly as planned. For Soviet citizens, such a thought would not likely have occurred anyway, because censorship in the Soviet Union carefully suppressed every reference to announcements of forthcoming launches in America.

However, the *"space jump"* had the effect of encouraging the political decision makers on the U.S. side of the Space Race. On the same day that Shepard received his award at the White House, Vice President Johnson presented a report on America's options in space. Three weeks later, on May 25, 1961, President Kennedy presented his renowned address to a joint session of the U.S. Congress and ignited the riskiest and most fascinating race of the Twentieth Century.

KENNEDY LAYS DOWN THE GAUNTLET. THE SOVIET REACTION

Forty days after the flight of Gagarin and twenty days after Shepard's Mercury-3 mission, Kennedy sounded the charge and addressed his *"Special Message to the Congress on Urgent National Needs"*. He outlined the military and political measures that were to be implemented to counter the threat of communism and announced the dramatic and ambitious goal of sending an American safely to the Moon and back before the end of the decade. The response of the audience tended more towards the polite than the enthusiastic. Many felt that the president's vision was totally unrealistic.

The news also drew a response in the Soviet Union. Khrushchev challenged his experts and scientists about the possibility of a Soviet landing on the Moon and it was Academician Valentin Glushko who responded with a report which explained that the only possible plan would be the *"Wernher Von Braun project"*. [3] According to Von Braun, a flight to the Moon would firstly require the construction of a large orbital space station, or platform, which would involve putting about 70 powerful rockets into orbit along with crews and equipment. Once the space-platform had been constructed, it would be possible to assemble the rocket

---

[2] Some objected to this description, calling Gagarin's flight an *"incomplete orbit"*, since he had landed west of his take-off point and, strictly speaking, had not completed one full orbit.

112 The Space Race changes direction

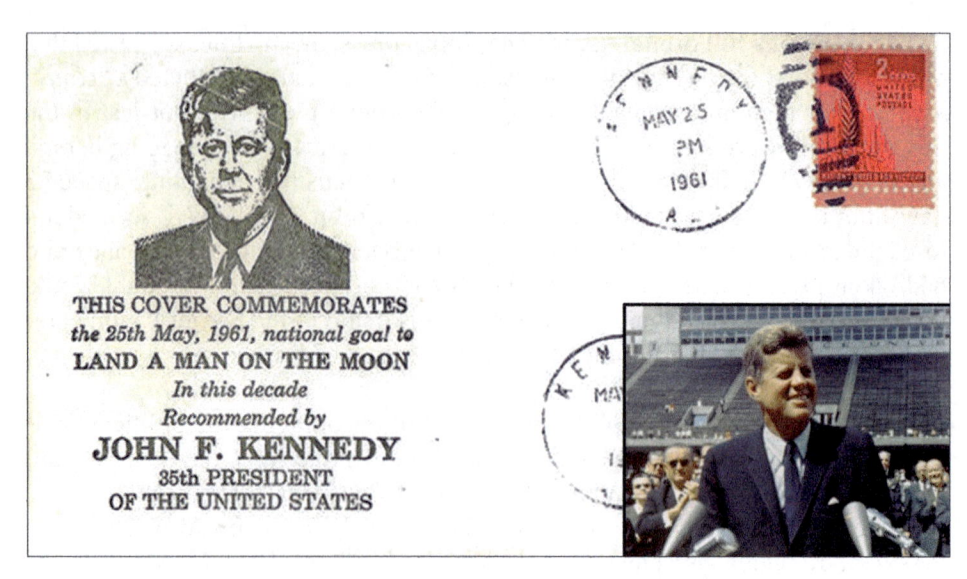

Figure 3.4 (left): Cover commemorating the national goal to "*land a man on the Moon before this decade is out*", recommended by President John F. Kennedy on May 25, 1961. (From the collection of Steve Durst, USA). (inset) Kennedy speaks at Rice University, Houston, Texas on September 12, 1962. (Courtesy NASA.)

in space and then launch it to the Moon. Reassured by this assessment, Khrushchev denounced Kennedy's goals as "*propaganda*"[3].

Aware of Glushko's report, a Moscow engineer, Yuri Khlebtsevich, wrote a letter to the Soviet Academy of Sciences, saying that the Moon could be reached more quickly and easily by other means. He based his views on a book, *The Conquest of Interplanetary Space*, written in 1929 by the Ukrainian scientist Yuri Kondratyuk[4]. According to Kondratyuk, it was not necessary for the entire

---

[3] Even in the USSR, space scientists approached Kennedy's Moon challenge warily. Boris Chertok, the brilliant engineer who designed most of Sergei Korolev's guidance systems, reported in his book "*Our first thought was 'Is this a bluff, or is he serious?' Our impression was that it was not a bluff. The feeling within the engineering community was that it was realistic and could be done. And it gave us satisfaction, because we were sure it would force our leadership to pay more attention to the Soviet space program.*" (Chertok [2009], p.250)

[4] See *Yuriy Kondratiuk. Space conquering*, in http://see-you.in.ua (accessed in March 2018). Yuri Vasilievich Kondratyuk (real name Aleksander Ignatyevich Shargey) was born on June 21, 1897, in Poltava, Ukraine. In 1914, the 17-year-old apprentice began a fundamental work, *To Those Who Will Read in Order to Build*, dealing with rocket flight basics, without any previous knowledge of world science achievements in this field. The 104-page manuscript, full of new astronautical-related ideas, with strong scientific and technical background, was first published only in 1964 by the Institute of Natural History and Technology (USSR Academy of Sciences).

Shargey was unable to graduate from the Petrograd Polytechnical Institute as he was enlisted to fight during the First World War. During the Civil War in Ukraine after the 1917 revolution,

spacecraft to land on the Moon. The best way to reach the Moon would be to put a rocket into lunar orbit and then send a small *"excursion cabin"* down from that orbit. This would be the solution that was adopted by the Americans for the Apollo program. Khlebtsevich's report was unacceptable to Gluskho, who sent back an abrupt refusal. He had already reported to Khrushchev and felt that an Academician must be true to his word. He had no intention of reopening a closed issue.

As a major expert in the field of electronics and the author of many inventions, Khlebtsevich had access to foreign technical publications and had a good idea of the state of American work on space projects. Appealing to the common sense of the academicians, he urged them to understand that by refusing to pay attention to Kondratyuk's ideas, they would be condemning the Soviet Union to defeat in the Moon race with the Americans. As a loyal Russian patriot devoted to his country, Khlebtsevich came up with another proposal, suggesting using the available technology to send a *"tankette-laboratory"* to the Moon. This would be a self-propelled trolley of modest dimensions but equipped with scientific instruments, which would allow the Russians to demonstrate their own way of exploring the Moon, even without the participation of men. This time, the letter rejecting his proposal was extremely abrupt and even threatening. The engineer was advised to mind his own business and not keep offering unwanted advice.

---

he was first enrolled in the White Guard and then to the Denikin Army that fought against Bolsheviks. As a pacifist, Shargey actually fought against no one but was still considered an enemy of the new communist power. In order to escape political persecution, Shargey changed his name to Yuri Kondratyuk. In 1927, he went to Novosibirsk (Russia) as a hoist builder. Two years earlier, he had sent his work *"The Conquest of Interplanetary Space"* to Moscow where, despite the enthusiastic appraisal of many scientists, publication of his work was suspended. Kondratyuk published 2000 copies at his own expense, doing much of the typesetting and operating the press himself, not only to save costs but also because the equations in the book posed problems for the printer. Kondratyuk's discoveries were made independently of Konstantin Tsiolkovsky, the acknowledged 'Grandfather of Spaceflight' who was working on the same issues at that time. The two never met.

Applying his engineering skills to local problems, Kondratyuk designed a huge, 13,000-ton grain elevator (quickly nicknamed 'Mastodon'), built of wood and without a single nail, since metal was in short supply in Siberia at that time. His ingenuity would come back to haunt him when he was investigated as a saboteur by the NKVD in 1930. The lack of nails in the structure was used as 'evidence' that he had planned it to collapse. Convicted of anti-Soviet activity, he was arrested and sentenced to three years in a gulag. During his imprisonment, Kondratyuk learned of a competition to design a large wind-power generator for the Crimea and, without any previous experience in this field, submitted a design for a concrete tower capable of generating up to 12,000 KW. His project was acknowledged as the best, ahead of two other projects worked out by two specialized scientific schools. As a result, Kondratyuk was transferred to Kharkiv to proceed with the project. In Moscow, after meeting with Sergei Korolev, who was impressed by his research, Kondratyuk was offered a position as chief theoretician in a secret 'Jet Lab' but turned it down, mindful of his now-hidden past. He never saw his giant wind-electric power generator in operation as its construction was stopped and never finished. Disillusioned and completely disappointed, Kondratyuk joined the Soviet army soon after the beginning of WWII and was killed in battle in February 1942.

Figure 3.5: A Ukrainian stamp featuring Yuti Kondratyuk was issued on June 21, 1997, to celebrate the 100th anniversary of his birth (enlarged in inset).

Surprisingly, under Soviet conditions at that time, Khlebtsevich refused to give up. He began writing articles to newspapers and reviews and made a short amateur film about his 'tankette-laboratory' which was shown in clubs where he delivered lectures on the subject. At this point, the First Section of the Academy of Sciences (the secret police department) intervened. Khlebtsevich was summoned to the police and warned that rocketry and space technology were secret subjects and that he could expect serious consequences if he continued to show a film that *"misleads the population concerning the prospects of exploring outer space."* There was no option left for Khlebsevich other than to give up the struggle and return to work, until he was thrown out of his job. Nine years after this *"mistaken"* idea was rejected, Khlebtsevich's 'tankette-laboratory' was implemented and sent to the Moon as Lunokhod, without involving the man who had conceived the idea. [4]

GUS GRISSOM: AMERICA PUTS THE THIRD MAN IN SPACE

On July 21, 1961, Virgil I. 'Gus' Grissom repeated Shepard's flight aboard his own Mercury capsule, *Liberty Bell 7*. There were some improvements over Shepard's spacecraft, including a large window with a periscope and an escape hatch that could be blown using a detonating plunger to give the occupant a better chance of survival in an emergency. After Shepard's sanitation problems, Grissom was fitted with a urine collection system, later reporting that it *"worked as advertised"*. [5]

Grissom's Mercury mission, MR-4, performed a parabola of 15 minutes and 37 seconds, including five minutes of weightlessness. This would be the last use of a Redstone launcher on a Mercury mission and all subsequent Mercury flights would carry the acronym 'MA' because they would be launched using an Atlas rocket.

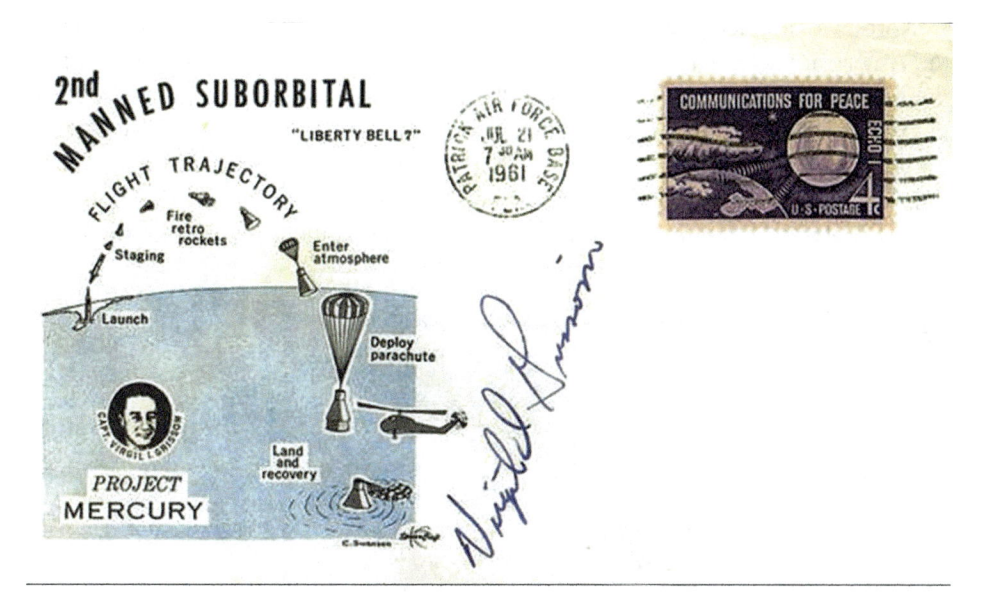

Figure 3.6: Cover commemorating the second flight of the Mercury program, with Gus Grissom aboard *Liberty Bell 7*.

After the splashdown, the escape hatch of *Liberty Bell 7* blew off and water began flooding into the capsule. The spacecraft took on too much water and sank into the deep ocean, 300 nautical miles (560 km) east-southeast of Cape Canaveral. The same fate almost befell Grissom while the recovery team were attempting to grab the capsule. Procedures would quickly be changed to ensure that recovery of the astronaut took priority on subsequent missions. Inflatable floatation gear was also added to the suit kit.

Grissom became the only man ever to lose his spacecraft. After several unsuccessful attempts, *Liberty Bell 7* was finally found and recovered from a depth of nearly 16,000 feet (4.9 km) on July 20, 1999, the 30th anniversary of the Apollo 11 lunar landing. Ironically, another exit hatch would contribute to the death of Grissom in the Apollo 1 fire of January 1967.

116   The Space Race changes direction

A NEW RUSSIAN FIRST: AN ENTIRE DAY IN ORBIT

Sergei Korolev, always well informed about American plans, knew that after the suborbital flight of Grissom, there were no more U.S. launches scheduled until the following year. Not in good health, Korolev decided to take a short vacation on the Black Sea coast with his wife and daughter. But in the middle of July, Khrushchev caught up to him even there and summoned Korolev to his summer palace on the Black Sea, under the pretext of awarding him another gold medal. Khrushchev's real objective was to instruct Korolev that the next Soviet space flight should take place no later than the beginning of August – in just a few weeks.

Korolev explained to Khrushchev that there was no need to rush things and that the Americans would not be launching anything else with a man onboard before the end of the year. Khrushchev's motivation for this urgency remained unclear, but whatever his reason, Korolev found himself rushing back to Moscow to embark on yet another launch. On August 6, 1961, Gherman Titov, Gagarin's back up on the first flight, was successfully launched aboard Vostok 2 from the Baikonur Cosmodrome. A month short of 26 years old at launch, Titov was (and still remains) the youngest space traveler ever to fly.

Figure 3.7: Russian stamp commemorating the flight of Gherman Titov, the first to spend an entire day in orbit.

General Nikolai Kamanin, the cosmonaut team's 'mother hen' wanted to limit the mission to three orbits, as did the crew physicians, because the physiological effects of exposure to prolonged weightlessness were still unknown. The two dogs who had flown a sixteen-orbit mission on Korabl-Sputnik-2 nine months earlier had experienced convulsions. Korolev insisted that the flight would be a day-long mission, which would allow a full evaluation of functions such as eating, dealing with body wastes and sleeping. Some of the Soviet doctors also wondered whether it would even be possible to wake up someone who had fallen asleep in zero gravity. Korolev was confident and wanted to move ahead quickly, however, because even after three orbits, the flight would still have to complete a full 24 hours because of the requirement to land safely in Soviet territory. The spacecraft would only be in a suitable position for such a return after 17 orbits. Between the third and the

17th orbit, it would not be possible to land within Greater Russia, so it made sense to go for the full one-day mission. Due to the short time available, Korolev planned the flight according to the same model as the flight of the dogs: 17 orbits and return to Earth at the point of departure, even though two such flights had ended in failure. The designers and engineers worked day and night on the braking system, which had broken down after being in a vacuum for 24 hours. They tested the revised system dozens of times to ensure it worked as planned, which it did. [6]

The launch went perfectly and Titov's flight finally proved that humans could live and work in space. He became the first person to orbit the Earth multiple times and to spend more than a day in space, with quite a substantial workload to get through in orbit. Unlike Gagarin, Titov also took manual control of his capsule – for a short time.

Titov enjoyed the spectacular views and reported his experiences with a poetic touch, instilled in him by his father, who was a schoolmaster in the village of Polkonikovo where Titov grew up: *"I had the feeling that our Earth is a sand particle in the universe comparable to a particle of sand on the shore of the ocean. It was strange to have a black dome above me and our earthly blue sky below. The Earth flashed as a multi-faceted gem, an extraordinary array of vivid hues that were strangely gentle in their play across the receding surface of the world… framed in a brilliant, radiant border. The colors were extraordinary – vivid, yet tender – and the light streaming through the cabin carried a strange shade as if it were filtered through stained glass."*

Titov was the first photographer – and 'videographer' – in space. Incredibly, Yuri Gagarin had not carried a camera on the world's first space flight, and neither did Alan Shepard or Gus Grissom (the American astronauts were photographed during their missions, but only by automated cameras mounted in the Mercury capsule). Titov took pictures of the Earth with handheld Zritel cameras, partly to demonstrate – for military officers anxiously awaiting the Zenit spy satellites – the potential for spying purposes of human flight. Titov also ate his meals, which consisted mostly of purees in tubes, along with candies and chunks of bread and sausage. He reported that the food was *"joyless."*

Although Soviet sources insisted that Titov was in excellent health during the flight, they would eventually admit that his vestibular system had experienced *"some changes manifested in unpleasant feelings."* Titov was the first person to suffer from 'space sickness' and also the first to sleep in space.[5] Ground controllers called to wake him after the 13th orbit in space but he did not respond immediately, making them worry that their worst fears had come true. Titov eventually

---

[5] Chertok noted in his book: *"Sleep in space! If anything, that was one of the most important experiments. If a person could sleep in space in a spacesuit in weightlessness, without a comforter and pillows, that meant he could live and work! This is why Korolev had fought with Kamanin, arguing in favor of a 24-hour flight. For three orbits, one could forego sleep and all the other physiological needs, including a tasty dinner, until returning to Earth."* (Chertok [2009], p. 188).

118   **The Space Race changes direction**

roused and commented that he was feeling much better. The rest of the flight went well and Titov landed in the same region that Gagarin had done four months earlier. There was a nerve-wracking experience on the descent, however. Titov landed just a dozen meters from a railroad on which a train was travelling at the time. For the following flights, a Ministry of Railways representative would be included on the State Commission, to coordinate the railroad schedules with the launch programs. [7]

A team of doctors took Titov under their care and within 24 hours the cosmonaut was in good enough condition to be able to fly to Moscow and be received by Khrushchev. After that, he was hospitalized again.

The Soviet propaganda machine immediately got to work. In just a few days, the usual educational posters appeared, the first of which was produced immediately after the landing of Vostok 2. The inscription under Gagarin's portrait mentioned his status as Hero of the Soviet Union, while the inscription under Titov's portrait did not. It could be argued that this was a stock poster kept ready for such emergencies and that it was 'personalized' by adding the portraits of the two cosmonauts and sending it to the printer sometime between the day Vostok 2 landed (August 6) and the day Titov was proclaimed Hero of the Soviet Union (August 9).

Figure 3.8: The poster "To Space!", by R. Dementiev, was issued by Voyenizdat, Moscow, immediately after the flight of Titov and was widely distributed.

Figure 3.9: After a few days, the propaganda poster "Glory to the Sons of the Party!" by A. Ustinov was issued, featuring a smiling Nikita Khrushchev flanked by Titov and Gagarin on the tribune of Lenin's Mausoleum.

Figure 3.10: Other impressive posters of this period would make history, such as "Be Proud, Soviet, You Opened a Path from the Earth to the Stars!" designed by Mikhail Soloviev and published in Moscow by Izogiz.

A few days later, it became clear why Khrushchev was in such a hurry to have this launch. The space spectacular helped to draw the attention of the world away from the construction of the Berlin Wall, one of the high-water marks of the Cold War. [8] Khrushchev had ordered the East German government to begin work on the Wall on August 6, the same day that Titov was blasted into orbit.

After Vostok-2, Soviet manned launches were suspended for a year. The priority was to get the Zenit spy satellites operational. The Zenit was much like a Vostok and was launched by a Vostok-type booster. As there was only one facility available to launch such a booster, the crewed flights had to be suspended. Korolev had managed to sell his crewed space program as complementary to the space reconnaissance one, but that link between the two efforts was sometimes detrimental to Korolev's program.

At this time, both the Soviets and the Americans were conducting high-altitude nuclear tests. When the American Operation Fishbowl test was fired in July 1962, the Soviets used it as a convenient cover story for suspending their crewed flights, claiming that they were waiting for the radiation from the Fishbowl test to die down. [5]

By the end of 1961, the Americans had made up some ground. Out of the total of 60 launches that year, 49 of them were by the U.S., with 25 successes. In addition to the flights of Shepard and Grissom, the Americans also put meteorological and telecommunications satellites into orbit. In contrast, the Soviets achieved only 11 launches, with just six successes. Strangely, however, the Western media showed greater sympathy with the Soviets and only issued significant reports on American space missions when they failed, mentioning the successful ones only briefly. The anti-American stance of the French media was particularly crude. [9]

IT IS TIME TO THINK ABOUT THE MOON

The historic speech of President John F. Kennedy in May 1961 strongly accelerated the American space program. At that point, the Mercury program, which until then did not have a clearly defined purpose, became the first stage of the race to the Moon. From this point on, NASA had to step up its studies of human behavior in weightless conditions, while the debate about which was the best method to get to the Moon no longer remained an academic one.

NASA was studying two competing architectures for the lunar landing. The obvious solution, called 'Direct Ascent', envisioned lunar missions of a type now more often associated with science fiction: a giant Nova rocket taking the entire 'spaceship' – in one piece and of considerable mass – on a trans-lunar injection trajectory to arrive directly at the Moon. The entire technology was still to be conceived and developed at this point, however. The other architecture was called

'Earth Orbit Rendezvous', or EOR, and implied launching the lunar vehicle in sections which could then be assembled in space. This architecture, which had long been favored by Wernher Von Braun, would use less powerful launchers but would also require numerous launches and probably a space station. Building a space station did not fit into Kennedy's time frame.

The solution came in the form of a third option, called 'Lunar Orbit Rendezvous', or LOR, which was proposed by John Houbolt, a young engineer at NASA's Langley Research Center. [10] LOR called for the assembly of several light elements that would satisfy the various needs of the mission's different phases: not a single piece, but more segments with different functionality[6]. The project seemed cumbersome but would be achievable in a short time using technologies already available. It would only require an upgrade of the Saturn rocket (which was already under development) and the addition of a small lunar module. But the LOR concept initially met great resistance, not least from Von Braun, who insisted that his Saturn V had to be seen only as the intermediate stage of the giant rocket he had in mind.

Houbolt persisted, however, and went over the heads of his superiors, writing directly to NASA Associate Administrator, Dr. Robert Seamans, urging: "*Do we want to go to the Moon or not? Why is a much less grandiose scheme involving rendezvous ostracized or put on the defensive. I fully realize that contacting you in this manner is somewhat unorthodox, but the issues at stake are crucial enough to us all that an unusual course is warranted.*" It was more than a year after President Kennedy's challenge had been announced before NASA was persuaded that the challenges of LOR would be easier than the alternative methods. In the end, NASA liked the elegance and relative simplicity of the LOR solution and adopted it on June 7, 1962, despite opposition from the charismatic and renowned Wernher Von Braun. In hindsight, it must be said that were it not for Houbolt, it is unlikely that the United States would have met the deadline. His vision and tenacity helped NASA to achieve President Kennedy's goal of landing a man on the Moon by the end of the decade.

---

[6] Actually, the LOR concept had first been fully developed in December 1958 as 'Manned Lunar Landing and Return' (MALLAR) by Thomas Dolan, an American engineer working at Vought Astronautics. At the time, it was largely ignored by NASA administrators (Courtney [1979], p. 66). The fundamentals of the LOR concept had already been outlined years earlier, by Yuri Kondratyuk in 1916, by Hermann Oberth in 1923 and by the British scientist and Interplanetary Society member Harry E. Ross in 1948. It is likely that Houbolt had read the Kondratyuk manuscript "*To Those Who Will Read in Order to Build*" (that had been published in Russian in 1964) only when it had been translated into English in 1965 (David Sheridan: "*How an idea no one wanted grew up to be the LEM*", Life magazine Vol. 66 No. 10 [Mar 14, 1969], pp. 20–24).

122 The Space Race changes direction

Figure 3.11: John Houbolt explains the Lunar Orbit Rendezvous (LOR) method that NASA would eventually use to take Apollo to the Moon. Courtesy NASA.

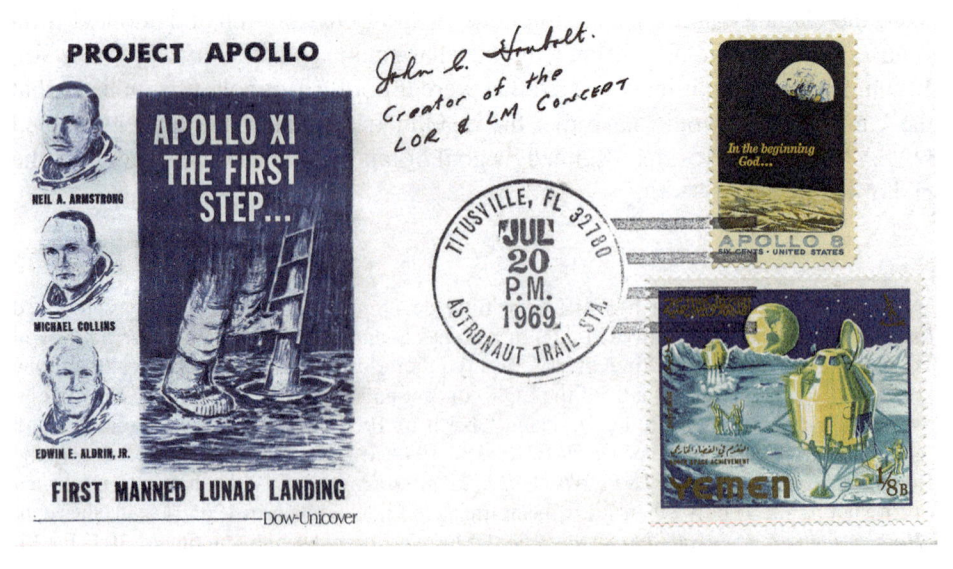

Figure 3.12: Cover with autograph of John Houbolt, "*creator of the LOR and LM concept.*"

Together with the LOR project, NASA also launched the new Gemini program, which would be sandwiched between Mercury and Apollo in order to develop techniques and experience that would be required for the new operating philosophy of space rendezvous and docking between spacecraft.

SOVIET LUNAR PROGRAM

The Moon was also becoming a topical issue in the Soviet Union. After monitoring, via the media, the debate taking place in the States, the Russians also made a choice, opting for a complex combination of EOR and LOR. Their model would require two vehicles to be launched. A gigantic rocket, which would eventually be called N-1, would leave its third stage in orbit carrying the unmanned lunar module. The crew of two would be launched by a Proton rocket a few hours later aboard a lunar spacecraft. After docking, the third stage engines would be reignited to take the whole spacecraft to circumlunar orbit. Once lunar orbit was achieved, a cosmonaut would enter the lunar module and descend to the surface of the Moon. He would plant the Soviet flag and collect some lunar samples before returning to the lunar module and lifting off again to rejoin his colleague in the other spacecraft in lunar orbit.

Figure 3.13: Artist's concept of a Soviet manned lunar landing. Courtesy Jean-Marie Le Cosperec.

The order to design the N-1 was given towards the end of 1961. N-1 would require sufficient power to launch 40–50 tons into orbit. Sergei Korolev would only take care of the conceptual phase, with Vladimir Chelomey, who had no experience of human spaceflight, entrusted with the main part of the lunar program: the LK-1 spacecraft required for circumnavigating the Moon. The following year, Khrushchev ordered Korolev to make the N-1 more powerful in order to launch the OS-1 or *Zvezda*, a 75-ton orbital station armed with nuclear weapons. No other use of the N-1 was planned or authorized. In an interview, Vasily Mishin, Korolev's deputy and then successor, would later complain: "*Superficial and contradictory decisions were sent down to us. Korolev was a resolute, independent-minded and far-sighted man, so he resisted such decisions, but that put him at odds with top leaders!*" [11]

Keeping an eye on the American plans to get to the Moon, Korolev realized that the Soviet Vostok craft was inadequate. They would need a lightweight, manageable three-seat spacecraft with docking capabilities. Korolev made his deputy Leonid Voskresensky responsible for drawing up the design of the Soyuz spacecraft, well aware that, due to the rivalry of Glushko and Chelomey, he would not be able to count on a more powerful booster than the "*monster*" with 21 engines that had taken Vostok into orbit. For the new project, he tried not to depart too much from the Vostok, and Soyuz was conceived as the natural and 'economical' evolution, preserving the essential elements such as the spherical capsule and the cylindrical last stage of the rocket. The whole structure of the Soyuz was only slightly larger than the Vostok, did not differ from it substantially in weight and had almost the same dimensions. [12] Korolev also planned a spacecraft to land a man on the Moon and bring him back to the waiting Soyuz in lunar orbit, but Soviet leaders rejected the plan and continued to support Chelomey's LK-1 project.

JOHN GLENN: THE FIRST AMERICAN IN EARTH ORBIT

The Americans were now proceeding with their finalized space program and mission MA-6, to which Robert Gilruth – the director of NASA's Manned Spacecraft Center – had assigned John Glenn with Deke Slayton as his backup, initially planned to launch before the end of December 1961. But Gilruth, who had a near-fanatical concern over the safety of the astronauts, delayed the launch several times. On January 27, 1962, Glenn sat in the Mercury capsule on top of the Atlas rocket for over five hours waiting for the clouds to break. In the end, amid disappointment and some friction, he was forced to step down once again. Finally, on February 20, 1962, the countdown reached zero and *Friendship 7* left the pad on a near-perfect launch, in front of 50,000 spectators watching from the beaches near the Cape and a hundred million TV viewers watching live around the world.

Figure 3.14 (above): A commemorative Swanson cover, cancelled at launch time at the military post office of Patrick Air Force Base. (below): First Day Cover for the stamp #1193, commemorating the *"Mercury Mission"*. On this machine cancellation, 'Cape Canaveral' appeared for the first time instead of the usual 'Port Canaveral'. The ArtCraft cover erroneously reproduced the Redstone rocket that had been used for the previous suborbital missions, rather than the Atlas which Glenn rode for this flight.

After a perfect launch, Glenn circled the globe three times during a flight lasting 4 hours 55 minutes and 23 seconds. On the second orbit, however, telemetry indicated that the heat shield was loose and there was significant concern that the capsule could burn up on reentry through the atmosphere. Flight Control decided to keep the retrorockets in place over the shield during reentry, in the hope that the pressures encountered would keep the shield in place. The fear that the unforeseen additional weight might modify the spacecraft's controls made all, including Glenn, hold their breath until the flight ended safely only 60 kilometers from the planned splashdown target. The capsule was taken promptly aboard the USS Noa. It was later revealed that there had been no problem with the heatshield and that the issue was with the flight telemetry, a conclusion that caused some friction between Glenn and Flight Director Chris Kraft.

126   The Space Race changes direction

Figure 3.15: Commemorative cover of the recovery of John Glenn's Mercury mission, cancelled aboard the USS Noa.

The secret Mercury stamp

Minutes after the successful splashdown of Mercury 6, an official statement was released by the Post Administration and postmasters were able to open the sealed parcels they had received marked 'top secret' to release the Project Mercury stamps. The stamp was to have been designed by Paul Calle, a renowned artist working for NASA, *"But he was out of the country and unable to begin design concepts"*. [13]

Figure 3.16: Charles Chickering's commemorative stamp for Project Mercury.

(continued)

> The task was therefore assigned to Charles R. Chickering, an employee at the Washington Bureau of Engraving and Printing. Chickering was known as the designer of dozens of stamps, including the *Fort Bliss Centennial* issued in 1948, which became the first 'space stamp' and featured a rocket designed after the German V-2 in the center. All operations for the production of this stamp were to be kept absolutely secret (if Glenn's flight had failed, the stamp would probably not have been issued).

Figure 3.17: Chickering's Fort Bliss Centennial stamp featured a rocket similar to the German V-2.

To keep the project quiet, Chickering worked from home while claiming to be away on vacation. It would be one of his last jobs before his retirement. The picture engraver, Richard Bower, also gave the impression he was on leave, but went into the Bureau at night. Howard Sharpless produced the lettering on weekends. Chickering produced the first two preliminary designs, both in portrait format to produce a sense of the endless height of deep space. They were returned to him with suggested revisions and he modeled the final design in the requested landscape format to emphasize the "*limitless width of space*".

NASA, however, was apparently unhappy with Chickering's original designs, and with others developed by Norman Todhunter of the Citizen's Stamp Advisory Committee. The agency wanted its own vision reflected in the stamp, so Charles de M. Barnes, an employee in NASA's Office of

(continued)

128 The Space Race changes direction

(continued)

Figure 3.18: Preliminary designs for Chickering's Project Mercury stamp

Educational Services, became involved in the project. Different solutions were worked out, but although recent studies have credited him with the horizontal sketch featuring *Friendship 7* while reentering the atmosphere, inter-agency agreements led to Barnes being ignored as the stamp's designer and Chickering was responsible for modeling the final stamp design. [14]

Figure 3.19: Concept designs for the Project Mercury stamp by Charles de M. Barnes.

The final stamp featured an image of the Mercury *Friendship 7* capsule circling the Earth against a field of stars. The face value of '4¢' featured alongside the wording "*U.S. MAN IN SPACE*" and the large name "*PROJECT MERCURY*". Unlike the usual stamps issued in the 1960s, there was no indication of "*U.S. Postage*", and rumors started to circulate that because of this, the stamp had no legal value. The Post Office immediately clarified that the wording had intentionally been omitted to focus on the exceptional space event.

John Glenn: The First American in Earth Orbit

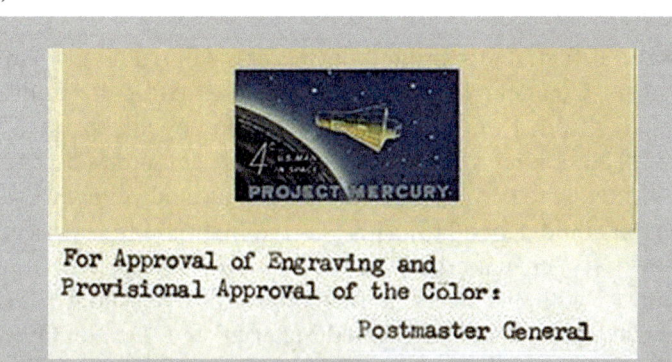

(continued)

Figure 3.20: The final draft of the Project Mercury stamp.

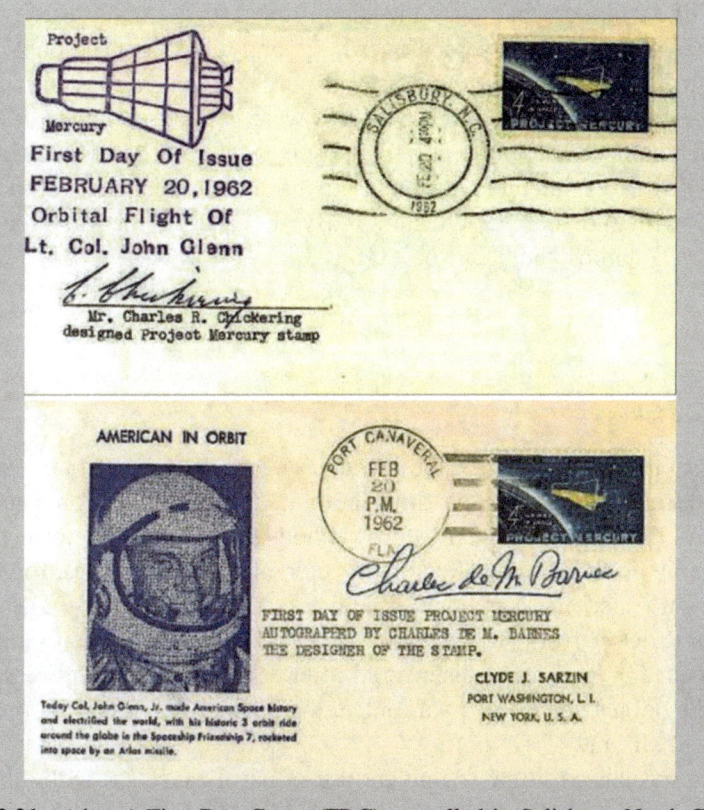

Figure 3.21: (above) First Day Cover (FDC) cancelled in Salisbury, North Carolina, one of the 305 postal offices that received the 'secret package'. The cover is signed by Charles R. Chickering. (below) FDC cancelled in Port Canaveral and signed by Charles de M. Barnes.

(continued)

> This was the first stamp printed on the new Giori Press – named after Italian inventor Gualtiero Giori – which permitted two-color or even three-color engraving from a single plate in one pass through the press. The secret was in the rubber inking rollers, whose surfaces were precisely cut to apply each ink selectively to parts of the same plate. Blue and yellow were used for this stamp, on white paper. [15] This was the first time in American history that a previously unannounced commemorative stamp had been issued simultaneously with the event it memorialized. The stamp sheets were secretly packed and marked 'Classified Material' and 'Do Not Open' before being sent to 305 postal offices, addressed for the attention of the postal inspectors to prevent curious postmasters from sneaking an early peek. Immediately after Glenn's safe return, the order arrived from Washington to remove the wrappers and place the stamps on sale.
>
> As news of the stamp spread over radio and television, the public began lining up at their nearest post office to purchase it. The popularity of the event meant that over 289 million stamps were issued, more than double the average quantity for commemorative postage stamps at the time. Many collectors drove considerable distances to have their blank envelopes postmarked with the unexpected stamp. By the end of the first day, some 10,290,850 stamps had been sold. [16]

The Riser fakes

As part of the enthusiasm for the successes of the space missions, space philately became a national pastime, both in the Soviet Union and in the USA. As the business began to grow (thousands of commemorative items were being produced for collectors to celebrate every mission), forged covers began to become a major problem for both superpowers. The case of Charles R. Riser from Bowie, Maryland, and his 1974 Federal Grand Jury indictment for mail fraud discovered after a *"Philatelic detective story"* is something of a legend. Riser was one of the greatest American astrophilately forgers of all time.

Riser specialized in the fraudulent use of naval cancels used by the seven U.S. Navy ships involved in recovering the Mercury missions. With considerable skill, he duplicated the cancels of USS Lake Champlain (CVS 39), USS

(continued)

(continued)

Decatur (DD 936), USS La Salle (LPD 3), USS L.F. Mason (DD 852), USS Noa (DD 841), USS Randolph (CVS 15) and USS Stormes (DD 780).[7] Charles Riser also created a number of different fake cancelling devices that imitated many of the post office cancels, including Port Canaveral and Wallops Island. Fake covers exist for many of the Mercury flight launches and recoveries, as well as Gemini and MOL recoveries.

In some cases, Riser typed additional information about the event on the cover. He was also very skilled at forging autographs of the astronauts, ship's captains and other personnel and therefore many of the Riser covers also contain questionable 'signatures'.

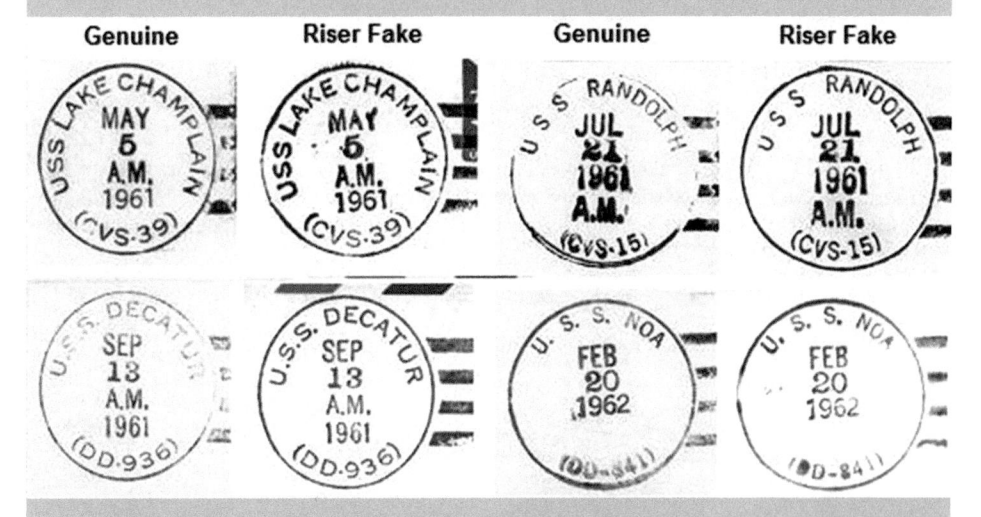

Figure 3.22: Small differences can be seen in the diameter of the cancellations and in the detail of the layout and shapes of the lettering.

Sometimes, the covers were also accompanied by a Certificate of Authentication from Charles R. Riser, Space Cover & Autograph Expertizing Service.

---

[7] More details may be found in the study by Paul C. Bulver, Dr. Reuben A. Ramkissoon and Lester E. Winick: *Study of Suspect Space Covers*, 2nd Edition, ATA Space Unit, 2001, which is the most comprehensive publication on this topic ever produced.

(continued)

(continued)

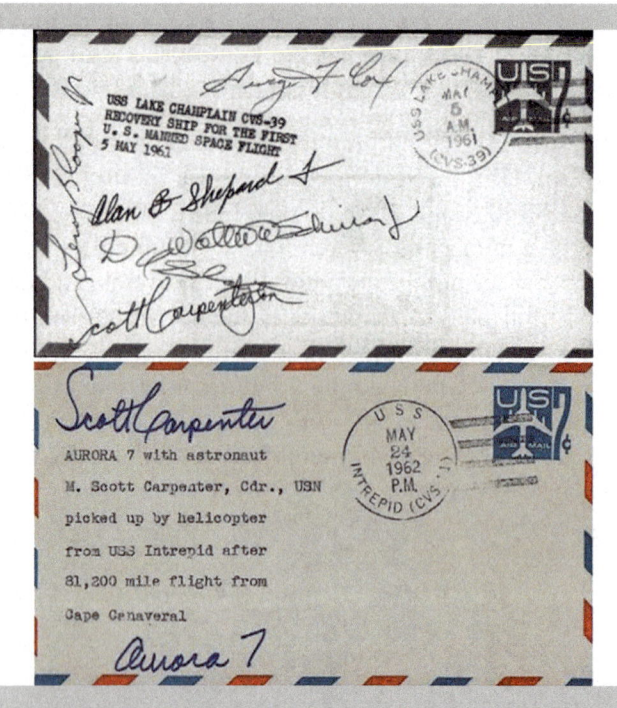

Figure 3.23 (above): Fake cover commemorating the recovery of Alan Shepard's *Freedom 7* Mercury mission. The cover bears the forged signatures of Shepard and other 'Original Seven' astronauts. (below) Forged cover for the recovery of Scott Carpenter's *Aurora 7* Mercury mission.

Figure 3.24: The Riser covers were often addressed to H. Flick (above) or F.C. Shade (below).

John Glenn: The First American in Earth Orbit   133

The backdated USS Noa covers

Another well-known controversy surrounding Glenn's Mercury mission was that of the 'backdated' Noa covers. The USS Noa was originally a secondary recovery ship in the Atlantic Ocean, not the prime recovery vessel. Until the actual recovery, nobody knew that the USS Noa would recover Glenn and the MA-6 capsule.

Like most ships, the USS Noa certainly carried a number of stamps in its onboard post office and somebody took the opportunity – while stocks lasted – to cancel the commemorative envelopes of the historic event. What aroused suspicion and sparked the investigation by the postal police were the stamped envelopes with the new Mercury stamp that were cancelled onboard the Noa on the first day of issue, coinciding with the day Glenn and the Mercury spacecraft were recovered. As mentioned earlier, the stamps had been delivered to

Figure 3.25: Two genuine USS Noa commemorative covers. Figure 3.34 (above): was postmarked aboard the Noa on the morning of the day that John Glenn was recovered with his Mercury capsule. (below) This cover was cancelled onboard when the ship docked at the harbor three days later and could take the new Mercury stamps onboard.

(continued)

134 The Space Race changes direction

(continued)

the post office the day before in sealed packages which were not to be opened until further notice, but they had not been delivered to the recovery ships as they had already been offshore for a few days by that point. The Noa could not have had this stamp onboard and the First Day cancellations onboard the Noa could therefore only have been backdated, postmarked at a later date when the ship was back in dock and could take the new stamp onboard. Conventionally, it was always assumed that only those covers franked with the older stamps (different from the new Mercury stamp) were authentic, as they were genuinely cancelled aboard the USS Noa on February 20.

Australian researcher Ross J. Smith has reexamined the entire story of the USS Noa envelopes with cachet featuring the recovery of the Mercury capsule. [17] First of all, it appears that these covers were professionally printed and that their quality was far superior to what was available aboard the USS

Figure 3.26: Two backdated Noa commemorative covers. The Mercury stamp was not available on the ship on the date of the postmark and was only delivered on February 23 once the ship had returned to the harbor.

John Glenn: The First American in Earth Orbit 135

(continued)

> Noa at the time. Thus, they must have been printed ashore. From the court records, we also know that when the covers were cancelled, the cachet on them was already printed. Smith concluded therefore that these covers were printed ashore between February 20 and 23 and delivered to the Noa when it docked in port. At that time, the indication 'PM' was no longer significant, and the postal clerk removed it from the postmark, before legitimately cancelling the 1,500 covers with the February 23 date. When he then backdated the 300 additional covers, he just changed the date to February 20, 1962, and did not bother with the PM time.
>
> It is likely that the postal clerk ran out of Project Mercury stamps and started using whatever was to hand, using them on backdated covers. Therefore, the type of stamp applied cannot be used as proof that a cover was backdated. Overturning popular opinion, Smith concluded that, irrespective

Figure 3.27: Two backdated Noa commemorative covers. When the postal clerk decided to backdate 300 more covers, he used whatever stamp was to hand as he had run out of the new Mercury stamp but did not use the 'PM' marker.

(continued)

(continued)

> of the attached stamp, only those covers bearing the 'PM' time designation (either blank or with a different cachet) would be guaranteed to be postmarked legitimately on the day of recovery. Irrespective of the cachet or stamp, *all* of the covers without a 'PM' time designation are, very likely, backdated.

SCOTT CARPENTER'S FLIGHT EQUALS THE SOVIETS

The Americans now had a long-term program – to get to the Moon before the "*Russians*" – as well as a short-term one of beating the Soviets who had so far accumulated one record after another. The next mission planned was Mercury-Atlas 7 for the second half of April 1962, for which Deke Slayton was assigned with Wally Schirra as his backup. But a minor heart issue, idiopathic atrial fibrillation, caused Slayton to be grounded in March, less than 10 weeks before launch.

Figure 3.28: Cover commemorating the flight of Scott Carpenter aboard *Aurora* 7. (From the collection of Renato Rega, Italy.)

The mission would have a flight plan similar to that of Glenn's MA-6 mission: three orbits around the Earth and a landing in the waters of the Caribbean Sea, with some additional experiments mainly aimed at putting the astronaut in a central role and measuring his ability to work in weightlessness. To everyone's surprise, the mission was not assigned to Schirra after Slayton's grounding, but instead went to Scott Carpenter, Glenn's backup on the previous flight. As the MA-7 mission was likely to be a repeat of Glenn's MA-6 flight, Carpenter was considered the best-prepared astronaut as he had already trained for a similar mission. The assignment disappointed all involved. Slayton was angry at having lost MA-7 but was mostly devastated by the medical news he had just received, while Schirra was angry at having been dropped in favor of Glenn's backup. Carpenter spent more time apologizing than training. [18]

During the flight of MA-7 *Aurora 7* (4 hours 54 minutes 9 seconds), Carpenter conduced a number of scientific experiments, including studying how liquids behave in weightlessness, observing the airglow layer of the atmosphere and photographing terrestrial features. He also ate solid food items for the first time (in the form of freeze-dried cubes in a plastic bag) rather than pastes squeezed out of a tube. By the end of his second orbit, Carpenter's fuel supply had dropped to worryingly low levels and while the astronaut did not seem concerned about this, Flight Director Chris Kraft certainly was.

While proceeding with scheduled experiments, Carpenter was so focused on changing the film in the camera for a last round of pictures, and on photographing the 'firefly' particles flying off the spacecraft, that he was late beginning his pre-retrofire checklist. Some people have argued that his mission was overloaded with unrealistic task lists but whatever the actual facts, *Aurora 7* missed the recovery area by 250 miles and, for a few heart-in-mouth moments, there was a very real risk that Carpenter would be America's first space fatality. [19] At the tracking station in Guaymas, Mexico, astronaut Gordon Cooper was monitoring the mission's progress. As he came over the horizon from Hawaii on his last pass, Carpenter was well behind the retrofire timeline and in the wrong attitude. Tears came into Cooper's eyes and he buried his head in his hands because he was certain his friend was going to be burned alive on reentry. CBS broadcaster Walter Cronkite announced to the TV audience in a choking voice: *"We may have lost an astronaut."*

But Carpenter made it down, descending into the Atlantic downrange of the targeted splashdown point. After learning that the recovery crew were at least an hour away, Carpenter decided to exit the cramped capsule. When the first recovery aircraft arrived on the scene an hour later, they found him blissfully at ease in his orange life raft, enjoying the sea breeze, his mind wandering, untroubled by thoughts of being alone. When the Navy divers arrived a few hours later and swam up to his raft, Carpenter nonchalantly offered them some food from his survival kit. He was eventually picked up by helicopter and taken to the aircraft carrier USS Intrepid. By convention, the Intrepid was therefore indicated as the recovery ship, despite that fact that the first ship to reach the scene was the destroyer USS Farragut. The Farragut was not equipped to recover the capsule but maintained close watch, recovering Carpenter's personal equipment and checking the floatation attachment on the spacecraft.

138 The Space Race changes direction

Max Faget, who designed the Mercury capsule, would later call Carpenter "*a better poet than an astronaut,*" while Chris Kraft would blame the mission's problems on Carpenter's poor performance, as the astronaut had ignored repeated instructions to conserve fuel and check his guidance instrumentation. Carpenter was the sixth human and the fourth American in space, but after his controversial flight he would never fly in space again. His flight brought the U.S. level with the Soviets for the number of orbital launches and a new American mission was announced for September, with Wally Schirra scheduled to launch on the Mercury-Atlas 8 (MA-8) mission.

Figure 3.29: Cover canceled aboard the recovery ship Intrepid at the end of Carpenter's Mercury mission.

THE FIRST SOVIET "GROUP FLIGHT"

By now, Soviet Premier Nikita Khrushchev was becoming angered by the thought that the Americans might somehow gain supremacy for the number of flights performed, or for the number of orbits or astronauts flown. On the suggestion of Dmitry Ustinov, who supervised the rocket industry, Khrushchev immediately ordered Korolev to organize something striking. Nikolai Kamanin was disgusted by this way of extemporizing without defined programs. He complained in his diary: "*Such is the style of our leadership. They've been doing nothing for almost half a year and now they ask us to prepare within 10 days an extremely complex mission, the program for which has not even been agreed upon.*" Korolev, a skilled organizer as well as a technical genius, did indeed invent something 'striking', compatible with the available technologies, to be exploited for propaganda purposes. To provide "*another proof for the entire world that the Americans are hopelessly behind the USSR,*" he decided to move up a flight that was already in planning and then added a second 'paired' flight. But the missions had to be pushed forward due to the priority of the Zenit photoreconnaissance satellite program, which required the same Vostok rocket and the same R-7 pad at Baikonur.

In May, the second Zenit launch ended with a first stage failure, the rocket exploding 300 meters above the pad and doing enough damage to put the launch complex out of operation for a month. By mid-July, the pad had been restored to use, but there was another delay when the United States carried out a high-altitude nuclear test, known as Starfish Prime, on July 9. The test had unexpected consequences, in that it released high levels of radiation into the upper atmosphere and space, knocking out several satellites and making any manned space launch unsafe for at least a month. By the second week of August, however, the radiation levels had diminished sufficiently for the Vostok 3/4 mission to proceed.

Vostok 3 was launched with Andrian Nikolayev onboard on August 11, 1962. After lengthy arguments between Korolev and Kamanin, his flight was intended to cover 64 orbits around the Earth in four days. Twenty-four hours later, Vostok 4 was launched with Pavel Popovich onboard, to perform the first "*group flight.*" On its 17th orbit, Vostok 3 was expected to return to the position it had been in on its first orbit, and its passage could be measured with extreme precision. It was then possible to calculate the exact moment to launch the second rocket, 24 hours after the first, so that Vostok 4 could be positioned close to its sister ship on orbit. NASA expected a rendezvous at this point, but intercepted flight communications indicated that the two spacecraft had not even attempted to do so. Vostok had no capability of maneuvering. It could be placed into a well-defined orbit but could not change its trajectory. The minimum distance between the two capsules was initially 6.5 kilometers, but neither Nikolayev nor Popovich were able to take an active role and change the orbital parameters of their respective capsules. One thing they could do was to unfasten their 'seatbelts', thus becoming the first humans to fly weightless in space.

The Vostok 3/4 missions marked the first time that more than one spacecraft had been in orbit at the same time. They not only appeared to be mastering the capability of 'formation flying' in space, but they also far exceeded the flight duration records of all the preceding Soviet and American flights combined. The missions were also the first to broadcast television images back to Earth, but the flights were cut short when the ground crew mistook a part of the conversation as a code-phrase requesting that the two cosmonauts be brought back early due to some problem[8].

Figure 3.30: Commemorative 'propaganda' card signed by Andrian Nikolayev (Vostok 3) and Pavel Popovich (Vostok 4). The autograph at the bottom is that of Yuri Gagarin.

Vostok 3 landed after 3 days 22 hours and 22 minutes near Karakalinsk in Kazakhstan. Vostok 4 landed just seven minutes later near Atas, south of Karaganda. In the subsequent historic press conference, these missions – which had proven the reliability of the rocket cluster-engines used by Korolev and the accurate calculation of the trajectory – were presented by the Soviet press as a special achievement and an amazing success, which *"leave the Americans far behind."* None were more willing to believe this than the Americans themselves. [20]

---

[8] The space motion sickness experienced by Gherman Titov during Vostok 2 indirectly led to the premature return of Pavel Popovich's Vostok 4. Prior to the Vostok 3/4 group flight, it had been decided that the cosmonauts would use the phrase *"observing thunderstorms"* to communicate to ground control that they were experiencing a serious attack of motion sickness and needed to land as soon as possible. Having launched on August 12, Popovich was supposed to return to Earth on August 16, one day after Andrian Nikolayev on Vostok 3, but at one point, Popovich reported *"observing thunderstorms,"* triggering the decision by ground controllers to bring him home early. It later turned out that this had been a misunderstanding. Popovich had genuinely been observing thunderstorms while flying over the Gulf of Mexico and was actually feeling excellent. The story is partly corroborated by excerpts from the diaries of Nikolai Kamanin published in 1991. (Hendrickx [1996], pp. 46–7.)

WALLY SCHIRRA: A "TEXTBOOK" FLIGHT

Wally Schirra's flight, which had been pre-announced for September and had triggered the launch of the Vostok 3 and 4 missions in response, was delayed until October 3. Schirra was launched aboard the *Sigma 7* spacecraft for mission Mercury-Atlas 8 (MA-8). The plan was to perform six orbits, and since the Mercury capsule had originally been designed to perform only three orbits it was upgraded with about 20 modifications. Apart from a steering malfunction on the Atlas booster which caused the big rocket to go through an unscripted slow roll after leaving the pad – and had everyone holding their breath – Schirra performed a *"textbook flight"* and the mission was a technical success, with all the demanding engineering objectives fully achieved. The Cuban Missile Crisis soon eclipsed the Space Race in the news, however.

Thanks to his professionalism on this flight, Schirra would fly into space again on the Gemini 6 and Apollo 7 missions, becoming the only American astronaut to participate in all three programs. During his six orbits of Earth on MA-8, lasting 9 hours 13 minutes 11 seconds, Schirra performed the first live American broadcast from space. MA-8 became the longest American manned orbital flight achieved at this point in the Space Race, though it remained well behind the multiple-day record held by Nikolayev's Vostok 3 mission. MA-8 would also mark the first time that a human flight had splashed down in the Pacific Ocean.

Figure 3.31: Commemorative cover for the mission of Wally Schirra aboard *Sigma* 7. (From the collection of Renato Rega, Italy.)

142   The Space Race changes direction

GORDON COOPER: A NEW AMERICAN RECORD FOR SPACE ENDURANCE

Once Schirra's mission was over, NASA officially announced the next flight for Gordon Cooper, the last active astronaut from the first group still awaiting his first space mission (Slayton was still grounded). The program for the new flight included a longer mission duration than the previous ones and meant that several systems in the Mercury capsule would have to be modified, adapted and reconfigured again to support the mission. These changes included the addition of extra batteries and oxygen tanks as well as greater reserves of water. To compensate for the additional weight, technical tools that had proven of little use were removed, including the periscope and a redundant set of thrusters. Given that the Earth would continue to rotate while the mission was in orbit, it was estimated that MA-9 would fly over the entire surface of the planet between the 33rd parallel north and the 33rd parallel south. Tracking station coverage would therefore have to be extended. On May 15, 1963, Cooper was launched on the Mercury-Atlas 9 mission (MA-9). His spacecraft, *Faith 7*, completed 22 orbits of Earth in 34 hours 19 minutes and 49 seconds.

Cooper performed 11 experiments as planned and also became the first American to sleep in space. He did not experience much of an appetite during the flight and only ate because it was scheduled. The food containers and water dispenser system proved unwieldy and he was unable to prepare the freeze-dried food packages properly, so he limited his consumption to cubed food – which he found largely unpalatable – and bite-sized sandwiches.

While the first 19 orbits of the MA-9 mission were mostly unremarkable, the final orbits severely tested Cooper's piloting skills because almost all of the

Figure 3.32: Commemorative cover for Gordon Cooper's *Faith 7* mission, the last and longest of the Mercury program.

onboard systems malfunctioned. The electrical system did not function, the environmental control system was saturated with carbon dioxide, the mission clock was inoperative, and temperatures in the spacecraft exceeded 130 degrees F. The system failures accumulated with each orbit. One of the failures was in the automatic flight control system, and Cooper was forced to reenter on manual control, thus shifting from passenger to pilot. Cooper had to align his spacecraft for retrofire using the horizon as reference and a watch for timing, then manually operate the reaction control system to counter dangerous spacecraft oscillations during the retro burn. But the astronaut seemed perfectly relaxed, taking all the emergencies in his stride and then concluding: *"Other than that, everything's fine!"*

His manual reentry was done perfectly by-the-book and, incredibly, he landed within five miles of the recovery ship USS Kearsage. Both Cooper and his capsule were recovered promptly, thus establishing the record for the most accurate landing in the Mercury program. After the precise landing made by Schirra on the previous mission, Air Force man Cooper enjoyed the idea of a *"blue suiter"* beating his Navy colleague to a pinpoint landing in the sea. The manual reentry had also disproven the derogatory *"spam in a can"* putdown Chuck Yeager had made about the Mercury program before it began.

Cooper restored the balance in the unproclaimed USSR/USA competition and established a new American endurance record in space at 34 hours. After the MA-9 mission, there was a debate about whether to fly one more Mercury mission. Mercury-Atlas 10 was proposed as a three-day, 48-orbit mission, to be flown by Alan Shepard in October 1963. Eventually, however, NASA officials decided that the Mercury program had fulfilled all of its goals and that it was time to move on to Project Gemini. The end of the Mercury program was announced, and MA-10 was cancelled.

VALERY BYKOVSKY PULVERIZES EVERY SOLO SPACE FLIGHT RECORD

While the Soviets were still well ahead with the solo space flight records established the previous year by Andrian Nikolayev and Pavel Popovich, on the orders of a grumpy Premier Khrushchev they promptly looked to establish Soviet leadership in space once again with a new set of space spectaculars: the launch of the first woman in space and a new 'rendezvous' of two spacecraft traversing the same orbit. The endeavor was presented to the press as a *"long-duration joint flight."* Khrushchev wanted three spacecraft to fly in space simultaneously (and at one point consideration was also given to sending Komarov), but there were only two Vostok spacecraft available and there was insufficient time to build another one.

144   The Space Race changes direction

Figure 3.33: Cover and stamp commemorating the flight of Valery Bykovsky aboard Vostok 5. His flight of almost five days remains the longest solo mission in history.

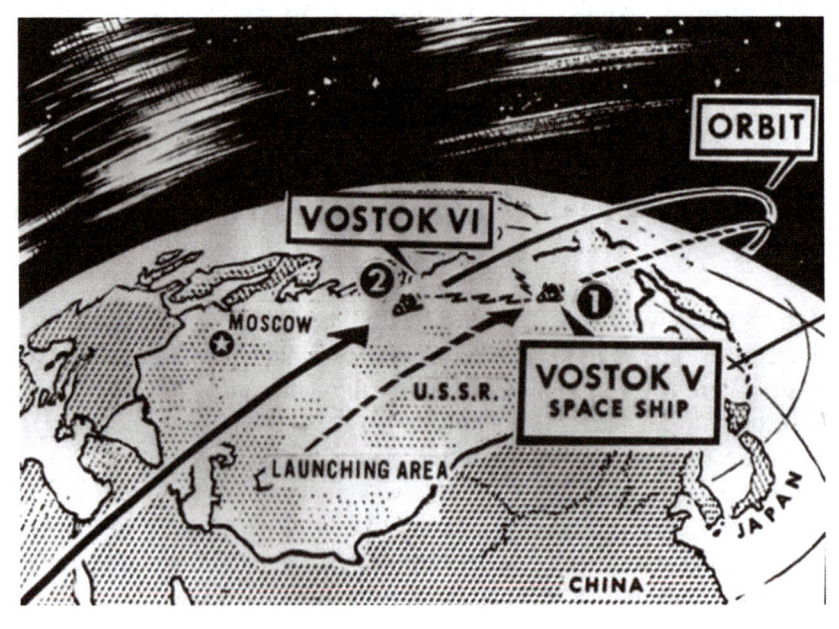

Figure 3.34: Drawing depicting the orbits of Bykovsky in Vostok 5 and Tereshkova in Vostok 6

The launch of Vostok 5, carrying Soviet Air Force Lieutenant Valery Fiodorovich Bykovsky, was initially scheduled for June 12, 1963, but because of the intense solar flare activity, the mission was delayed by two days. During the countdown on June 14, the Attitude Control Handle suddenly stopped working, while a cable detached and lodged under Bykovsky's seat. At the explicit request of the cosmonaut, both issues were fixed without interrupting the countdown and the mission, planned for a record eight days, began at 11:58:58 on June 14, 1963.

Shortly into the flight, the Control Center realized that errors in calculations had resulted in a flight path significantly lower than expected. In addition, the increased friction with the upper layers of the atmosphere began to affect the Vostok's perigee, dropping from 174 to 154 km. Temperatures in the service module began to rise. Onboard, Bykovsky experienced problems with the waste management system, making life in the capsule very uncomfortable. Troubles also arose with the survival system and the internal temperature decreased from 30 degrees C to 10 degrees C. It was decided to return the spacecraft to Earth after only five days in space. However, Bykovsky's record of 118 hours 56 minutes 41 seconds in space pulverized the solo spaceflight record, and the Vostok 5 mission holds the world duration record for a single-crew spacecraft to this day.

VALENTINA TERESHKOVA: THE FIRST WOMAN IN SPACE

Two days after Bykovsky, on June 16, 1963, Vostok 6 was launched carrying Valentina Vladimirovna Tereshkova, the first woman into space. [21] Her callsign for the flight was Chaika (in Russian Чайка, meaning 'Seagull').

On its first orbit, Vostok 6 came within about five kilometers of Vostok 5, the closest distance achieved during the flight, and Tereshkova established radio contact with Bykovsky. Over the subsequent orbits, the two spacecraft progressively drifted apart, with no possibility of the cosmonauts influencing their trajectories. After the second day, communications between the two spacecraft were only possible via the Control Center.

In an interview published in *Komsomolskaya Pravda* in 2007, Tereshkova reported that Vostok 6 had launched faultlessly and the flight went as planned until entering Earth orbit. After she had orbited Earth a few times, the Control Center realized that, orbit by orbit, her spacecraft was moving away from the Earth and this could make it impossible to reenter the atmosphere. The Control Center corrected the automatic orientation and the problem was solved. Information about this mistake remained classified and was only revealed to the newspapers forty years later. [22]

But Tereshkova's troubles were not over. She was ordered to remain strapped into her seat for all the 70 hours of her flight in the tiny spacecraft, wearing her spacesuit and helmet. After a few orbits in microgravity, she began to experience

146　The Space Race changes direction

Figure 3.35: The Soviet stamp issued in December 1963 featured a seagull flying around the Earth

the symptoms of Space Adaptation Syndrome (SAS, or space motion sickness), with nausea, physical discomfort and vomiting. On day two, she developed a cramp in her right leg and on day three the pain became unbearable. Her helmet also put pressure on her shoulders and a sensor on her head continually itched. As a consequence of the SAS, Tereshkova was physically unwell and uncomfortable, with no possibility of cleaning herself. During the landing procedure, she ejected at 7,000 meters, as planned, to land under her personal parachute. *"To my horror,"* she reported, *"I saw that I was heading for a splashdown in a large lake instead of on solid ground. We were trained for such a circumstance, but I doubted I would have had the strength to survive."* Fortunately, a high wind blew her over the shore, but also resulted in a heavy landing. She struck her nose on her helmet, making a dark bruise. Tereshkova was in pain, dirty and almost unconscious and was immediately hospitalized once the recovery team had found her. For propaganda purposes, and the honor of the Soviet Union, it was essential that the return of the first spacewoman should be seen as triumphal. After she recovered. Tereshkova was brought back to the landing site, cleaned up and given a pristine spacesuit, so that she could be filmed for the official news releases.

The idea of training female cosmonauts was initiated by Sergei Korolev in 1961, after the historic mission of Yuri Gagarin, even though Kamanin took the credit for it in his diaries. [23] Initially, the project was opposed by both

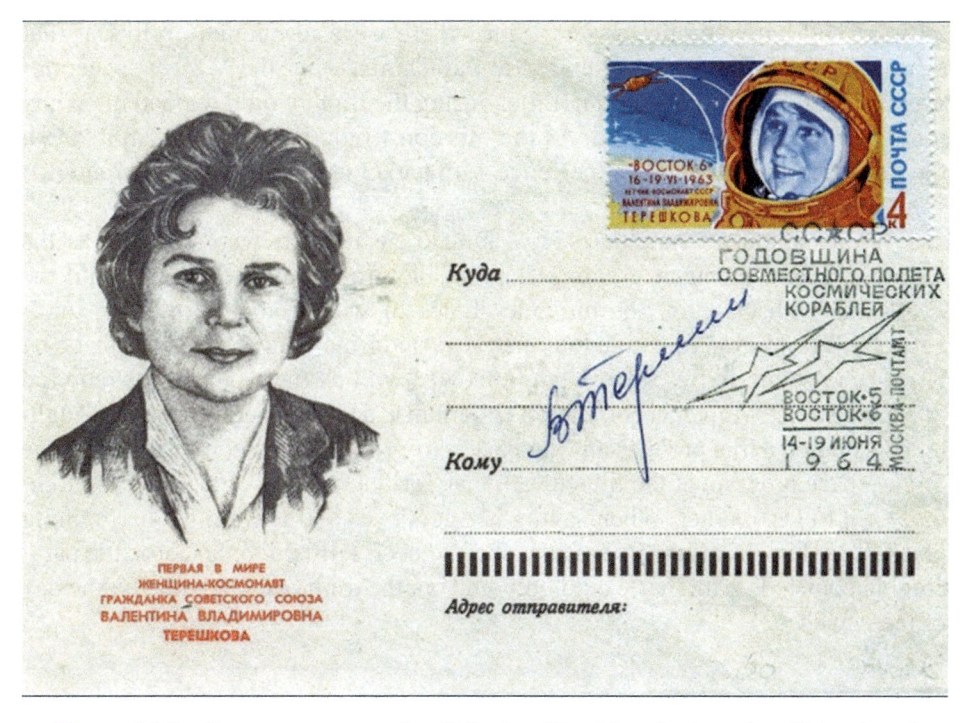

Figure 3.36: Cover commemorating Valentina Tereshkova's three-day flight aboard Vostok 6 in 1963.

Figure 3.37: The first woman in space. (Courtesy Chris Calle.)

148   **The Space Race changes direction**

military and bureaucratic powers. Since there were few female pilots at that time, the search had to be extended to parachutists, but the Vostok spacecraft was completely automatic and nothing during the flight would depend upon any input from the 'pilot' onboard. As the mission would be politically motivated, the only requirements were experience with parachutes and a strong communist spirit.

From an initial 58 candidates, five female cosmonauts were selected as the second USSR cosmonaut group. Details of the group remained secret until the 1980s. As with Gagarin, the ultimate choice of who would fly the mission – between Tereshkova, Valentina Ponomoryova and Irina Solovyova – was made by Khrushchev himself. Tereshkova was endowed with all the typical traits required of a New Soviet Woman: a committed communist, a textile factory worker, daughter of a soldier, a true proletarian – and a *"pretty girl."* The secrecy surrounding the Soviet space program at the time meant that Tereshkova was ordered to say nothing, even to her mother. When saying her goodbyes and leaving for her training phase, Tereshkova could only say that she had been selected for an acrobatic parachuting course. Her mother would only find out the truth over the radio on the day of the launch.

Figure 3.38:   As with Yuri Gagarin, Tereshkova was chosen for her flight by Khrushchev.

Two hours after Tereshkova had landed, and after a five-day mission, Bykovsky was ejected from his spacecraft at 7,000 meters (23,000 feet) above the ground and landed safely at 14:06 Moscow Time on June 19, 1963, some 540 kilometers from Karaganda in what is now Kazakhstan. A few dozen people living nearby rushed over to help him and brought him by car to his spacecraft – which had landed a further two kilometers away – in order to take the ritual pictures.

Figure 3.39: Cover commemorating the joint flights of Bykovsky and Tereshkova in 1963.

SOVIET SUPREMACY CONFIRMED

With the first woman in space, the Soviet Union gained immense esteem worldwide, which the Politburo was able to exploit in a masterly manner for propaganda purposes. The achievement was heralded as a triumph and a few days later, Tereshkova was awarded Pilot-Cosmonaut of the Soviet Union, Hero of the Soviet Union and Gold Star of the Order of Lenin honors. TASS emphatically declared that: "*Valentina Tereshkova was set forth in the same glorious path of progress already walked through by the most renowned Russian women of the past. The triumph of the first spacewoman is obliged to the extraordinary scientific success of Sophie Kovaleskova, the first female professor, and to Sophie Peroskaja, the Russian revolutionary who was executed by hanging after attempting assassination of the czar. Without social emancipation the path to science would be barred to women."*

Figure 3.40: Several Soviet stamps referred to Tereshkova and her historical achievement, both in the USSR and in the satellite countries. The stamps shown here were issued in 1963 and 1964.

Five months later, the Party's General Secretary Khrushchev announced another unexpected event worldwide: the first marriage of two cosmonauts, a *"space family."* Valentina Tereshkova and Andrian Nikolayev, the third cosmonaut in space and the only bachelor to have flown, were married on November 3, 1963. The marriage ceremony took place at the Moscow Wedding Palace and had a huge propaganda resonance. It was rumored that the marriage was at the insistence of Khrushchev. The couple were assigned a luxury apartment on the Kutuzovskij Prospekt. On June 8, 1964, Tereshkova gave birth to a daughter, Aljenka Andrianovna, but the marriage was not as idyllic as the media claimed. Nikolayev was quite a gruff man and the 'space family' fell apart within a few years. As with the American astronauts of that era, however, a divorce would have meant the end of their careers, so the couple remained together. The marriage only officially broke up in 1982.

With the Americans having announced the end of the Mercury program, the Soviet Vostok program also ended following Tereshkova's flight. The female cosmonaut program was disbanded shortly afterwards and no woman would fly into space again until twenty years later, in 1982, when Svetlana Savitskaya was assigned to fly aboard Soyuz T-7 and then Soyuz T-12, this time fully integrated into the cosmonaut corps and not just for propaganda purposes (other than to beat the first American woman, Sally Ride, into space). [24]

Figure 3.41: A Soviet 'space family'. Tereshkova married fellow cosmonaut Andrian Nikolayev in November 1963.

References

1. *Race to the Moon 1957–1975*, Greg Goebel, in vc.airvectors.net, Chapter 13 (accessed in February 2018).
2. **Rockets and People – Vol. 3: Hot Days of the Cold War**, Boris Chertok, NASA SP-4110, Washington D.C., 2009, p. 79.
3. **The Russian Space Bluff – The inside story of the Soviet drive to the Moon**, Leonid Vladimirov, Dial Press, London, 1973, p. 107.
4. Reference 3, pp. 174–78.
5. Reference 1, Chapter 14, (accessed in February 2018).
6. Reference 3, pp. 109–110.
7. Reference 2, p. 193.
8. Reference 3, p. 110.
9. Reference 3, p. 14.
10. *John Houbolt, pivotal figure in reaching the Moon*, Umberto Cavallaro, in AD*ASTRA, no. 21, June 2014, p. 15.
11. *Designer Mishin speaks on early Soviet space programmes and the manned lunar project*, interview with Vasiliy Mishin in *Spaceflight*, Vol. 32, March 1990, pp. 104–6.
12. Reference 3, pp. 121–2.
13. Personal communication between Chris Calle (Paul Calle's son) and the author. The full story can be found in *The Secret Mercury Stamp is Fifty Years Old*, Umberto Cavallaro, in AD*ASTRA, No. 12, March 2012, pp. 19–20.

152 The Space Race changes direction

14. *Charles R. Chickering – Cachetmaker – Part I*, Mark Lerner, in *First Days, Journal of the American First Day Covers Society*, #384' July 2010, pp. 15–16.
15. *The Giori Difference: The 'Top Secret' Project Mercury Stamp*, in postalmuseum.si.edu (accessed in February 2018).
16. *Search continues for secret stamp honoring John Glenn's historic spaceflight*, www.collectspace.com (accessed in February 2018)
17. *Was this Noa Cover Backdated?* Ross J. Smith, *Astrophile*, May 2007, pp. 108–111; see also *The Day the John Glenn and the U.S. Postal Service Shook the World*, Douglas A. Kelsey, *American Philatelist*, February 2002, pp. 130–143.
18. **Challenge to Apollo: The Soviet Union and the Space Race 1945–1974**, Asif Siddiqi, NASA SP-4408, Washington D.C., 2000, p. 243.
19. **The All-American Boys**, Walt Cunningham, iBooks, New York, 2003, p. 101.
20. Reference 3, pp. 115–117.
21. **Women Spacefarers: Sixty Different Paths to Space**, Umberto Cavallaro, Springer Praxis Books, New York, 2017, pp. 1–8.
22. *Un bluff le immagini del rientro dallo spazio*, Fabrizio Dragosei, *Corriere della Sera*, March 7, 2007, p. 24 (in Italian).
23. *Korolev Facts and Myths*, Bart Hendrickx, *Spaceflight* Vol. 38, February 1996, pp 44–48.
24. Reference 21, pp. 9–16.

4

The astronaut sits in the driver's seat

THE GEMINI PROGRAM ANNOUNCED

The Mercury program had achieved all that was asked of it. It was a fast-track project, a makeshift solution intended to catch up with the Soviets, but the successful conclusion of the program helped the Americans to regain confidence. Now that they knew that man was able to survive in space and, if required, could manage emergencies, the Americans began to feel that they could compete and set themselves a precise goal: to beat the Soviets and arrive at the Moon before them. Having chosen the LOR solution, the path had been charted. What was required next was to develop and test a rocket powerful enough and to build a steerable spacecraft.

The LOR approach chosen by NASA required the instigation of another manned program between the two already established, Mercury and Apollo[1]. This third, bridging program would be used to develop certain spaceflight capabilities in support of Apollo. It was called Gemini[2].

---

[1] The Apollo program was designed in the last years of the Eisenhower administration as an evolution of the Mercury program. The name 'Apollo', that of the Greek god of music and light, was chosen by Abe Silverstein, the head of NASA's Saturn Vehicle Evaluation Committee (later known as the Silverstein Committee). *"Apollo riding his chariot across the sun was appropriate to the grand scale of the proposed program,"* Silverstein said. (See *What's in a Name?* Emily Kennard, www.nasa.gov [May 15, 2009], accessed in February 2018.)

[2] The name 'Gemini', in Latin 'twin' or 'double', referring to the third constellation of the zodiac with its twin stars Castor and Pollux, was chosen for a spacecraft capable of carrying an astronaut crew of two.

154 The astronaut sits in the driver's seat

Before starting the Apollo program, aimed at the Moon, the Americans would need to test the techniques of rendezvous and docking, and other orbital maneuvers, which would be essential to allow the Lunar Module (LM) to separate and re-dock after the lunar mission. Extravehicular activities (EVA, or spacewalking) would have to be practiced, atmospheric reentry would need perfecting and the precision of the landing at a target point in the ocean needed to be improved. In contrast to the Mercury capsule, Gemini was designed to be able to alter its orbit and dock with another craft, the first of these being the Agena target vehicle which also had a powerful rocket engine that could be used to perform large orbital changes. Gemini was the first American manned spacecraft to be equipped with an onboard computer to control mission maneuvers.

One of the main ingredients in this new program would be the human factor. On April 18, 1962, NASA decided to hire a second group of astronauts, with advanced engineering backgrounds and extensive experience as test pilots. This new selection, which would be known as the 'New Nine', was selected from 253 candidates and was officially presented as 'Group Two' in Houston, Texas, on September 17, 1962.

Figure 4.1: Cover commemorating the official presentation of 'The New Nine' astronauts selected in 1962.

Figure 4.2: Official picture of the first two groups of NASA astronauts. Sitting in the front row are the 'Original Seven'. From left to right: L. Gordon Cooper Jr., Virgil I. Grissom, M. Scott Carpenter, Walter M. Schirra Jr., John H. Glenn Jr., Alan B. Shepard Jr., and Donald K. Slayton. Standing in the second row are the 'New Nine'. From left to right: Edward H. White II, James A. McDivitt, John W. Young, Elliot M. See Jr., Charles Conrad Jr., Frank Borman, Neil A. Armstrong, Thomas P. Stafford, and James A. Lovell Jr. (Courtesy NASA.)

GEMINI SLOWLY TAKES SHAPE

An early issue with the new program was one of misconception, with Gemini initially considered a simple upgrade of the Mercury program, so that when it was announced on December 7, 1961, Robert Gilruth called it *"Mercury Mark II"*, a sort of *"two-man Mercury."* It was only later that NASA realized that it would not be possible just to leverage off Mercury technologies and that a radically new spacecraft would be required, incorporating several risky cutting-edge technologies.

Other problems came with the qualification of the launch vehicle. NASA selected the Titan II intercontinental ballistic missile (ICBM) as the Gemini booster, which fit perfectly with the program's requirements, not least because its total thrust capability was some two-and-a-half times more powerful than the Mercury-Atlas rocket. The only drawback was that Titan II had yet to fly. To qualify the rocket as an ICBM and to test it for use in the Gemini program, the U.S. Air Force (USAF) planned a series of 32 launches of Titan II throughout 1962 and 1963. Both Cape Canaveral and Vandenberg Air Force Base would be used for the launches. Man-rating the Gemini-Titan would require electrical and hydraulic redundant systems to be implemented, security systems to be developed so that telemetry could be controlled during the flight, and possible malfunctions promptly identified.

156   The astronaut sits in the driver's seat

The contract with Martin Marietta was signed by NASA two months after the merger between Martin Company (which designed the Titan II missile) and American-Marietta. When the program entered its operational phase, the USAF was already working on the development of the new Titan III version and was quite reluctant to invest in the previous model, but Titan III would prove to be unreliable in the short term. On the other hand, Titan II showed a nasty tendency of 'pogo oscillation' (excessive longitudinal vibrations in the first stage. The nickname for this vibration problem, which bore similarities to the action of a pogo stick, was invented by NASA engineers). While this was of little concern to the USAF for military use, it greatly worried NASA officials, as the phenomenon would be harmful to astronauts on a manned Gemini flight. Analysis of Titan I data revealed that the problem had always been there, but it had been ignored since it was not a major issue when the payload was a nuclear warhead. It would take more than a year, and a dozen test runs, to identify the causes and remedy them.

Figure 4.3, Figure 4.4: Fine tuning and testing of the Titan II rocket would require over a year. (From the collection of Steve Durst, USA.)

Figure 4.3, Figure 4.4: (continued)

At the insistence of the USAF, a new method of return from reentry, by way of gliding to a landing strip, was also trialed. This would eliminate the need to deploy an expensive fleet of naval vessels to recover the spacecraft[3], as well as removing the risk of sinking after landing. After reentering the atmosphere, the spacecraft would turn from a capsule with limited control into a pilotable vehicle by deploying an inflatable 'flexible wing'. This had been invented by Francis Rogallo, an amateur kite flyer who eventually became a NACA aeronautics project engineer responsible for the wind tunnel, and then a researcher at NASA's Langley Research Laboratory. [1]

North American Aviation was given the contract to develop the 'Parawing' (as NASA renamed it) but the company, distracted by its Apollo work, did not have the resources to do a good job on it and, after poor test results, rising costs and delays – in the race against time – the 'Parawing' would be abandoned in 1964 in favor of the old-fashioned parachute and splashdown return profile.

---

[3] For John Glenn's 1962 mission, there were 23 ships in the Atlantic Ocean and one in the Pacific. Wally Schirra's flight in 1963 employed the largest naval fleet, with 21 ships stationed in the Atlantic and another six in the Pacific.

Figure 4.5: 25 flights of the 'Parawing' were performed in 1964 to test the new landing system. (From the collection of Steve Durst, USA.)

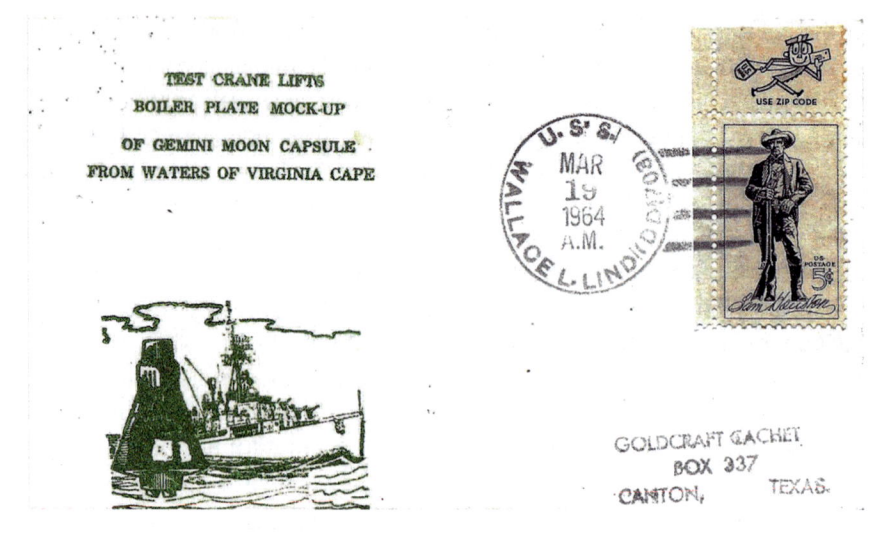

Figure 4.6: The USS Wallace L. Lind, practicing the recovery of the Gemini module off Norfolk, Virginia. Cover commemorating the test performed on March 19, 1964. This rarely seen cover was cancelled aboard the USS Lind. (From the collection of Pietro Della Maddalena, Italy.)

NASA GETS BACK ON TRACK

President John F. Kennedy had set NASA a tough schedule for reaching the Moon, and the whole Gemini program was in serious trouble, with costs starting to climb out of control and schedules definitely behind planning and increasingly in doubt. The brilliant project manager, Jim Chamberlain, was replaced in 1963 by NACA old-timer Chuck Mathews, who tried to get things back on track. NASA's Administrator, James Webb, realized that strong leadership and direction would be critical to achieving success with the extraordinary goals of the multiple concurrent programs that were being developed in parallel, including Gemini, Saturn and Apollo. He coopted George Mueller, who had already shown his determination in handling critical projects such as the USAF Atlas, Titan Minuteman and Thor ballistic missile programs, as well as in the development of other space-related projects such as Explorer VI and Pioneer V.

Mueller joined NASA as Associate Administrator in 1963. He soon realized that *"there isn't any management system in existence."* To improve efficiency and to put Gemini (as well as Saturn and Apollo) back on track and solve the problems with the Bureau of the Budget, Mueller began a complete restructuring of NASA, from the top down. He created the Office of Manned Space Flight (OMSF) at NASA Headquarters (now the Human Operations and Exploration Mission Directorate) and had the three NASA centers devoted to manned spaceflight reporting to him directly. Those three were: The Manned Space Center (MSC – which would later become the Johnson Space Center, JSC – headed by Robert Gilruth) that was developing the Gemini and Apollo craft in Houston; the Marshall Space Flight Center (MSFC, directed by Wernher Von Braun) that was developing the Saturn V rocket in Huntsville, Alabama; and Cape Canaveral in Florida (which would shortly become the Kennedy Space Center) that was developing launch structures. [2]

During his six years of service at NASA, Mueller would introduce a remarkable series of management changes within the Agency, including the winning 'all-up testing' approach (see Chapter 5). On November 1, 1963, 'Program' replaced 'Project' in the title of the office that directed Gemini. The change reflected its responsibility for the program as a whole and not merely for the spacecraft. [3]

THE DYNA-SOAR BECOMES EXTINCT

The excessive attention of the U.S. Air Force in the Gemini program triggered a strong reaction from NASA, which claimed autonomy when it came to space programs. A review of the whole situation led to the USAF resizing its 'space' projects,

160 The astronaut sits in the driver's seat

with the X-20 Dyna-Soar[4] program bearing the brunt of the changes. The hypersonic glider program, already in doubt in the eyes of many, was still in its 'project' phase and had thus far cost $660 million. It was intended as a reusable spaceplane, capable of attacking a target nearly anywhere in the world at the speed of an ICBM[5] (achieved by traveling into space and reentering the atmosphere) and was designed to glide to Earth again like an aircraft under the control of a pilot[6]. Supported by a small but powerful USAF lobby, Dyna-Soar had very vague military purposes, ranging from reconnaissance from space (that was already being carried out adequately by the Corona project) to bombardment from space (for which intercontinental rockets had been developed), to space rescue (which overlapped NASA's remit).

Figure 4.7: Cover commemorating the first experimental launch of Dyna-Soar, postmarked at Cape Canaveral on September 18, 1963. (From the collection of Stefano Matteassi, Italy.)

The Dyna-Soar program was finally cancelled in December 1963. [4] Some would ironically comment that calling the project something that sounded like 'dinosaurs' could only lead to its inevitable extinction. [5]

At the end of 1963, after a delay of 11 months, the Americans finally announced the beginning of two programs: Gemini and Saturn.

---

[4] The U.S. Air Force described the X-20's gliding landing style as "*DYNAmic SOARing.*"

[5] The X-20 was the culmination of concepts that had begun with Eugen Sänger in 1928 and had progressed in Germany during WWII as a way to bomb America from Europe.

[6] The experience gained in the design and development of Dyna-Soar would be useful later in the design of the Space Shuttle.

THE ASTRONAUTS OF GROUP 3: FRESH FORCES ARRIVE

Despite the delays accumulated by the Gemini program, NASA still had to look at the Apollo program and to achieving President Kennedy's target, so they put out a call for more new astronaut candidates. The third group of 14 new astronauts was presented to the public on October 18, 1963. They were: Edwin E. 'Buzz' Aldrin, William A. Anders, Charles A. Bassett, Alan L. Bean, Eugene A. Cernan, Roger B. Chaffee, Michael Collins, R. Walter Cunningham, Donn F. Eisele, Theodore C. Freeman, Richard F. Gordon Jr., Russell L. 'Rusty' Schweickart, David R. Scott, and Clifton C. Williams.

Figure 4.8: Cover commemorating the announcement of 14 new NASA astronauts in October 1963.

THE GEMINI PROGRAM LIFTS OFF

On April 8, 1964, a Titan II rocket finally put the Gemini-Titan 1 (GT-1) unmanned test vehicle into orbit, the first flight of the Gemini program. Its main objective was to test the structural integrity of the new spacecraft and the modified Titan II ICBM. Six minutes after launch, the spacecraft achieved orbit, although thanks to a little excessive speed provided by the launcher, it was a higher orbit than planned. Gemini 1 performed almost 64 revolutions, until the orbit decayed due to atmospheric drag. Although only the first three orbits were considered to be part of the mission, the Mission Control Center kept monitoring the spacecraft

162 The astronaut sits in the driver's seat

until its destructive atmospheric reentry – according to the plan – on April 12 over the south Atlantic. There was no intention to recover the spacecraft and in fact, four large holes had been drilled through its heat shield to ensure that it would not survive reentry. The spacecraft was the first to carry a computer for guidance.

The Gemini 1 test flight confirmed that the 'pogo effect' had been corrected, which meant that plans for a Gemini manned flight could proceed. NASA announced the prime crew for the first manned mission as Gus Grissom and John Young, with Wally Schirra and Tom Stafford as their backup crew. The mission was scheduled for early December 1964.

Figure 4.9: Covers commemorating the launch of Gemini 1 and the announcement of the crew for the first manned flight, Gemini 3. (From the collection of Steve Durst, USA.)

VOSKHOD 1: THE MOST ABSURD ADVENTURE IN SPACE EVER

When the Americans announced their new Gemini program at the end of 1963, which would launch two astronauts into space, Nikita Khrushchev summoned Sergei Korolev and ordered him to fly not two, but three cosmonauts before November 7, 1964, the anniversary of the October Revolution: "*If the Americans have a vehicle for two, we prove our superiority by flying three men.*" [6] Korolev tried to explain that it would be impossible to prepare a new spacecraft – and especially a new rocket that was powerful enough – in just a few months, but to no avail. Khrushchev was not interested in such tedious technical details[7].

The most important thing as far as the Soviet leader was concerned was the spectacular headline-making potential of such a flight: "*to launch into space three Soviet citizens before the Americans can launch their two astronauts.*" The threat was clear: If Korolev was unable to fulfill the "*task entrusted to him by the Party and the Government,*" then that task would be passed to somebody else who would be ready to do the job. Once again, the specter of Vladimir Chelomey hung over Korolev's shoulder. [7] His biographies would later say that 'S.P.' (the nickname of Korolev among his colleagues and employees, from the initials of his first two names, 'Sergei Pavlovich'), already known for his bad temper, would become intractable during this period. His deputy, Leonid Voskresensky, wanted Korolev to try to convince Khrushchev that it would be better to direct Soviet manned spaceflight programs towards the construction of new scientific space stations, but Russian experts knew that pursuing this path would cost them the Space Race. Unable to deal with the stresses of this latest situation, Voskresensky collapsed and was hospitalized.

Korolev was already working on a new spacecraft, but it was still far from completion. The only spacecraft available for Khrushchev's mission was the old single-seat Vostok, whose internal diameter was less than two meters at its widest. But with so little time available, the only possibility left was to reconfigure the internal space, by removing all the scientific equipment and minimizing the survival reserves and safety systems. Vostok was renamed Voskhod, in order to deceive the outside world into thinking that the Soviets had constructed a brand new spacecraft in just a few months. It was absolutely forbidden to disseminate any information or drawings about the 'new' vehicle. [8] However, despite the refit, it soon became clear that it would be impossible to squeeze three people within the Voskhod module, even if they chose the smallest of the cosmonauts.

---

[7]This section was first published as "*50 Jahre Wos-chod 1: Die absurdeste Raumfahrtmission aller Zeiten*", in WeltraumPhilatelie no. 256 (Winter 2014), pp. 20–23 [in German]; "*Voschod 1 - neiabsurdnéisi kosmickv let vsech dob, 50. vvroci letu*", in Kosmos no. 3/2014, pp. 112–115 [in Czech]; "*Voskhod-1. La più assurda avventura spaziale di tutti i tempi*" in AD*ASTRA no. 22 (October 2014), pp. 13–15 [in Italian]; "*Voskhod-1. The most absurd adventure in space ever*", in ORBIT (quarterly Journal of the ASSS - UK) no. 104 (January 2015) pp. 21–22 [in English].

164   The astronaut sits in the driver's seat

Figure 4.10: Commemorative Mission Cover featuring Konstantin Feoktistov, with a special postmark and the signatures of the three Voskhod 1 cosmonauts (from left: Boris Yegorov, Feoktistov and Vladimir Komarov).

Then, Konstantin Feoktistov, the chief engineer of the department for return equipment, made a daring and risky proposal bordering on madness: fitting the three cosmonauts into the Voskhod without spacesuits, relying entirely on the hermetic sealing of the capsule. When Korolev questioned *"Who on Earth would [be willing to] fly without his spacesuit?"* Feoktistov answered: *"Me, to begin with!"* Thus, the eminent engineer, a man in less than robust health and with obvious sight problems, suddenly became a cosmonaut. In addition to Feoktistov, the chosen crew was Boris Yegorov, a young and diminutive physician who became the first doctor in space, and Vladimir Komarov, the best of the Cosmonaut Corps at that time[8].

---

[8] The three-man Voskhod 1 spacecraft not only enabled Khrushchev to outshine America's two-man Gemini spacecraft, it also provided Korolev with the opportunity to break the Air Force's monopoly in the cosmonaut team. (Hendrickx [1996], p. 47.)

With surprising speed, a chair was manufactured for each cosmonaut, perfectly adapted to their individual shape. Under Korolev's ruthless direction, three dummies were rapidly constructed to fit the seats. It quickly became apparent that it was impossible to fit all three into the capsule conventionally. A triangular arrangement of the three seats was also tried but this did not work either. Eventually, after several trials, a less obvious but viable solution was found: packing the cosmonauts into the capsule like sardines in a tin. As the smallest of the crew, Yegorov was placed in the front seat – raised with respect to the other two. Komarov, the 'pilot' cosmonaut, who would have no role in piloting the vehicle, sat crouched beneath him, with engineer Feoktistov alongside.

It would be fatal for the three cosmonauts to eject from the capsule without their spacesuits, so Korolev had to work out how to bring the whole capsule down under its own parachutes. Among the items removed in trying to reduce the weight of the capsule were many of the explosive bolts. This meant that there would be no emergency system to rescue the three cosmonauts during the first 27 seconds of the flight. Three months before the launch, Voskhod 1 still weighed 220 pounds more than the 11,700-pound launch capability of the existing rockets. The race to eliminate even more weight necessitated the oddest of tricks and included putting the three cosmonauts on a strict diet.

Figure 4.11: First Day Cover, serviced on October 19, the day the four stamps were issued. One stamp is devoted to each of the three crewmembers and the other (the blue one used to frank the cover) devoted to the mission. The cover also bears the stamp of Komarov, signed by the cosmonaut.

Figure 4.12: Stamp depicting the crew of Voskhod 1; Komarov, Feoktistov and Yegorov.

The first test flight of an unmanned Voskhod spacecraft was launched on October 6, 1964, under the name of Cosmos 47 and carrying a dog and the mannequin 'Ivan Ivanovich'. This unique test was considered sufficient for this high-risk venture and the first capsule carrying three cosmonauts was launched a few days later.

On the morning of October 12, 1964, a bus carried the three cosmonauts to the foot of the 38-meter-high rocket. The cosmonauts were wearing light jackets. After the official greeting, Sergei Korolev went up to the cosmonauts and embraced each of them one by one, something he had never done before. The official explanation for this unusual behavior was that it would have been impossible on previous flights because the cosmonauts were wearing spacesuits. [9] The cosmonauts squeezed themselves into the capsule one at a time and, for the first time, the hatch was closed and hermetically sealed from the inside.

The launch went without a hitch and Voskhod completed 17 orbits around the Earth, according to the well-tested routine. Crammed into the spherical capsule, the three cosmonauts suffered space sickness and, for the first time in a space mission, they had no tasks to perform. "*It was a circus act*," as Korolev's deputy Vasiliy Mishin put it, "*for three people couldn't do any useful work in space.*" [10] During the flight, they greeted the athletes competing at the Olympics in Tokyo and, in keeping with tradition, spoke with Khrushchev over the phone, connected from his dacha on the Black Sea.

Korolev's biographers have pointed out that despite its triviality ("*Yes, Nikita Sergejevich… You are right Nikita Sergejevich… At your command Nikita Sergejevich… Thanks, Nikita Sergejevich!*"), this conversation is noteworthy because it was the last public conversation by Khrushchev. Voskhod landed in Central Asia on October 13, 1964 and the following day, Khrushchev was suddenly summoned to Moscow. Escorted from the runway where he landed, the "*irresponsible voluntarist*" Khrushchev was taken directly into the building of the Central Committee of the CPSU, where he was relieved of all Party and government positions. Unfortunately, it was soon realized that with Khrushchev overthrown, the Soviet space program had lost its main sponsor.

Figure 4.13: Official commemorative First Day Cover (FDC) issued by Kniga on October 19, 1964, on the day of issue of the three stamps. The cover is signed by (left to right) Komarov, Feoktistov and Yegorov. (Courtesy signedfdc.blogspot.com)

The cosmonauts were due to be welcomed in Moscow on October 15, but they did not arrive there on that day, nor the following few days, remaining back in the steppes of Central Asia pending new orders. They would be received by new leaders Leonid Brezhnev and Alexei Kosygin during the pair's first public appearance a week later.

Before the program had begun, Chuck Yeager had derided the American astronauts and their Mercury capsule as "*spam in a can*." The Voskhod and its three-man crew would be similarly referred to as "*sardines in a tin*." [11] But the mission did achieve another first. Even if the three-man Voskhod followed in the Russian tradition of a 'Potemkin village' (an impressive façade disguising a shabby building) in space, its propaganda value was in emphasizing that the Soviets had already been able to send three cosmonauts into space while the Americans had announced – but were yet to implement – a program for only two astronauts. Unusually, a 24-page brochure with 15 black-and-white photographs was widely distributed, showing the cosmonauts' training, the flight, the glorious return and the parades through the cheering crowds. The fact that only one of the three cosmonauts was a trained pilot was presented as a clear demonstration of the absolute reliability of the new

168　The astronaut sits in the driver's seat

Voskhod spacecraft. The cosmonauts had flown in a 'shirt-sleeve environment', without spacesuits, because Russian spaceships were now so safe that the suits were no longer necessary. [12] The Western world was left to react to the new Soviet space spectacular, wondering what kind of revolutionary new spacecraft the Russians had been able to develop[9]. However, Korolev asked that the propaganda should stop referring to the 'unlucky' light jackets and was eventually heeded.

Voskhod 'French' fakes

Figure 4.14: 'French' fakes commemorating the missions of Voskhod 1 and 2. As usual, they bear the fraudulent Baikonur-Karaganda cancel (a Post Office was only opened in Baikonur in April 1975, just prior to the joint USSR-USA Apollo-Soyuz mission).

---

[9] Years later, Alexei Leonov would recall in his book: "*the spacecraft completed a flight of 16 orbits with three cosmonauts aboard, prompting both envy and admiration of the West.*" (Scott-Leonov [2004], p. 95.)

GT-2: HALTED BY A LIGHTNING STRIKE

A second unmanned Gemini flight, Gemini-Titan 2 (GT-2), was launched on January 19, 1965, three months after the Soviets had gained the front pages of the newspapers again with their triumphal three-man Voskhod flight. The suborbital Gemini 2 focused on testing the thermal shield and validating the recovery system. The mission had been delayed several times, firstly when a lightning strike knocked out power to Launch Complex 19 in August 1964. The launch vehicle then had to be dismantled and stored in a safe place to protect it from Hurricane Cleo, which passed over Cape Canaveral in August, and then two other hurricanes (Dora and Ethel) which arrived a few weeks later. On December 9, 1964, the launch countdown reached zero and the first stage engines were ignited, but the Malfunction Detection System detected a loss of hydraulic pressure and shut the engines down again barely a second after ignition. By this point, NASA had already given up any hope of putting the first manned Gemini mission into orbit in 1964.

Shortly after the launch of Gemini 2, the Mission Control Center suffered a power outage and did not get its systems back online until the flight was almost over. Control of the mission was transferred to a tracking ship. It was later discovered that the power outage had been due to an overload of the electrical system, caused by the network television equipment used to cover the launch. Despite this, most of the mission's goals were satisfactorily achieved, with the heat shield and retrorockets functioning as expected. The Gemini 2 reentry module was recovered by the aircraft carrier USS Lake Champlain and would later fly again, on November 3, 1966, on a test flight for the USAF Manned Orbiting Laboratory (MOL) project. It would become the first vehicle ever to be reused in space.

Figure 4.15: Scarce 'really run' cover sent by Ensign David Sorenson to his father of the day of the recovery of Gemini 2. (From the collection of Steve Durst, USA.)

THE LAST DRAMATIC CHANCE TO BEAT THE AMERICANS

Following the fall of Khrushchev, a much more hopeful Korolev asked his staff to prepare a detailed report for Brezhnev and Kosygin. The report explained the current state-of-the-art of the space programs in the USSR and the U.S., and bluntly stated that the Soviet attempts to reach the Moon were being conducted not with any precise scientific program, but only with the ambition to *"beat the Americans at any cost"* and snatch from them any new space 'firsts'. This criterion could no longer be followed because the Americans were now much more advanced with their rocket motors and electronic instruments. The paper also disclosed the story behind the preparations of the Voskhod 1 mission which – it was later reported – both impressed and horrified Brezhnev and Kosygin[10].

Figure 4.16: In the mid-1960s, lunar landscapes often appeared on the Kniga cachets, as if to indicate that the Moon was the next target of the Soviet space program.

Korolev's report suggested several proposals for the future Soviet space program and recommended dropping the idea of landing on the Moon, limiting the research

---

[10]This section was published as *"Die letzte Gelegenheit die USA zu schlagen"*, in WeltraumPhilatelie no. 257 (Spring 2015), pp. 13–17 [in German]; and *"Voskhod-2: L'ultima drammatica corsa per arrivare prima"* in AD*ASTRA no. 24 (March 2015), pp. 6–8 [in Italian].

to activities that would actually be achievable such as sending automated probes. It suggested no longer panicking in response to American spaceflight announcements, with emergency launches aimed solely at 'going one better', and recommended addressing efforts towards studies of both a suitably powerful rocket-carrier and the design of orbital scientific stations. These proposals were apparently favorably accepted. Encouraged by their attitude, Korolev promised his new 'bosses' he would do everything possible one final time to "*put one over on the Americans*" who, as part of their Gemini program, had announced plans to send an astronaut outside the spacecraft on a spacewalk. Korolev began to prepare a last flight of the Vostok/Voskhod spherical capsule and arranged the mission of Voskhod 2.

At the suggestion of Voskresensky, it was decided not to depressurize the cabin in order for a cosmonaut to step outside, but instead to place an airlock between the cockpit and the exit. The preparations for the mission were interrupted by Voskresensky's ongoing health problems. His years of imprisonment, the ceaseless and exhausting toil, and the continuous stress had led to him suffering serious heart disease. He was hospitalized several times in his last years but recovered and kept returning to work.

Later in the year, on December 14, 1965, the Voskresenskys went with friends to a concert in Tchaikovsky Hall. After the concert, they dropped in on their friends and it was here that Voskresensky himself asked for an ambulance. By the time it arrived, he was already dead. He was just 52. [13] The death of his associate deeply affected Korolev. On his own tombstone, it says that without Voskresensky, it would not have been possible to launch Sputnik before the Americans. Physically in poor health himself, Korolev was forced to resume work just a day after the funeral of his friend.

VOSKHOD 2: THE FIRST SPACEWALK

For Voskhod 2, although he no longer had Khrushchev harassing him with telephone calls from the Kremlin, Korolev was determined, as always, to move heaven and Earth to beat the Americans one more time. Alexei Leonov and Pavel Belyayev were selected as the crew for the flight, but this mission also came with a great many weighty problems to solve and it was now that Korolev really began to miss the regular input from the ailing Voskresensky, supporting him with his stream of ideas. There was still room to reduce the amount of onboard subsistence carried, but two cosmonauts in spacesuits would weigh no less than three in jackets. There was also the new airlock to consider. At the last moment – after the food had already been loaded together with "*a small portion of kharcho, a spicy Georgian soup of rice, meat, onion and garlic*" that he had requested, Leonov had a change of mind, as if he had had a premonition. He recalled: "*At the last moment, I ordered most of this food to be replaced with extra ammunition for my pistol. What use would we have for food in a mission expected to last only twenty-four hours?*

Much better to carry more cartridges for self-defense, in case our spacecraft landed in an area with wild animals." [14]

Figure 4.17: Soviet Post celebrated the new space first of Voskhod 2 on March 19, 1965, by issuing a non-perforated stamp designed by the Lesegri team. Under the watchful eye of KGB censorship, they had to invent a fancy, stylized spaceship, totally different from the real one, which it was strictly prohibited to reproduce. A perforated version of this stamp was issued few days later, on March 23. The stamp is signed by Alexei Leonov.

The mission was launched on March 18, 1965 and was announced by Radio Moscow shortly after the launch, without providing any details. It was only after the end of the Cold War and the dissolution of the Soviet Union that the world would learn just how close this mission had come to tragedy.

After 12 minutes spent in open space, Leonov's space suit had inflated, expanding like a balloon as a result of the so-called 'football bladder effect'. [15] His suit had stiffened so much that Leonov was not even able to activate the shutter on his chest-mounted camera to photograph the Voskhod, nor retrieve the camera that had immortalized his spacewalk.

More serious problems arose when Leonov tried to reenter the capsule through the inflatable airlock, as he later recalled: *"As I edged closer to the airlock's entrance, I realized I had a very serious problem. My spacesuit had ballooned in vacuum to such a degree that my feet had pulled away from my boots and my fingers no longer reached the gloves attached to my sleeves. No engineer had been able to foresee this... Now the suit was so misshapen that it would be impossible for me to enter the airlock feet first as I had in training. I simply couldn't do it. I had to find another way of getting back inside the spacecraft, and quickly. The only way it seemed possible was by squeezing head first into the airlock."* [16]

After trying for eight minutes, Leonov finally managed to enter the airlock, but got stuck sideways. His exertions had raised his body temperature and after many attempts to get fully inside, he was becoming exhausted. Eventually, he decided to depressurize his suit by opening the valve that discharged the air, running the risk of decompression sickness and blood embolism: *"The only solution was to reduce the pressure in my suit by opening the pressure valve and letting out a little oxygen at a time as I tried to inch inside the airlock. At first, I thought of reporting what I planned to do to Mission Control. But I decided against it. I did not want to create nervousness on the ground. And anyway, I was the only one who could bring the situation under control."* [17]

Figure 4.18: Postcard signed and designed by Alexei Leonov, who is also a fine artist.

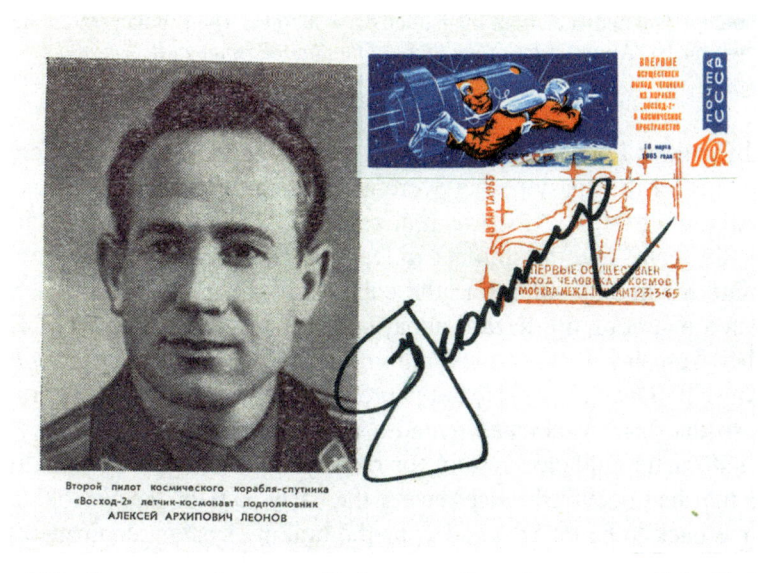

Figure 4.19: Commemorative cover with the non-perforated stamp, cancelled with the pictorial red-ink postmark used on March 23, 1965 in the Moscow International Post Office, as indicated in the bottom line of text. The Russian text reads: *"For the first time a man has gone out in the cosmos."* On the same day, a similar pictorial postmark – with a few differences in the design – was also used at the Moscow Main Post Office, with black ink.

Leonov would later confide that his helmet included a suicide pill, to use in the event that things went wrong and Belyayev was forced to abandon him in space. [18] He also recognized that if his training had not been so intensive, he would never have been able to perform the complicated maneuvers that had saved his life. [19] *"From the moment our mission looked to be in jeopardy,"* Leonov recalled, *"transmissions from our spacecraft, which had been broadcast on both radio and television, were suddenly suspended without explanation. In their place, Mozart's Requiem was played again and again on state radio."* [20]

Figure 4.20: A fixed-date pictorial postmark was used on March 18, 1966 to commemorate the first anniversary of the launch of Voskhod 2. The Russian text in the external crown reads: *"Anniversary of the flight of the Sputnik spacecraft Voskhod 2."*

The difficulties Leonov experienced in reentering the spacecraft were not the last problems for the mission; they were just the start of a series of dire emergencies. Firstly, after the EVA, it was realized that the Voskhod exit hatch was not hermetically sealed and the ship began leaking air. The automatic system tried to compensate for this by saturating the cabin atmosphere with oxygen, which in turn created a serious fire hazard aboard Voskhod 2. In the words of Korolev's biographer Yaroslav Golovanov: *"the tragic shadow of Valentin Bondarenko loomed over Voskhod 2."* [21] Next, the pressure inside the cabin began to increase steadily so that the cosmonauts feared a massive explosion. Then they realized that the automatic guidance system for reentry was not working. Finally, an old problem that had occurred twice during the early tests of the Vostok capsule in 1960 came back to haunt Voskhod 2, as the braking system, controlled from the ground, failed to work on orbit 17. Korolev ordered the cosmonauts to operate the system by hand on the next orbit, but going into an 18th revolution meant that they could no longer land in southern Russia. All department aircraft stationed in Siberia and in the Arctic area were alerted. For four hours at the Command Center, there was no direct contact with the spacecraft and it was unclear what had happened to it. Mission Control had no idea where they were, or whether they had survived.

Finally, a report came in from a civil helicopter that had discovered a red parachute, some 30 kilometers southwest of the town of Bereznyaka in the Northern Ural forest, approximately 1,500 km (930 miles) west of where they were supposed to land. They had also spotted the cosmonauts but had no idea how to rescue them. They tried to toss a rope ladder down but, as Leonov recalled: *"We would have had to be circus acrobats [to use it]. It was a flimsy, unreliable ladder and our spacesuits were too heavy and stiff to allow us the agility of scaling its rungs. As news of our whereabouts was relayed from pilot to pilot in the area, more aircraft started to circle above us. There were so many at one point that we worried there would be a serious accident if one collided with another."* [22]

The dense, snow-covered and inhospitable Siberian pine forest prevented the helicopter from landing near the cosmonauts, and there were no populated areas nearby. *"We were only too aware,"* Leonov commented, *"that the taiga where we had landed was the natural habitat of bears and wolves. It was spring, the mating season, when both animals are at their most aggressive. We had one pistol aboard our spacecraft, the firearm I had stowed away at the last moment, but we had plenty of ammunition."* [23] The TP-82 shotgun would remain a routine part of the crew's survival kit until it was officially abolished in 2007 following the ISS Expedition 16 mission.

The following day, a small advanced rescue team – including two doctors and a cameraman – reached the cosmonauts on skis.

It would be another two days before Leonov and Belyayev were rescued; the time it took to clear two patches of forest large enough to land rescue helicopters. Overall, the recovery and transfer of the two cosmonauts took twice as long as the duration of their mission in space. [24] It would be almost four years before the Soviets attempted another EVA.

The real tragedy, however, was that Voskhod had taken up three critical years that should have been devoted to the development of Soyuz, for the sake of space spectaculars. Voskhod 2 would be the last mission headed by Korolev and the last prestigious record achieved by the Soviets. Indeed, it would be the last Soviet manned space mission for some time. After the success (and near-tragedy) of Voskhod 2, the Soviets unexpectedly suspended all space activities.

Figure 4.21: This picture – from the collection of Walter Hopferwieser, Austria – flew aboard Voskhod 2 with Leonov and Belyayev and was presented to Vladimir Belyayev (no relation) who was the first to arrive on skis at the landing site. On the front was written *"To Volodya Belyayev"* and it was signed by both cosmonauts. On the reverse, it was signed by 15 leaders of the Soviet space program, including Sergei Korolev and Nikolai Kamanin; Soviet Air Force Marshals Konstantin Vershinin and Roman Rudenko; and cosmonauts Yuri Gagarin, Gherman Titov, Valery Bykovsky, Valentina Tereshkova and Sergei Anochin.

GT-3: THE FIRST SPACECRAFT 'FLOWN' IN SPACE UNDER PILOT CONTROL

While the Soviets were grabbing the headlines with the first three-man crew on Voskhod 1 and the historic first spacewalk by Alexei Leonov on Voskhod 2, the Americans seemed to be marking time. But they had been moving forward

GT-3: The First Spacecraft 'Flown' in Space Under Pilot Control 177

Figure 4.22: Official FDC serviced by Kniga for the commemorative stamps issued on May 23, 1965. Both of the cosmonauts have signed this cover; Leonov on the left and Belyayev on the right.

methodically and NASA would now begin to step up the pace, launching ten next-generation manned spacecraft over the next 20 months and performing increasingly complex tasks with each new mission. This was probably one of the reasons why the Voskhod flights were suddenly discontinued. The success of the Gemini missions over the next two years would far exceed all of the accomplishments of Vostok and Voskhod. The Soviets would have to come up with something better to upstage the Americans in space.

The first manned Gemini flight, Gemini-Titan 3 (GT-3) was launched on March 23, 1965, just after the conclusion of the Soviet Voskhod 2 mission. Its crew was made up of veteran astronaut Gus Grissom and rookie John Young, the first member of the second astronaut group to fly in space. Grissom would become the first human to fly two space missions (although Joe Walker had been the first to *reach* space twice, on suborbital flights of the X-15 in 1963). Following the well-established habit of the Mercury flights, NASA allowed Grissom to name his Gemini spacecraft. Recalling how *Liberty Bell 7* had sunk in the ocean after his Mercury 4 mission, Grissom chose the name *Molly Brown* – after the popular Broadway musical of the time, "*The Unsinkable Molly Brown.*" The name was intended to bring good luck to his first manned Gemini crew, but was not appreciated by NASA managers, who asked him to come up with an alternative. Grissom then ironically suggested *Titanic*, so *Molly Brown* was retained for the mission. NASA would call a halt to the

Figure 4.23: Commemorative covers cancelled in Perm (above) and Volgograd (below). To commemorate the first anniversary of the flight, a pictorial postmark was used on March 18, 1966, in six different postal facilities throughout the Soviet Union: Moscow, Moscow International, Kemerovo, Volgograd, Perm and Vologda. The Russian external text reads "*Anniversary of the flight of P. Belyayev and A. Leonov in the spacecraft Voskhod 2.*" The post office where the postmark was used is identified just below the date. Among the most interesting are the covers cancelled in Perm and in Volgograd.
Perm, in the Ural Mountains, was the rural settlement with a small post office that was nearest to the landing place (about 75 kilometers, 47 miles away). The inhospitable forest is featured in the cover cachet and green ink was used for the postmark. This was the first space-related postmark used during the Soviet Era in Volgograd, the largest city near to the Kapustin Yar Cosmodrome, which, at that time, was still secret.

tradition after the flight of Gemini 3 and the custom of naming spacecraft would only restart, for operational reasons, with Apollo 9 in 1969.

Gemini 3 was simply a technological validation mission to ensure that the spacecraft would perform as designed and could support its crew. The mission orbited only three times over the course of less than five hours. Grissom and Young also performed some basic scientific experiments and tested the new spacecraft's

GT-3: The First Spacecraft 'Flown' in Space Under Pilot Control

Figure 4.24: Mission emblem for The Gemini 3 flight of Grissom and Young aboard *Molly Brown*.

Figure 4.25: Covers commemorating the launch of GT-3. As highlighted on the cover, the goal of the mission was to flight-test the Gemini spacecraft.

180   The astronaut sits in the driver's seat

maneuverability in orbit. For the first time ever, a crewed spacecraft was actually 'flown' in space under pilot control, thus paving the way for manned orbital rendezvous tests. The mission is also remembered for the corned beef sandwich that John Young 'smuggled' aboard the spacecraft with the help of Wally Schirra, who had acquired a reputation as a practical joker. While in orbit, Young offered the sandwich to his crewmate Grissom in zero gravity, spreading breadcrumbs around the capsule. Bread would prove problematic as a space food. The episode stirred up a hornets' nest at NASA and in Congress, where some members were looking for any excuse to cut agency funding and were very keen to come down on the frivolity and apparent ill-discipline on a mission that had cost substantial taxpayer dollars.

NASA promised to take steps to prevent the recurrence of such an episode on future flights. This episode did not help to maintain good relations between Chris Kraft (Mission Control) and Deke Slayton (Chief of the Astronaut Office). Relations between the two remained strained in the following years, even during the 'Apollo Era', with Kraft continuing to show Slayton and the Astronaut Corps a lack of respect and camaraderie and often directing his ire towards them.

Regardless of all this, Gemini had now proven to be a reliable system and to make up for lost time, NASA's top brass decided to schedule the remaining launches of the program at two-month intervals instead of three-months as originally planned.

The 'Plugged-9' Covers

Figure 4.26: Cover for the mission of Gemini 3, featuring the well-known 'Plugged-9' fake cancel (shown in more detail bottom right).

(continued)

> With the first manned Gemini mission – which coincided with the introduction of the postal ZIP Code (Zonal Improvement Program) in the Cape Canaveral hand cancel – a new forged postmark appeared on the American market: the well-known 'Plugged-9' cancel (see Figure 4.26). The origin of this fake cancel is not entirely clear, although its discovery in 1972 was soon followed by the arrest of William Ronson, who was found guilty at his trial in New York.
>
> The forged cancel device was never found and initially, the whole episode seemed to be about an 'inappropriate' use of an official Post Office stamp that the alleged perpetrator and cover servicer had used for his cacheted envelopes, which were marketed under the trade name 'Orbit Covers'. This shocking episode marked the end of production of 'Orbit Covers' although after his release from prison, Ronson again tried to reenter the covers 'business' over the next few years, with little success as the popularity of such covers was on the wane. 'Plugged-9' fake covers first appeared in the mid-1960s. They exist for all of the Gemini missions from GT-3 to GT-12 and for the missions of Apollo I, II, III, 8, 9, 10 and 12. Subsequent studies of the various cancels suggested the existence of a false cancel device that Ronson had probably produced by cloning a Cape Canaveral Post Office device. Whatever its origin, these fakes are unfortunately quite common.

GT-4: THE AMERICANS ALSO WALK IN SPACE

The first 'true' Gemini mission was Gemini 4 (GT-4), which was launched on June 3, 1965 and carried two rookie astronauts into orbit, Jim McDivitt and Ed White. During the lift-off, the mission was directed, tracked and supported as usual by the Mission Control Center at Cape Canaveral, but at the conclusion of the launch phase, control was taken over for the first time by the new, improved Mission Control Center (MCC) operating at the Manned Space Center in Houston, Texas.

The mission did not start well, with the MCC fearing the worst as the Titan's old problem of pogoing returned for a while during the ascent, causing the astronauts to stutter over the communications link. Fortunately, the booster then smoothed out its flight. Gemini 4 was originally supposed to last for seven days, putting the U.S. within reach of Soviet capabilities. The crew attempted a rendezvous with the upper stage of the Titan II that had carried them into orbit, but after fruitlessly consuming much of their fuel, the crew had to abandon the attempt. NASA came to realize that maneuvering a spacecraft in orbit was not as simple as had been assumed and would require more specific research to understand.

A second objective for this mission was the first American Extra-Vehicular Activity (EVA). White exited the spacecraft and floated around in space, using a compressed-air gun, the Hand-Held Maneuvering Unit (HHMU), to push himself

182 The astronaut sits in the driver's seat

Figure 4.27: (above) Commemorative GT-4 launch cover. Some astrophilately 'purists' prefer this kind of plain cover, with no cachet, that can be recognized and appreciated only by experts familiar with dates and places who can relate the cover and mission. (below) Commemorative cover by Treyco that marks the first operational use of the new Mission Control Center in Houston, Texas. The two postmarks (launch and splashdown) draw attention to the new duration record set by the Gemini 4 crew, who spent four days in space. (From the collection of Steve Durst, USA.)

out and control his walk in space. His EVA lasted 23 minutes and Ed White became the second man, and first American, to conduct a spacewalk. The crew also tested an improved David Clark Company G-4C spacesuit. The primary change was the addition of a thermal/bumper garment consisting of a layer of felt, to provide micrometeorite protection and retain heat.

The excess fuel and battery power consumption necessitated shortening the mission to four days, but even this shortened duration helped to fine-tune the crew's diet and health techniques that would be adopted for the later long-duration missions. The spacecraft splashed down in the Atlantic Ocean after 4 days 1 hour 52 minutes and was recovered by the USS Wasp.

While White's spacewalk was not a novelty after Alexei Leonov's earlier EVA, the fact that it was announced in advance and broadcast live on TV thrilled both the international and domestic television audiences. "*Many press reports in the West,*" recalled Leonov, "*later claimed that White had been the first to perform a spacewalk and that mine had been a fake; that the film of me outside the Voskhod 2 had been staged in a laboratory. The reports were taken so seriously that the Guinness Book of Records, for instance, for some time recorded White as the first man to walk in space. NASA did nothing to contradict the false claims."* [25]

Figure 4.28: Fake 'Plugged-9' covers exist for all the Gemini missions.

Gemini Celebrated in the USA with a 'Space Twins' Stamp

Ed White's spacewalk during Gemini 4 would be featured in the twin stamps issued by the U.S. Postal Service on September 29, 1967 to celebrate the successful conclusion of the Gemini program. U.S. space stamps were particularly rare at the time. Apart from the one issued in 1948 for the Fort Bliss Centenary (which depicted one of the V-2 rockets acquired by America as spoils of war, together with the scientists who had designed and built them), only three other U.S. space stamps existed: The Echo-1 stamp of 1960; the Mercury stamp of 1962; and the Robert Goddard stamp of 1964.

The assignment to design the Gemini stamp was entrusted to Paul Calle, a well-known artist who worked in the NASA Fine Art Program with the aim of artistically recording the early steps of American Space Exploration. In 1962, Calle had to decline the invitation to design the Mercury stamp due to too many concomitant obligations[11]. This was Calle's first experience with stamps and it

---

[11] In a private communication with the author in March 2011, Chris Calle, Paul's son, recalled: *"In 1962, my father was asked to design the Mercury stamp, but he was out of the country. He was on assignment in Bermuda – I believe an Air Force Art Program assignment – and he was unable to begin design concepts. That's what I remember… just one of many things I wish I could still ask my dad. Can you imagine if he had designed the Mercury, Gemini Twin AND the 1969 First Man on the Moon stamps!"*

(continued)

(continued)

> fascinated him: "*The Gemini work really was the beginning of a new career for me,*" he explained. [26] "*Designing stamps is truly a unique experience! The subject matter is chosen for its national significance, usually of historic importance and the conception of the design must be thought of in terms of art in miniature form. Rather than 'think big', the designer must 'think small'!*" [27]
>
> The stamp had to be a 5¢ first-class rate stamp and, as the name 'Gemini' suggested, nothing less than a pair of twin stamps could express the success of the Gemini program. It would be the first twin stamp issue in the history of U.S. philately. "*The design of the 'Twin Space Stamp' presented me with a challenge unique in postal stamp design at the time,*" Paul Calle recalled. "*The assignment was to design a commemorative issue that would symbolize the successful conclusion of the NASA Project Gemini Program of Space Exploration. The unique aspect of the challenge was to conceive a design that graphically would be pleasing when used as a double stamp, and, when one twin was separated from the other, the design of the single stamp had to be a complete composition of its own.*" [28]

Figures 4.29–4.31: Sketches drafted by Paul Calle for the study of the Gemini twin stamp. (Courtesy Chris Calle.)

(continued)

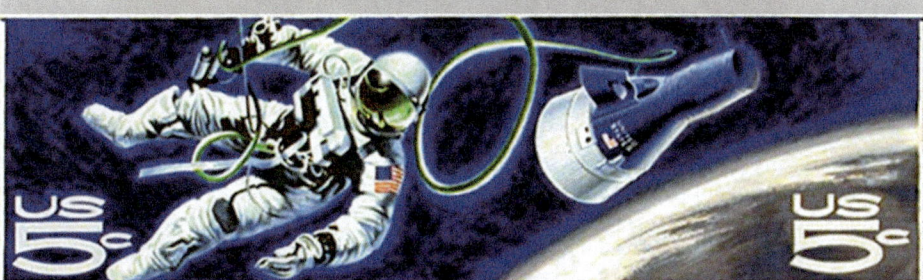

Figures 4.29–4.31: (continued)

(continued)

186 The astronaut sits in the driver's seat

(continued)

> Calle refined his preliminary 'thinking' sketches and submitted them to Stevan Dohanos, who headed the Citizens' Stamp Advisory Committee in charge of selecting art to appear on United States postage stamps. "*Soon we realized,*" Calle recalled, "*that the first U.S. spacewalk had a great visual impact and was well suited to symbolically express the success of the Gemini program.*" [28] The two stamps were connected graphically by the umbilical cord that provided oxygen, communication and a tether between the astronaut and his spacecraft. [29] "*My wife Olga and I attended the first day ceremony,*" Calle remembered, "*and I was seated next to Mike Collins at the luncheon. I was really excited to be sitting next to a real astronaut and a spacewalker on Gemini 10. I asked Mike what it was like to walk in space. He replied: 'Frankly, Paul, it was so cramped in the spacecraft, I could hardly wait to open it up and get out of that thing'.*" [29]

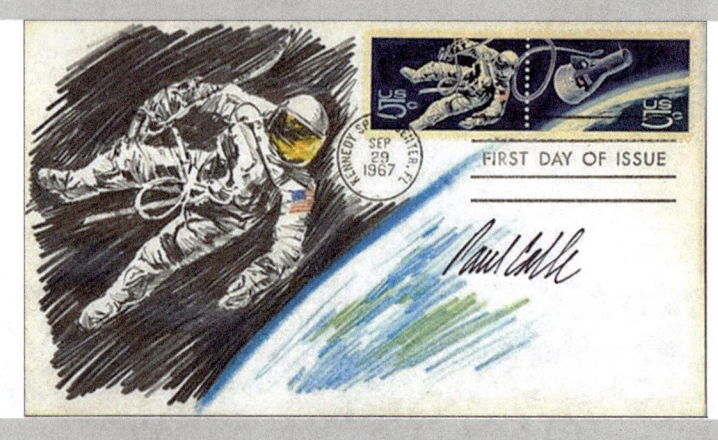

Figure 4.32: Hand-painted cover by Paul Calle. (From the personal collection of Chris Calle, USA.)

GT-5: THE USA SURPASSES SOVIET RECORDS

Two months later, the record for endurance in space passed into American hands for the first time, thanks to the Gemini-Titan 5 (GT-5) mission of Gordon Cooper and Pete Conrad, which surpassed the previous record of five days held by the Soviets with Vostok 5. The GT-5 mission lifted off on August 21, 1965 and lasted about 191 hours (nearly eight full days).

GT-6 & 7: A New Endurance Record and the First Rendezvous in Space

Despite NASA forbidding the astronauts to name their mission, following the *Molly Brown* saga of GT-3, Cooper's emblem design for the GT-5 mission included the motto 'Eight Days or Bust'. In the end, NASA allowed Cooper's mission patch design, but insisted that the motto must be covered up, in case they 'busted' before making it to eight days. With this mission, Cooper became the first man to *orbit* the Earth for the second time. Gus Grissom had been the first to achieve two space missions, but the first of those – on Mercury-4 – had only been a suborbital flight.

Figure 4.33: Mission emblem for Gemini 5. Gordon Cooper's "*8 Days or Bust*" motto was rejected by NASA and removed from the mission patch

For the first time on this mission, because of its longer duration, new fuel cells were used rather than the short-lived conventional storage batteries, although there were still some issues. Track changes were also attempted on this flight, in order to practice maneuvering the spacecraft in orbit. The astronauts did not seem to have much of an appetite during the mission and consumed on average about 1000 calories per day, well below the intended 2,700 calories.

GT-6 & 7: A NEW ENDURANCE RECORD AND THE FIRST RENDEZVOUS IN SPACE

Gemini 6 was originally intended to be the first mission to rendezvous and dock with a modified Agena upper stage, launched specifically for that purpose, but the Agena target failed to reach orbit and blew up over the Atlantic

188   The astronaut sits in the driver's seat

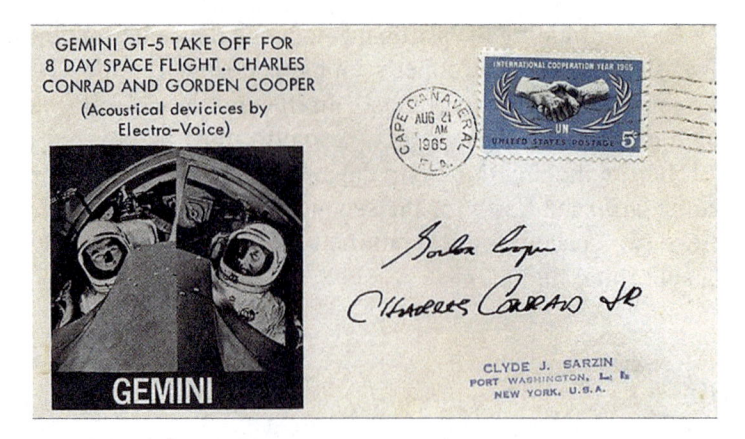

Figure 4.34: Cover commemorating the launch of GT-5. This was also one of the missions for which a number of 'Plugged-9' fake covers were produced.

Ocean. The countdown for the launch of Gemini 6, scheduled for 90 minutes after the Agena launch, was immediately halted, much to the frustration of the two astronauts, Wally Schirra and Tom Stafford, sitting in the capsule on Pad 19.

McDonnell's Gemini Spacecraft Chief Walter Burke then suggested that NASA simply replace the Agena target with a new one – the Gemini 7 mission that was in advanced preparation – and have the two spacecraft rendezvous with each other. Senior NASA management at first rejected this idea, which seemed unrealistic given that there was only one launch pad at Cape Canaveral that could support a Gemini-Titan launch and it was not set up for two launches in sequence. Moreover, it was felt that Mission Control could only handle communications with one orbiting crewed spacecraft at a time. However, the idea was enthusiastically supported by the astronauts and NASA mission personnel, who found viable solutions to make the proposal work. In a few days, all the pieces came together and Gemini 7 was launched on December 4, 1965, carrying Frank Borman and Jim Lovell. Once they had reached orbit, they would await the arrival of Gemini 6, which had been rescheduled for launch immediately after Gemini 7. At 14 days, the Gemini 7 mission would set a new record for the longest spaceflight, which it held until the flight of Soyuz 9 in June 1970.

Onboard Gemini 6A was the same crew as in the original Gemini 6 mission, but the amended name was an indication that Schirra and Stafford's mission goals had been modified. The launch had to be scrubbed again when an umbilical cable dropped a second too soon, causing the Titan's engines to shut down after ignition.

GT-6 & 7: A New Endurance Record and the First Rendezvous in Space

Figure 4.35: Cover commemorating the flight of Gemini 7, the 14-day mission of Frank Borman and Jim Lovell.

With the booster's fuel tanks fully pressurized, there was a danger of the Titan exploding like a bomb. Schirra, the commander of the mission, was aware of the failure and had to decide instantly whether the two astronauts should eject. As the Titan hadn't budged off the pad, the level-headed Schirra evaluated that there was no need to eject, thus making it possible to reschedule the mission. Had he chosen to abort, the dual mission would have had to be cancelled. The crew were fortuitous, in a sense, as the later inspection of the booster revealed that someone had neglected to remove a dust cover during engine assembly, an oversight that might have had disastrous consequences. [30]

After three days of intense work, Gemini 6 was finally launched successfully on December 15, 1965, and the two spacecraft made the planned rendezvous, or close approach in space, a little over seven hours after launch. The two Geminis maneuvered with each other for several orbits, coming as close as one foot (30 cm) apart. Although the Soviet Union had twice previously launched pairs of Vostok spacecraft, they could only passively approach and establish radio contact with each other (coming no closer than several kilometers apart and in different orbital planes). Gemini 6 and 7, however, were the first to move into close proximity, directly piloted by the astronauts, and could have docked if they had been so equipped.

190   The astronaut sits in the driver's seat

Figure 4.36: Commemorative covers for the launch of GT-6 (above) and for its recovery by USS Wasp (below).

The astronauts were highly impressed with the fine-control maneuverability of their spacecraft. Velocity inputs as low as 0.03 meters per second (0.10 fps) provided very precise maneuvering. After attaining all their mission's goals, Gemini 6A reentered the atmosphere after only 26 hours, splashing down within 18 kilometers (11 miles) of the planned site, northeast of the Turks and Caicos Islands in the Atlantic Ocean, where it was recovered by the USS Wasp. This was the first truly accurate reentry, and the first recovery to be televised live. Two days later, on December 18, GT-7 also splashed down and was recovered by the same USS Wasp. Lovell later described the flight as: *"something like sitting in a latrine for two weeks without access to a shower."*

In their post-flight debriefing, Borman and Lovell noted that the food rations had generally been of good quality, but they strongly disliked the freeze-dried protein bites and advised against them being included on future missions. They also suggested that more breakfast items would have been nice, that NASA should avoid including bite-sized food which readily produced crumbs that floated loose in the cabin, and that the packaging of some items needed improving. Because of the mission's duration, Gemini 7 carried a significantly larger supply of food than previous flights, and the astronauts often found it difficult to remove the tightly-packed food containers, some of which had also not been stowed in the correct order for the day of the flight they were intended to be eaten.

GT-8: AN EMERGENCY IN SPACE

The sixth Gemini mission, Gemini-Titan 8 (GT-8) was scheduled as another attempt to perform the Agena docking originally planned for mission GT-6. Neil Armstrong (a civilian test pilot with long experience in the X-15 rocket research aircraft program)[12], was in command of the mission, with Dave Scott (the first astronaut of the third group to fly in space) as the pilot. It was the first mission for both astronauts. In the Mission Control Center, Walt Cunningham and Jim Lovell served as 'CAPsule COMmunicators', or Capcoms. The mission launched on March 16, 1966 and after just five orbits, GT-8 reached the Agena target, which had been launched on the same day and was now in the planned orbit and oriented in the proper attitude for docking. The docking proceeded perfectly, achieving another space first for the Americans: the first docking of two spacecraft in orbit. However, the euphoria would not last long.

Immediately after the docking with the Agena, a runaway thruster began rotating the joined vehicles. Armstrong fired the thrusters to compensate and correct the deviation but it soon started again. The two astronauts assumed that the fault was with the Agena and shut down its attitude control system, but the rotation continued, now in two axes, and GT-8 had to undock from the Agena. [31] Alarmingly, without the added mass of their target, the spin rate increased to about once per second: "*We were in an uncontrollable tumble in space. And it was about to get much worse*," Scott later recalled. [32]

Armstrong and Scott were in serious trouble. If the spin continued, they might have blacked out and would not have recovered. To get the spacecraft under control, they disengaged the maneuvering thrusters (including the faulty one) and fired the Re-entry Control System (RCS) thrusters to stop the spin, using almost 75 percent of the reentry maneuvering fuel. Mission rules dictated that the flight had to be aborted if the RCS was fired for any reason. Less than eight hours after lift-off, the crew of GT-8 were in the sea, floating in a raft in the Pacific waiting for a destroyer to pick them up from the emergency recovery zone off the Japanese

---

[12] Neil Armstrong was the first civilian U.S. astronaut. All those who had flown before were military pilots.

Figure 4.37: Covers commemorating the launch of GT-8 and the "emergency in space."

island of Okinawa. They were recovered by the USS Boxer after 40 minutes. It was the first time that a NASA mission had been abandoned due to an emergency situation.

VOSKHOD 3: IN SEARCH OF AN UNLIKELY NEW SOVIET SPACE SPECTACULAR

The Gemini program had begun scoring success after success and the Americans were not just catching up with the USSR. With each flight, they were overhauling the Soviets and carrying out new, more difficult tasks. The news that Gemini 5 had surpassed the USSR's record for endurance in space annoyed the Soviets, who were busy further modifying the Voskhod capsule in preparation for the flight of Voskhod 3. The new mission was intended as a manned long-duration flight, with cosmonauts Boris Volynov and Georgy Shonin, that would far surpass the record

Voskhod 3: In Search of an Unlikely New Soviet Space Spectacular

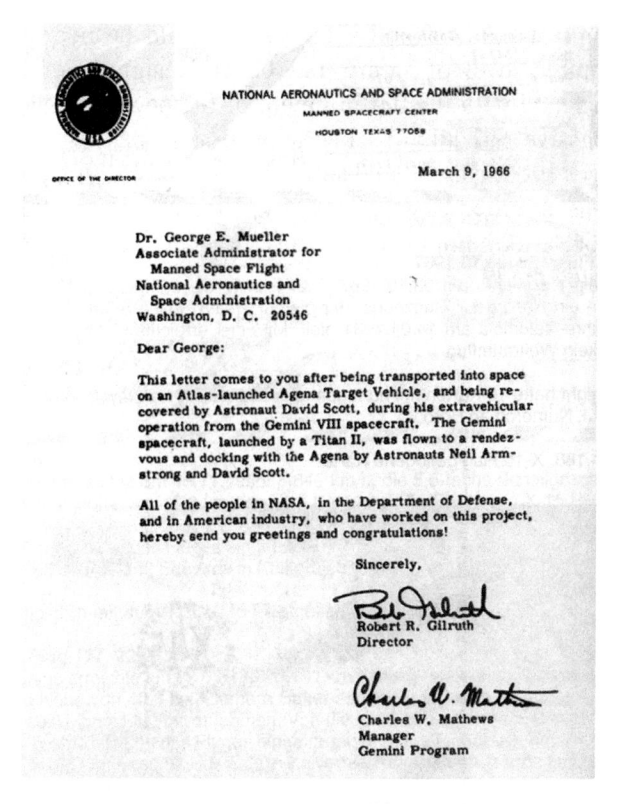

Figure 4.38: No philatelic covers were carried aboard the Gemini missions. This letter, addressed to George E. Mueller by NASA MSC Director Robert Gilruth and Gemini PM Charles W. Mathews, was microfilmed and sent into space aboard the Agena 8 satellite but, due to the emergency, the 11 × 8 mm microfilm could not be recovered by Dave Scott during the Gemini 8 mission as planned. It was retrieved four months later by Michael Collins, during his Gemini 10 EVA. (Courtesy of Walter Hopferwieser, Austria.) [33]

achieved by Gemini 5 (a mission that *Red Star*, the newspaper of the Soviet Defense Ministry had called "*A spy in the sky*"). The subsequent Voskhod 4 mission would be a scientific flight, including artificial gravity experiments, with test pilot cosmonaut Georgy Beregovoy and scientist Georgy Katys, while Voskhod 5 would be a military mission that would include cosmonaut Vladimir Shatalov.

Voskhod 3 was scheduled for launch in autumn 1965 and was due to set a new space endurance record of twelve days, but the Soviets quickly realized that the deadline could not be met due to serious problems with ripped parachutes and the environmental control system. The problems were compounded by rivalry between Soviet space medicine institutions. [34] The mission was rescheduled to coincide with the 23rd Congress of the Communist Party in early March 1966.

194 **The astronaut sits in the driver's seat**

A major redesign of the Voskhod subsystems would be necessary in order to reach an acceptable level of safety, but Chief Designer Sergei Korolev was already overburdened with the development of the new Soyuz ('Union') spacecraft and the massive N-1 lunar rocket, as well as plans to soft-land a probe on the Moon in early 1966. Unwell and suffering from a bleeding polyp in his intestine, Korolev was admitted to hospital early in the new year. Only days later, on January 14, 1966, after complications arose in what should have been a routine operation, Korolev died[13].

Figure 4.39: The dogs Veterok and Ugolyok on a Soviet stamp issued in 1966.

A test mission was launched on February 22, 1966, disguised under the cover name of Cosmos 110, completing a 21-day flight with two dogs onboard. Veterok and Ugolyok survived the flight, despite its orbital path taking them through the Van Allen Belts, but they returned in dreadful condition, with muscle wasting, dehydration, calcium loss and problems walking. The Voskhod 3 mission itself was delayed further and further. A failure of the R-7 necessitated another postponement and the mission was rescheduled for May. Meanwhile, the Americans were obtaining a sequence of successes, including the endurance record of 14 days (GT-7), the first rendezvous of two spacecraft (GT-7 and GT-6A) and the first docking in Earth orbit (GT-8). Leonid Smirnov, chairman of the Military-Industrial Commission, realized that the flight of Voskhod 3 would now serve no purpose for the Soviet government: even if it achieved a new record duration, it would not be spectacular enough to have an impact on world public opinion.

---

[13] See Chapter 5, page 215: The Sudden Passing of Korolev.

GT-9 Lifts-Off with the Backup Crew   195

Additionally, after the death of Korolev, his successor Vasiliy Mishin simply had no desire to begin his tenure with a now-obsolete spacecraft that had only a small margin of safety for the cosmonauts. Flight testing of the new Soyuz would begin with the launch of Cosmos 133 on November 28, 1966. Interest in another Voskhod mission diminished and while Voskhod 3 was never formally cancelled, it simply faded away for the lack of a spectacular reason to fly it.

GT-9 LIFTS-OFF WITH THE BACKUP CREW

The third attempt to dock with an Agena was performed by the Gemini-Titan 9 (GT-9) mission. The crew originally assigned to the mission were Elliott See and Charlie Bassett[14], but the two were killed when their T-38 jet crashed on February 28, 1966, as they reached the McDonnell factory in St. Louis where they were due to undertake some simulator training. The weather conditions were poor, with rain, snow and low visibility. Tom Stafford, who was piloting a second T-38 with Gene Cernan, decided to abort his approach and go around for another try. In the first aircraft, pilot Elliott See decided instead to try to land and the T-38 hit the roof of the factory building where the Gemini was being assembled, tearing off a wing and crashing into the parking lot at the rear. Both See and Bassett were killed instantly but fortunately there were no other fatalities, just 14 others injured, though none seriously. It could have been much worse. If See had flown a little lower, the T-38 would have slammed directly into the factory building, probably with considerably more fatalities and certainly destroying the Gemini spacecraft under construction there, setting the entire program seriously behind schedule. Stafford and Cernan remained circling around in the murk, not knowing what had happened. For a while, there was confusion and some assumed that it was Stafford and Cernan that had been killed.

For the first time in American space history, a mission would be performed by the backup crew, while Stafford would also become the first astronaut to fly twice in the Gemini program. GT-9 was launched on June 3, 1966 and all went well until the approach to the Agena target. During the rendezvous maneuvers, the astronauts realized that the nose fairing of the satellite had failed to eject (it would emerge that this was due to a launch preparation error on the ground), making it impossible for them to dock. The crew used the fuel to practice approaching from below the target.

---

[14] Elliott See and Charlie Bassett, like Neil Armstrong, were true civilian test pilots in the astronaut corps at the time. Whereas Armstrong had come from the X-15 program and was regarded as one of the best pilots of the corps, See had come from performing engine flight tests for General Electric, and was considered one of the weakest.

196   The astronaut sits in the driver's seat

Figure 4.40: (above) Cover cancelled in St. Louis on February 28, 1966, to commemorate the accident in which Bassett and See were killed (from the collection of Antoni Rigo, Spain). (below) Cover commemorating the launch of mission GT-9, signed by Bassett and See and then by Stafford and Cernan, who became the prime crew. (From the collection of David Ball, USA.)

Back at the Cape, backup astronaut Buzz Aldrin suggested that Cernan should cut the spring-loaded lanyards with surgical scissors from the equipment pack. Ground controllers were against this idea due to the possibility of puncturing Cernan's suit and also because of the risks involved in managing the explosive bolts that, for unknown reasons, hadn't worked. There were some frank exchanges and some friction with the crew over this but in the end, the idea was shelved. It has been reported that after this episode, Bob Gilruth (Director of the Manned Spaceflight Center in Houston) suggested to Deke Slayton (Chief of the Astronaut Office) that Aldrin should be put on the sidelines for a while.

Figure 4.41: Commemorative covers for the launch of Mission GT-9 (above) and its recovery (below), signed by Gene Cernan. (From the collection of Pietro Dalla Maddalena, Italy.)

On the third day, Cernan left the spacecraft and became the third man to make a spacewalk. But once again, he discovered that EVA was far from an easy and relaxing activity. After a huge effort, Cernan was sweating profusely and quickly becoming exhausted. His visor began to fog and he was unable to see, and Stafford feared the worst. The spacewalk was cut short after 128 minutes, without completing all the planned activities, but Cernan had at least achieved a new EVA duration record. The GT-9 mission ended on June 6 in the Atlantic Ocean, just 700 meters from the target splashdown site, where the capsule was once again recovered by the USS Wasp. The splashdown was the most accurate achieved so far, but there were disappointments with the mission thanks to the failed docking and problems with the EVA. Both procedures were reassigned to the Gemini 10 mission six weeks later.

GT-10: A RENDEZVOUS WITH TWO SATELLITES

The docking with an Agena target vehicle became the primary objective for the next mission, the Gemini-Titan 10 (GT-10) flight of command pilot John Young, who had previously flown on GT-3, and rookie pilot Mike Collins. The mission plan also included a rendezvous with the Agena target of the Gemini 8 mission, plus two EVA excursions and 15 scientific, technological and medical experiments. Gemini 10 launched on July 18, 1966 at 5:20 pm, exactly as scheduled despite tropical storm Celia threatening the Cape. The mission's own Agena target had been launched first from Pad 14. Once again, a substantial out-of-plane error in the initial orbit meant that GT-10 had to use up 60 percent of its fuel for the rendezvous. However, the docking with the Gemini Agena Target Vehicle 10 (GATV-10) was successful and Mission Control developed an alternative flight plan to enable the mission to fulfill its objectives. GT-10 would remain docked with the Agena as long as possible, utilizing the target's fuel for attitude control. Additional docking attempts were scrubbed.

On the second day, Collins performed the first 'Stand-Up EVA', opening the hatch and standing on his seat to photograph stellar UV radiation around the Southern Milky Way. The ultraviolet camera he used would not have worked on Earth due to the filtering effect of the atmosphere.

The following day, using the GATV propulsion system, Gemini 10 changed its orbit and climbed to 412.4 nautical miles (763.8 km) to meet with the dormant Agena GATV-8 left over from the Gemini 8 flight aborted four months previously. This maneuver set a new altitude record, surpassing the 475 km of the Soviet Voskhod 2 from March 1965. It was also the first time that a rendezvous had been conducted with two different spacecraft in the same flight. As the older Agena had no electricity onboard, the radar could not be used and the docking had to be accomplished visually. Later, attached to a 50-foot (15 m) tether, Collins performed his second EVA to traverse to the dormant Agena and retrieve the micrometeoroid detection panel from the side of the GATV-8, as well as the 11 x 8 mm microfilm with the letter to George Mueller (shown on page 193). Collins became the first person to go to another spacecraft in orbit, but like Cernan on Gemini-9, he found that all his tasks took longer and more effort than expected. Once again, his experience confirmed that despite its delights, EVA was dangerous, difficult and deceptive.

The Gemini 10 mission made a significant contribution in the race to the Moon, expanding NASA's capabilities. It demonstrated the use of a fueled satellite to provide propulsion for a docked spacecraft and the capability of an astronaut to travel to another spacecraft and back. The GT-10 spacecraft reentered on July 21, 1966, and splashed down in the Atlantic Ocean within sight of the prime recovery ship the USS Guadalcanal, only 5.6 kilometers from the intended target.

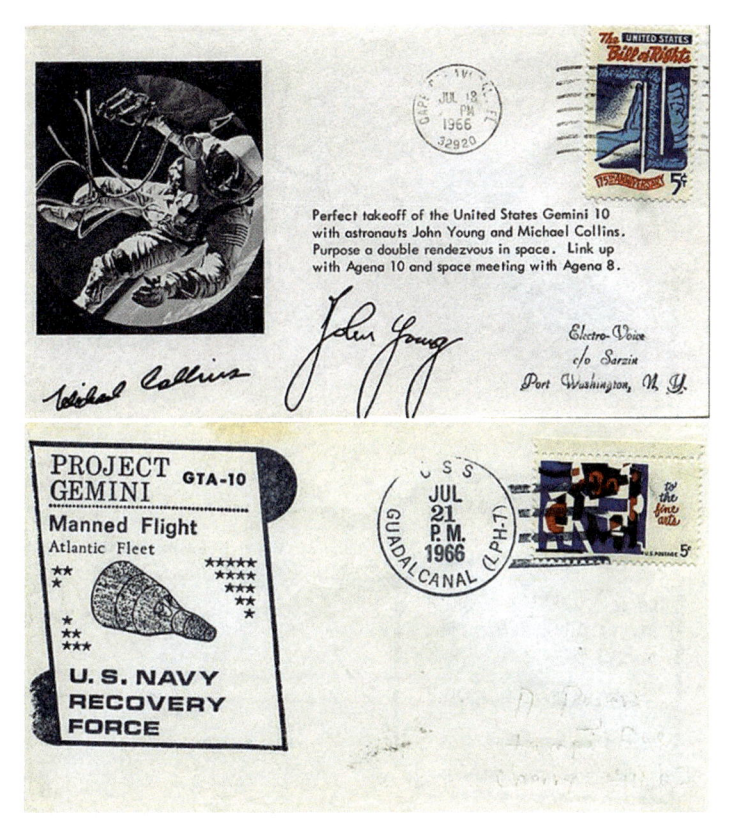

Figure 4.42: Commemorative covers issued for the launch of the GT-10 mission and its recovery by the USS Guadalcanal.

GT-11: AN UNPRECEDENTED RECORD ALTITUDE FLIGHT

Gemini-Titan 11 (GT-11) was launched on September 12, 1966, the crew consisting of Charles 'Pete' Conrad, the veteran of Gemini 5, and Richard 'Dick' Gordon who was undertaking his first spaceflight. The first goal of the mission was to dock with its Agena target on the first orbit. In order to achieve this, Gemini 11 had to be launched within an unbelievably short two-second launch window, the shortest in the history of the Gemini program and one that did not allow room for even the smallest mistake or delay. This would also simulate the future departure of the Lunar Module from the Moon to dock with the Command Module in lunar orbit. The launch worked flawlessly, within the first half-second of the two-second window.

Figure 4.43: (above) Postal stationary commemorating the launch of Gemini 11, The rubber stamp 'NASA cachet' is highly sought after by collectors. (below) Cover commemorating the recovery by the USS Guam.

At 11:16:42, just 94 minutes after lift-off, Gemini 11 successfully docked with the Agena while the two spacecraft were passing over the southern United States. Very little propellant was consumed and this allowed the astronauts to practice the docking and undocking procedure repeatedly. Once again on this mission, however, the EVA astronaut was unable to complete the planned spacewalk, with Gordon tiring very quickly simply trying to remain in place while he worked. His efforts caused his suit to overheat and blinded him with sweat, and Conrad called him back inside after just 38 minutes.

Pete Conrad had been taken with the notion of sending a Gemini on a trip round the Moon[15] and while that possibility did not come to fruition, he was able to use

---

[15] In its early days, there was talk of using the Gemini program to send men to the Moon. The proposal resurfaced in 1964, just months before manned flights began. The program looked so promising and Apollo, which started just a year after Gemini, was so troubled with problems that

GT-11: An Unprecedented Record Altitude Flight 201

the rocket on the Agena vehicle to raise Gemini 11 to a height of almost 1,374 km (nearly 740 nautical miles) above the Earth. That record for an orbiting manned spacecraft has yet to be broken, as the only manned flights to have achieved greater distances from Earth since then were the Apollo lunar landing missions[16].

Figure 4.44: India (left) and Texas (right) seen from Gemini 11 (courtesy NASA).

Gemini mission planners considered extending the program to include a lunar flight. A lunar Gemini mission would not have landed on the Moon, because it would not have had the right hardware, but even going into lunar orbit or just swinging around the far side of the Moon would have been a huge step for NASA in its race with the Soviet Union. Pete Conrad was intrigued by this proposal and went to Congress to argue his case, seeking support for at least one Gemini mission to the Moon. NASA's top leaders, James Webb and Robert Seamans, did not agree, contending that Apollo did not need a competitor. If Congress wanted to appropriate additional funds, Webb argued, it would be better to spend them on the program that had been designed to go to the Moon from the start. Even after Webb dismissed the scheme, Conrad still wanted to take Gemini as far as it would go, and began a small crusade to convince NASA management that there were good reasons for going really high. There were concerns about passing through regions of intense ionizing radiation called the Van Allen Radiation Belts (VARB). With the help of Bill Anders – who would be on the first crew to go to the Moon during Apollo 8 – Conrad argued that a high orbit would not pose any risks to the Gemini spacecraft and would be able to look into the radiation belts to help devise ways to minimize risks. Finally, NASA gave him a 'go' and apogee excursion became part of Gemini 11. The mission would demonstrate that trajectories through the VARB were not only survivable, but that the radiation doses received were inconsequential, suggesting that the Van Allen Belts were not constant about the planet, being denser in some regions than in others, and global radiation dosage was comparable to a chest x-ray. This discovery would be important for the Apollo missions to the Moon.

[16] The International Space Station orbits the Earth at an altitude of about 230 miles, while the Space Shuttles, over their combined 135 missions, usually orbited around 200 miles above the planet. Their peak operational limit was around 600 miles, or 970 km, during the Hubble Servicing Missions.

From their unique position, the astronauts also took over 300 photographs as part of the different science experiments, some of the most striking images of the Earth from orbit in NASA's history. Gemini 11 was one of the shortest missions in the program, returning to Earth on September 15, just 72 hours after lift-off. The flight ended with a splashdown in the Atlantic Ocean, within 4.6 kilometers of the USS Guam, the prime recovery ship. The splashdown was closer to the carrier than any previous Gemini mission.

MOL: THE SECRET AMERICAN MILITARY SPACE STATION

The Cold War was in full swing and espionage was an obsession for both superpowers. After the demise of their X-20 Dyna-Soar program in 1963, the U.S Air Force (USAF) began a top-secret program that would use a suitably modified Gemini capsule, 'Gemini B', to create a military presence in the sky by implementing espionage from an orbital platform. The Manned Orbiting Laboratory (MOL) was intended to conduct surveillance on both the USSR and China. The secret program, known within the National Reconnaissance Organization (NRO) by the code name 'Dorian', would take satellite photos, study life in space and perform other duties which have only recently been declassified. [35]

In 1965, the USAF appointed its first group of secret *"aerospace research pilots,"* who would soon be joined by another 10 chosen from among the best military pilots. Five (or more) two-man MOL flights in polar orbit would begin in 1968. While the NASA astronauts were making the headlines in the front pages of newspapers and giving interviews and autographs, the MOL astronauts were surrounded by the strictest secrecy.

On November 3, 1966, a Titan IIIC rocket placed the Gemini B capsule (refurbished after it was last used in the Gemini-Titan 2 test on January 19, 1965) into orbit for the Manned Orbiting Laboratory-Heat Shield Qualification test (MOL-HSQ) to analyze the aerodynamic configuration of the Gemini B. This was an unmanned mission and would be the only test or flight of the MOL program. Even though it had support from the military and the President, MOL was seldom fully funded. With growing pressure from the expansion of the Vietnam War, the perceived duplication of effort with NASA programs and the improved performance of unmanned spy satellites, MOL was cancelled in June 1969, together with the USAF's last chance to develop its own manned space flight program. MOL operated for five-and-a-half years and cost $1.56 billion, without launching a single manned vehicle into space[17].

---

[17] After the cancellation of the program, the small corps of military astronaut-spies was disbanded, with NASA hiring seven of them: Robert Crippen (later to become the pilot of the first Space Shuttle, STS-1); 'Dick' Truly (pilot of STS-2, commander of STS-8 and eventually Director of NASA from 1989 to 1992); Karol Bobko (who became a member of the support

Figure 4.45: Commemorative covers for the launch of the Gemini B test on November 3, 1966.

Figure 4.46: At least the MOL astronauts got a patch. The only public display of the original MOL patch is at the USAF Museum in Dayton, Ohio.

204   The astronaut sits in the driver's seat

Figure 4.47: The 'Magnificent Seven' secret military astronauts Group 1, selected in 1965 to work with the Manned Orbiting Laboratory (MOL). Michael J. Adams, Albert H. Crews, John L. Finley, Richard E. Lawyer, Lachlan Macleay, Francis G. Neubeck, James M. Taylor and Richard H. Truly were among the best military pilots at the time.

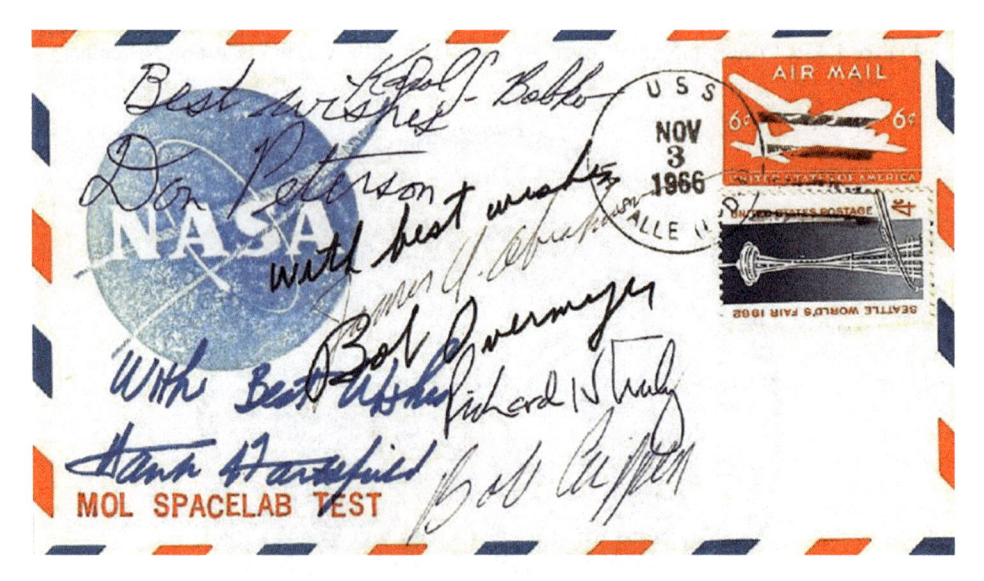

Figure 4.48: Even for this 'secret' program, there was a 'Riser fake' cover (see page 129) with the usual airmail envelope. In addition to the fake postmark of the recovery ship (Riser was a specialist in counterfeiting recovery ship cancellations), the envelope also carries forged signatures of the military astronauts, whose names were top secret at the time.

In response to MOL, Vladimir Chelomey would promote the highly secretive Soviet military espionage program, 'Almaz'[18].

GT-12: THE GRAND FINALE

Commanded by veteran James A. Lovell (who had set the endurance record in space aboard Gemini 7) and piloted by Edwin 'Buzz' Aldrin (known as "*Doctor Rendezvous*" because of his obsession with rendezvous in space, the subject of his PhD thesis at MIT), Gemini 12 was a successful conclusion to the Gemini program and addressed the final issues raised by the earlier flights so that Apollo could proceed. Among its goals, Gemini 12 included the fifth rendezvous and fourth docking with an Agena target vehicle, but most importantly it was intended to resolve some of the most pressing issues remaining with EVA, which were still of great concern to Apollo mission planners.

The 'EVA approach' was completely redesigned by NASA engineers, in terms of planning, equipment and training. Instead of the two EVAs performed on each of the previous three missions, a total of three spacewalks were planned for Gemini 12, in order to determine if the latest improvements could resolve the issues. After a first Stand-Up EVA, Aldrin performed a 'true' spacewalk of 2 hours 20 minutes, during which he performed several tasks on both the Gemini spacecraft and the Agena target. The latest Agena had been modified before launch and was equipped with an extensive set of handrails and footholds. Aldrin completed a final 55-minute Stand-Up EVA on November 14, with the three spacewalks totaling five-and-a-half hours. Eventually, Aldrin would say, in his usual style, that it was "*a piece of cake.*" A few years later, Gene Cernan would comment in his book: "*In true Buzz fashion, he would openly claim in later years that he had personally solved all the problems of EVA and that his spacewalk went smoothly because he was better prepared than the rest of us.*" [36]

---

crew for the Shuttle Approach and Landing Tests (ALT) and eventually the pilot of STS-6 and commander of STS-51D and STS-51J); Gordon Fullerton (first involved in ALT and then pilot of STS-3 and commander of STS-51F/Spacelab-2); Henry Hartsfield (pilot of STS-4 and commander of STS-41D and STS-51A); Robert Overmyer (first involved in ALT and then pilot of STS-5 and commander of STS-51B/Spacelab-3); and Donald Peterson (mission specialist on STS-6).

[18] Three Almaz space stations were launched between 1973 and 1976. To cover up the military nature of the program, the three stations were designated as civilian Salyuts. Salyut 2 (or Almaz 1) would fail shortly after achieving orbit and decay in the atmosphere without ever being occupied. Five crewed Soyuz expeditions would fly to Salyut 3 (Almaz 2) and Salyut 5 (Almaz 3), with three reaching their stations and only two of the missions being considered fully successful at that time.

206 The astronaut sits in the driver's seat

Gemini 12 splashed down into the Atlantic Ocean where it was recovered by the USS Wasp, the same ship that had recovered Gemini 4, Gemini 6, Gemini 7 and Gemini 9. The Gemini program had finally validated long-duration missions, rendezvous *and* spacewalking.

Figure 4.49: (above) Cover commemorating the launch of the GT-12 mission, with the highly sought-after 'NASA cachet'. (below): Cover commemorating the recovery by the USS Wasp.

One sad side story to the Gemini program concerned the end of the career of John 'Shorty' Powers as the astronauts' mouthpiece. A USAF lieutenant-colonel and war veteran, Powers served as a mission commentator during the Mercury missions and was known as the *"voice of the astronauts"* and the *"eighth astronaut."* He continued this role during Gemini. Powers had always been a hard drinker and his drinking had been getting progressively worse. It was discovered that he had been leaking inside information on the Gemini program to a reporter in return for bottles of whiskey and he was immediately removed from his prestigious position. He died on the last day of 1979, having effectively drunk himself to death.

References

1. **On the Shoulders of Titans: A History of Project Gemini**, Barton C. Hacker and James M. Grimwood, NASA Sp-4001, Washington D.C., 1974, pp. 18–20.
2. **Before This Decade is Out...**, Glen E. Swanson (Ed.), NASA SP-4223, Washington D.C., 1999, XVI, pp. 101–116.
3. Reference 1, p. 165.
4. **American X-Vehicles: An Inventory – X-1 to X-50**, Dennis R. Jenkins, Tony Landis and Jay Miller, NASA SP-4531, Washington D.C., 2003, p. 27; *Race to the Moon 1957-1975*, Greg Goebel, vc.airvectors.net, Chapter 16.6
5. Reference 1, pp. 118.
6. *Race to the Moon 1957-1975*, Greg Goebel, vc.airvectors.net, Ch. 16.5; **Challenge to Apollo: The Soviet Union and the Space Race, 1945-1974**, Asif Siddiqi, NASA SP-4408, Washington D.C., 2000, p. 384; and **Rockets and People – Vol. 3: Hot Days of the Cold War**, Boris Chertok, NASA SP-4110, Washington D.C., 2009, p. 230.
7. **The Russian Space Bluff – The inside story of the Soviet drive to the Moon**, Leonid Vladimirov, Dial Press, London, 1973, pp. 120–139.
8. Reference 7, p. 128.
9. Reference 7, p. 135.
10. *Race to the Moon 1957-1975*, Greg Goebel, vc.airvectors.net, Ch. 16.5.
11. Reference 7, p. 129; *Race to the Moon 1957-1975*, Greg Goebel, vc.airvectors.net, Ch. 16.5.
12. Reference 7, p. 138.
13. **Rockets and People – Vol. 3: Hot Days of the Cold War**, Boris Chertok, NASA SP-4110, Washington D.C., 2009, p. 451.
14. **Two Sides of the Moon: Our Story of the Cold War Space Race**, David Scott and Alexei Leonov, St. Martin's Press, New York, 2004, p. 101.
15. Reference 7, pp. 141–143.
16. Reference 14, p. 3, 108.
17. Reference 14, p. 109; also *Learning to Spacewalk. A cosmonaut remembers the exhilaration and terror of his first space mission*, in *Air & Space Magazine*, January 2005.
18. **Walking to Olympus: An EVA Chronology**, David S.F. Portree and Robert C. Treviño, NASA Monographs in Aerospace History Series #7, Washington D.C., 1997, p. 2.
19. Reference 14, p. 110.
20. Reference 14, p. 109.
21. *Korolev Facts and Myths*, Bart Hendrickx, *Spaceflight* Vol. 38, February 1996, p. 47
22. Reference 14, p. 117.
23. Reference 14, p. 116.
24. Reference 13, pp. 266–69; and Reference 14, pp. 110–122.
25. Reference 14, p. 129.
26. **Celebrating Apollo 11 – The Artwork of Paul Calle**, Chris Calle, AeroGraphics Inc., Bradenton, FL, 2009, p. 20.
27. **The Pencil**, Paul Calle, published by North Light Publishers, Westport, Conn., 1974, distributed by Watson-Guptill Publications, New York, p. 121.
28. Reference 27, p. 122.
29. Reference 26, p. 21.
30. *Race to the Moon 1957-1975*, Greg Goebel, vc.airvectors.net, Ch. 17.3.
31. **The All-American Boys**, Walt Cunningham, iBooks, New York, 2003, pp. 109-110.
32. Reference 14, p. 6.

33. *Pioneerraketenpost und kosmiche Post*, Walter Hopferwieser, Austria Netto Katalog Verlag, Vienna, Austria, 2016, p. 120.
34. *Encyclopedia Astronautica, "Voskhod 3"*, Mark Wade, www.astronautix.com.
35. *Declassified Manned Orbiting Laboratory/DORIAN Illustrations*, www.nro.gov (accessed in February 2018).
36. **The Last Man on the Moon: Astronaut Eugene Cernan and America's Race in Space**, Eugene Cernan and Donald A. Davis, St. Martin's Griffin, New York, 1999, p. 157.

5

Two tragedies block the race in space

NASA: STRONG DIRECTION TO ACHIEVE SUCCESS

With the last mission of the Gemini program successfully completed, NASA could now concentrate all of its efforts towards President John F. Kennedy's goal of landing a man on the Moon before the end of the decade. George Mueller, brought in to facilitate concurrent development of the many needed systems, had begun to introduce radical management changes.

One of his most remarkable achievements was the introduction of his controversial 'all-up testing' philosophy for the Saturn V launch vehicle, with each flight using the full number of live stages. This approach, which had already been used successfully on the Titan II and Minuteman programs, was strongly opposed by Wernher Von Braun's engineering concepts. The conservative test plan vigorously defended by Von Braun and his team for developing the Saturn V rocket called for a building block approach, where each stage would be flight-tested before adding the next one. Mueller's alternative meant that all the engines and systems would be tested on the ground and then each test flight would be of the full Saturn V stack. The flight tests of the Apollo prototypes would be launched using small rockets such as Little Joe and Saturn I. By instigating this approach, Mueller was able to bring both the costs and the timetable of the entire project under control[1].

---

[1] Mueller's approach would turn out to be a winning one. After the first two unmanned tests of Apollo 4 (SA-501) on November 9, 1967 and Apollo 6 (SA-502) on April 4, 1968, the third Saturn V would launch Apollo 8 on its lunar orbital mission at Christmas 1968, and the sixth would send Apollo 11 on its way to the first Moon landing. Von Braun would later acknowledge that without Mueller's approach, NASA would not have landed astronauts on the Moon before Kennedy's deadline.

210 Two tragedies block the race in space

After winning this battle, Mueller realized that he needed the right people in managerial roles and, with Administrator James Webb's consent, co-opted U.S. Air Force (USAF) General Sam Phillips, former director of the Minuteman program, to join NASA as the director of the Apollo program. Phillip, in turn, would bring over 160 people from the USAF.

Figure 5.1: George Mueller co-opted USAF General Sam Phillips as director of the Apollo Program, who in turn would bring over 160 others from the USAF, including Huston Balender and Al Vett.

Under Mueller, NASA manned spaceflight became a seven-day-a-week job. Meetings were called early in the morning to avoid having them disrupted by telephone calls, while important decision-making meetings were scheduled for Saturday or Sunday when there was less chance of disruption.

Another battle would take place behind the scenes, with NASA's Office of Space Science (OSS), against which Mueller did not hesitate to use his right of veto. The OSS, pushed by the lobbying of the American Agency of Geological Studies (USGS), wanted to use the Apollo missions to perform extensive geological research on the Moon. One of their strongest advocates was Gene Shoemaker, a brilliant scientist who could not accept that Apollo would be spending so much time on the Moon without taking the opportunity to study its geology. In this clash between the objectives of science and technology, many in the scientific community were appalled that the opportunity for scientific contribution afforded by Apollo was not proportional to the program's huge costs.

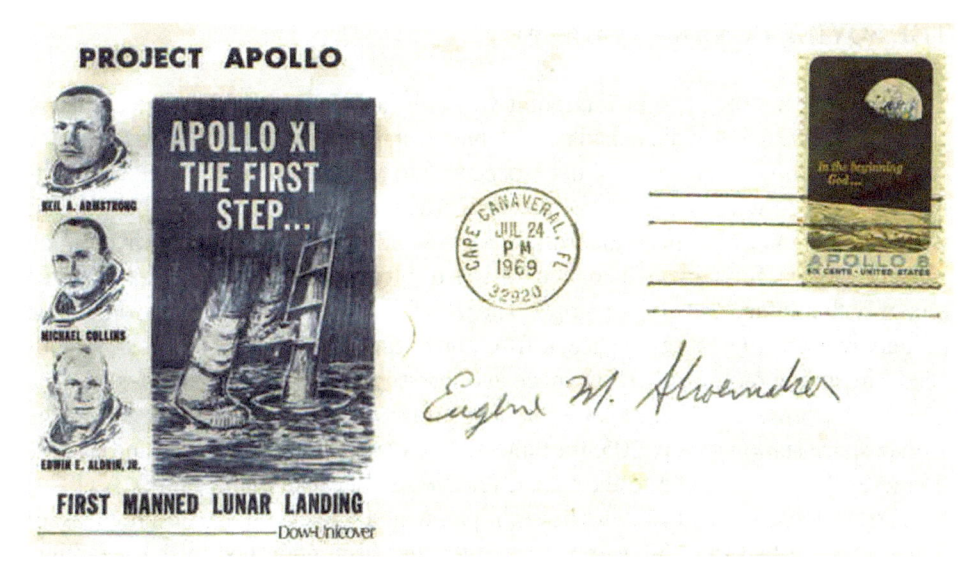

Figure 5.2: Apollo 11 commemorative cover, signed by Eugene M. Shoemaker.

The argument between the National Academy of Sciences and the Astronaut Office (notably Deke Slayton and Alan Shepard) grew quite heated on the subject. In 1965, NASA agreed to allow six scientist-astronauts to join the astronaut corps (thus inaugurating the age of the 'hyphenated astronauts'): Duane Graveline (a medical doctor)[2]; Joseph P. 'Joe' Kerwin (a navy flight surgeon); Owen K. Garriott, Edward G. Gibson and F. Curtis Michel (all physical scientists); and Harrison 'Jack' Schmitt (a geologist). They would be employed in the space program as 'sops' to the scientific community, *"in much the same spirit,"* according to Walt Cunningham, *"[that] the early Russians threw one another off the back of the sleigh to slow down the wolves."* [1]

With Mueller's methods bringing the Apollo project back on track and with the successes of the Gemini program still fresh, space programs were back in vogue. Despite the war in Vietnam and growing opposition in Congress, funding for the space program continued to be approved and in 1966, NASA obtained a record $2.967 billion for Apollo.

---

[2] Having not really wanted the scientist-astronauts in the corps in the first place, Deke Slayton was quick to take advantage of the fact that Graveline was filing a divorce case at the time of his selection. *"Worried that the divorce procedure could distract him from his new commitments,"* Slayton did not hesitate to remove Graveline from the corps, acting so quickly that he did not even have time to appear in the first group photo. The scientist-astronauts, in *"John Wayne's space frontier"* would have bigger handicaps to overcome than just the usual new-guy-on-the-block syndrome (Cunningham [2003], p. 298). The first of the remaining five to fly in space would be Jack Schmitt, on the last Apollo mission, Apollo 17.

THE SOVIET LUNAR PROGRAM

The Soviet space program at this point was still lacking a suitable spacecraft but, above all, the decision-makers had no real interest in it. Since the early 1960s, Korolev had been thinking of a new 'modular' spacecraft to replace the Vostok, but it was difficult to succeed with this when the proposal was not deemed useful by the military, who were not interested in conquering space beyond Earth orbit. To circumvent such military reserve, Soyuz was therefore designed to be a 'universal spacecraft' with a mix of both military and civilian goals. The Chief Designer attempted to promote the military version of his new spacecraft, emphasizing its ability to address various needs, including monitoring from space and interception of enemy satellites. For the time being, however, the task of espionage from space had been entrusted to the Almaz space station (see p. 205, footnote 18), while that of interception and inspection (ASAT) was assigned to the maneuverable satellite Istrebitel Sputnik A (IS-A, see p. 61). However, the IS-A was never implemented for several reasons, not least of which was the fact that the enemy's satellites had been equipped in the meantime with self-destruct systems to prevent such orbital inspection. More or less the same fate happened to Korolev's giant N-1 rocket, which he had designed to be a 'universal launch vehicle' capable of performing missions ranging from launching large space stations to manned missions to the Moon, or from exploring deep space with scientific probes to lifting massive military satellites into Earth orbit.

Seen from the outside, the Soviet system seemed monolithic and efficient, but there was no equivalent space organization to NASA in the USSR. Nor was there a coherent strategy for space exploration and therefore, as a consequence, no economic development program. The whole setup was characterized by internal rivalries, personal grievances and political plots. In ruthless competition, three different bureaus – OKB-1 of Sergei Korolev, OKB-52 of Vladimir Chelomey and OKB-5 of Mikhail Yangel – sought, through personal relations, the support of the Academy of Sciences to enable them to acquire more resources than their rivals in order to carry out duplicate programs. Any rise in the Kremlin's graces by a given official, friend or admirer of one or the other of the OKBs (or, indeed, any fall into disgrace) had such consequences that any given program fluctuated between continuous struggle, uncertainty, rethinks and course corrections.

At the same time, Mstislav Keldysh, the head of the Academy of Sciences, was much more concerned with isolating the pacifist Andrei Sakharov and expelling him from the Academy than he was with conquering the Moon, while the government's Military-Industrial Commission (or Voyenno-Promyshlennaya Komissiya, VPK), led by Dmitriy F. Ustinov which should have overseen space projects, was faithful to the Stalinist principle of promoting competition and did not discourage duplicated efforts. In the typical atmosphere of secrecy prevalent in the Soviet Union at that time, the rival and parallel projects flourished according to the 'Dog in the Manger' principle, of spitefully preventing others from having something for which you have no use yourself.

Figure 5.3: 1981 Soviet stamp featuring Korolev and the N-1 rocket.

On August 3, 1964, the Central Committee of the Communist Party and the USSR Council of Ministers passed Resolution No 655-268 (which would be declassified only in June 2003) and approved the lunar program that had been prepared for Nikita Khrushchev by Korolev and his team. [2] That program listed, among the ambitious (and probably unrealistic) schedules, a manned circumlunar flight in 1966, and invited *"the Soviet masses to engage with enthusiasm the purpose of bringing socialism to the Moon by 1967."*

Figure 5.4: A stamp from the early 1960s hinting that a circumlunar flight was one of the goals of the Soviet space program.

The projects and goals were described in detail in the first appendix to the decree, where the four main lunar projects of the 'Study and Mastery of the Moon' section (the most important part of the decree) were divided between Korolev's OKB-1 and Chelomey's OKB-52. The manned circumlunar flight was assigned to Chelomey and the manned lunar landing to Korolev. [3] After many ups and downs, Korolev was finally authorized to build Soyuz for the crewed N1-L3 lunar landing complex, his response to the American challenge to fly a man to the Moon.

The L3 would have used the LOK (Lunniy Orbitalny Korabl – Lunar Orbital Spacecraft) to carry two cosmonauts into orbit around the Moon. After docking, the LOK would have gone into circumlunar orbit, acting as the 'Command Module' or 'Mother Ship' for the piloted LK lander (Lunniy Korabl – Lunar Spacecraft). One of the two cosmonauts would have passed into the LK module via an EVA (because there would be no direct connection between the two modules) and would have landed on the lunar surface for a stay of approximately four hours, including a two-hour walk. [4]

Chelomey started to develop his own 7K-L1 (Zond) module for a crewed circumnavigation of the Moon, in competition with the N1-L3 'complex' designed by Korolev. Thanks to his bureau's political support – the Premier's son Sergei Khrushchev worked in Chelomey's OKB-52 – Chelomey succeeded in persuading the Kremlin that his suggested direct flight to the Moon was a much quicker and neater method than Korolev's cumbersome N1-L3 which, like the American solution, would require complicated docking operations and rendezvous. [4] Partially contradicting Resolution 655-268, Khrushchev entrusted OKB-52 with the task of providing the "*socialist response*" to the Americans and, very optimistically, the Premier urged Chelomey to put his L1 capsule into lunar orbit by October 1967, the 50th anniversary of the Bolshevik Revolution. In hindsight, Korolev's successor Vasiliy Mishin recalled in an interview: "*We should have had a long-term space research program. Unfortunately, we received a great deal of isolated instructions which pursued political objectives or were designed to boost our prestige... We were constantly pushed ahead. Superficial and contradictory decisions were sent down to us.*" [5]

The situation worsened after the dismissal of Khrushchev, who had at least been hungry for space 'spectacular' firsts to be exploited for propaganda purposes, even if they were meaningless or not enough to build a Soviet Space Program around. When Khrushchev was ousted on October 14, 1964, Chelomey lost his patron. Korolev was quick to take advantage of this and began working on getting the program removed from Chelomey's grasp. The new Leonid Brezhnev-Alexei Kosygin administration noted that while Chelomey's design office had received more funding over the years, it had recorded the most disappointing results, so they decreed the cancellation of existing orders and assigned the entire lunar program to Korolev's bureau. The target for the launch of the L1 capsule for

The Sudden Passing of Korolev

the mission around the Moon remained scheduled for October 1967, but while all these changes were taking place, the Americans had stolen a march and forged ahead, by more than three years. [6]

THE SUDDEN PASSING OF KOROLEV

While the situation with Chelomey seemed to have been resolved, Korolev's N-1 was also behind schedule and the Chief Designer was faced with the problem of increasing both the power and the range of the rocket. [4] The Soviets did not have any alloys that could withstand the high temperatures developed by large diameter engines and Soviet engineers had not been able to come up with adequate methods to reduce the heat. Korolev's solution was the development of the so-called 'cluster', or group, of small rockets, with their nozzles lying side-by-side and gathered together in a bundle to replace each single large motor. [7] The initial plan was to install 26 engines in the N-1 first stage, but this soon increased to 30 to develop a thrust of 4,590 tons. While the Americans were able to test an entire assembled engine module on their test stands and then install it on the launch vehicle, the Soviets were only able to test in pieces and then assembled those pieces, with no guarantee that they would properly interact. The N-1 booster did have a built-in safety failure control system (called KORD) which controlled each engine in each stage. If four engines in the first stage failed, the rocket would still work. Likewise, if a pair of engines in the second stage, or one engine in the third stage failed, the N-1 would still be capable of reaching low Earth orbit.

But the booster also suffered serious problems with pogo oscillations and vibrations. Identifying their origin was very difficult and required a great deal of experience and a lot of time, and only Korolev himself was able to do so. However, despite working 18 to 20 hours a day, he simply did not have enough time. His close collaborators reported that he was so tired by the end of the day at this point, and in such bad shape, that he did not even have the strength to climb the stairs to his apartment.

Korolev was declining fast at the end of 1965. In December, he had a series of medical checks and – as already seen (p.194) – was diagnosed with a colon polyp. At the beginning of January 1966, he checked into the Kremlin hospital and underwent a 'routine', simple operation, *"less complicated than an appendectomy,"* which nobody was concerned about. He was operated on not just by a famous surgeon, but by none other than the Soviet Minister of Health, the cardiac surgeon Boris Vasilyevich Petrovsky. In an interview released in late 1991, Korolev's daughter Natasha Koroleva, herself a surgeon and a professor at the Moscow Medical Academy, revealed further details of what happened. The operation discovered that, in addition to multiple polyps, Korolev also had a sarcoma.

216   Two tragedies block the race in space

Petrovsky, who had excellent Party credentials but poor medical skills since he had always dealt with administrative and political roles throughout his life, decided to remove the sarcoma and provoked an uncontrollable hemorrhage. He was unprepared for the complications that arose and, instead of trying to address the cause of the bleeding, decided to continue with the operation, which lasted over four hours. There was also a problem for the anesthetists. Needing to set up artificial respiration quickly, they found they were unable to intubate their patient. Korolev's jaw had been broken when he had been tortured in prison in 1938 and it was difficult for him to open his mouth wide. The problem was further exacerbated by his unusually short neck. The operating team had not thought of, or had not prepared for, this eventuality. Nor had the cardio-pulmonary bypass machine been prepared in advance. When Petrovsky finally realized the situation, he summoned his colleague, Soviet Army Head Surgeon Professor Alexander Aleksandrovich Vishnevsky, out of desperation. But it was too late and Korolev suffered heart failure. When Vishnevsky arrived, there was nothing he could do and it has been reported that he declared: *"I do not operate on corpses."* Korolev died on Petrovsky's operating table on January 14, 1966.

It would later be discovered that Korolev's sarcoma was completely encysted, there were no metastases and everything was clean. If Petrovsky had not decided to remove the sarcoma, Korolev could have lived with it. [8] After his death, there was no longer any need to keep Korolev's identity a secret and *Pravda* announced that the Chief Designer had died of a *"metabolic disorder,"* running a two-page

Figure 5.5: Korolev's place of honor in the Kremlin Wall.

obituary including a photograph. For the first time, the Soviet people learned who the Chief Designer was. Korolev was given a state funeral and his remains were buried in the Kremlin wall, the highest honor for a Soviet citizen. His image would become one of the most representative symbols of the Soviet space program and state propaganda exploited it, depicting him several times on USSR stamps.

Figure 5.6: Following his death, Korolev's image was frequently featured on Soviet stamps.

Many Western newspapers did not immediately grasp the importance of Korolev's accomplishments, however. The *New York Times* reported the news of his passing on page 82 of its Sunday January 16 edition. Even as late as 1968, some newspapers would still claim that Academician Leonid Sedov was the "*father of Sputnik.*" [9] When Korolev passed away, the Soviet manned lunar program died with him.

THE SOVIETS TAKE THE FIRST PICTURES FROM THE MOON

Two weeks after the loss of Korolev, the Soviets managed to achieve one more impressive 'first' over the Americans. On January 31, 1966, Luna 9 – the last creation of the Chief Designer – became the first man-made probe to make a soft landing on a planetary body other than Earth when it touched down on the Moon. It was almost as if the Soviets wanted to tell the world that the race to the Moon was ongoing while, once again, the equivalent American project (called Surveyor) had fallen two years behind schedule. A few minutes after landing in the Oceanus Procellarum (Ocean of Storms), the four petals that covered the top half of the Luna 9 spacecraft opened outwards for increased stability and the television camera began a photographic survey of the lunar surface.

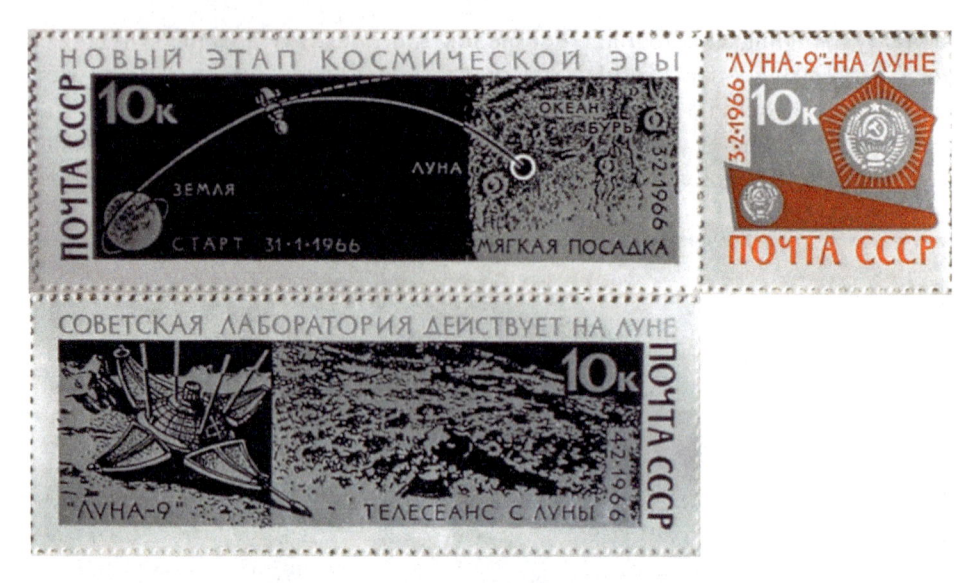

Figure 5.7: Soviet stamps commemorating the mission of Luna 9.

Figure 5.8: The poster "Report: Mission Accomplished!" was promptly issued and circulated after Luna 9 had landed, as if to confirm that the race to the Moon was still ongoing.

A set of pictures was transmitted back to Earth before the probe's batteries became exhausted on February 6. The frames were reworked at the Moscow control center and assembled to produce a panoramic view of the landing site. These were the first photographic data received from the surface of another celestial body. Perhaps one of the most important discoveries of the mission, however, was that a spacecraft could land on the Moon without sinking into the lunar dust, and that the lunar ground could support a lander.

After the loss of a probe launched on March 1, which failed to achieve its lunar trajectory and fell back into Earth's atmosphere to disintegrate (eventually declassified with the name Cosmos 111), the Soviets would set another new record with the launch of Luna 10 on March 31. The probe entered lunar orbit on April 3, 1966, to become the first artificial satellite of the Moon. It completed its first orbit in three hours. Luna 10 was a makeshift solution launched mainly to prevent the far more advanced American Lunar Orbiter from getting there first. It carried no cameras but transmitted a synthesized rendering of the '*Internationale*' to cheering Communist Party delegates in Moscow, who had assembled for the first Congress under Leonid Brezhnev's leadership. At the start of the Congress, mission controllers discovered a missing note in the hymn and decided to play the previous night's tape to the unsuspecting delegates, claiming that it was a live broadcast from the Moon.

Figure 5.9: Soviet stamp issued in recognition of the successful Luna missions.

Scientific instruments carried on the probe permitted extensive research to be conducted in lunar orbit, including measurements of the electrical, magnetic and radiation fields in near-lunar space infrared emissions from the Moon, as well as gravitational waves. The battery-powered satellite had a limited lifetime and radio transmissions ceased after 57 days, on May 30, after 460 orbits of the Moon and 219 active data transmissions.

Figure 5.10: Covers commemorating Soviet successes at the Moon through the Luna probes.

THE AMERICAN LANDER IS BEHIND SCHEDULE

The bold plan for an Apollo mission based on LOR was under preparation in the United States, but NASA still had a great deal more to learn about the destination if they wanted to land on the Moon by 1969 as promised. Telescopes had revealed little about the nature of the lunar surface and to obtain the necessary information about the lunar dust, craters, crevices and jagged rocks on the lunar surface, NASA planned to send out automated probes to take a closer look. But the Surveyor project, entrusted to the Jet Propulsion Laboratory (JPL) – which was still deeply involved in the Ranger project – was suffering serious schedule delays. [10] The Americans finally succeeded in launching their first lunar lander, Surveyor 1, only on May 30, 1966. On June 2, the probe descended into a 'ghost crater' of the Oceanus Procellarum. However, although the Soviet Luna 9 mission had soft-landed on the lunar surface three months earlier, it was only after a few failed attempts. With Surveyor 1, the Americans achieved a soft landing on their first try. [11]

The American Lander is Behind Schedule 221

The probe carried over 100 sensors onboard for studying the surface to find possible landing sites for future human missions and to verify that such sites were safe for human landings. Using its television camera, a film-based system which developed the fine-grained film onboard the spacecraft, as well as sophisticated radio-telemetry, Surveyor 1 scanned the pictures and transmitted 11,237 still photos of the lunar surface back to Earth. The images showed rocks on the Moon in amazing detail, documented the physical conditions of the lunar surface for Apollo and identified a myriad of craters, from thousands of kilometers in diameter down to tens of microns. Surveyor's instruments permitted the first chemical analyses of the materials of the Moon's surface and provided abundant information on local environmental, temperature and engineering data, right through to January 7, 1967.

Between May 1966 and January 1968, six Surveyor spacecraft made successful soft landings at predetermined points on the lunar surface. From the touchdown dynamics, surface-bearing strength measurements and eye-level television scanning of the local surface conditions, NASA learned that the Moon could easily support the impact and weight of a small lander and obtained the photographs, scientific and technological information necessary for the Apollo manned landing program. [12]

Figure 5.11: Covers commemorating the unmanned Surveyor program which gathered crucial information about the Moon for NASA prior to the Apollo missions.

222 Two tragedies block the race in space

The most well-known of these missions is Surveyor 3, launched on April 17, 1967, which landed on the Oceanus Procellarum on April 20. It transmitted 6,300 still images and would later be visited by the crew of Apollo 12 when astronauts Pete Conrad and Alan Bean piloted the Lunar Module *Intrepid* to a landing within 200 meters of the probe on November 19, 1969. Landing within walking distance of the (by then) inactive robotic lander proved Apollo's pinpoint landing capability and allowed the astronauts to return parts from the Surveyor to Earth for engineering assessment.

Figure 5.12: Astronaut Alan Bean inspects Surveyor 3, with the LM *Intrepid* in the background. (Courtesy NASA.)

"It was a real challenge," remarked astronaut Walt Cunningham, *"to determine the precise location of Surveyor from a distance equal to ten times the circumference of the Earth and then to land within walking distance of that spot. It would be comparable to dropping a grape into a Coke bottle from the top of the Empire State Building."* He added: *"The Russian space, military, and political establishments probably viewed our accomplishment on Apollo Twelve – pinpointing an object and then flying to it from a quarter of a million miles away – as second only to a successful lunar landing itself. The military implications of such guidance capability were obvious."* [13]

APOLLO ON THE LAUNCH PAD

On February 26, 1966, the Americans tested a Saturn IB in a suborbital flight for the first time, which – following Mueller's 'all-up testing' approach – carried an Apollo CSM. The Apollo-Saturn mission, officially named as AS-201, is sometimes incorrectly referred to as 'Apollo 1'[3].

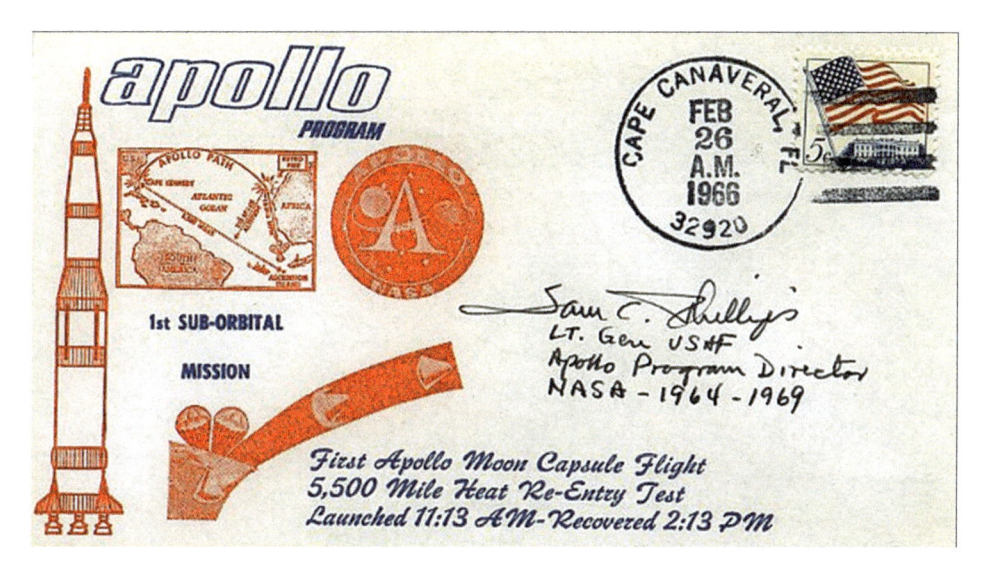

Figure 5.13: Cover commemorating the first suborbital test flight of the Apollo program, AS-201.

An orbital flight followed on July 5 (AS-203) and a second suborbital flight on August 25 (AS-202). These two missions are also improperly referred to as 'Apollo 2' and 'Apollo 3', as can be seen in the commemorative covers in Figure 5.14[4].

The Apollo program was now proceeding with great strides and within a year, the giant Saturn V – which would take humans to the Moon – had made it to the launch pad. The new rocket was to be launched on November 7, 1967 and there was much concern. It was the first time that the 'monster' had appeared on the launch pad and, traditionally, new rockets had never proven themselves on first launch. There was also a new complex procedure at Mission Control that required testing.

---

[3] Officially, 'Apollo 1' was used retrospectively to designate the unfortunate mission of Grissom, White and Chaffee, at the request of Grissom's widow Betty in 1967.

[4] Officially, no Apollo 2 and Apollo 3 missions exist.

224 **Two tragedies block the race in space**

Figure 5.14: Covers commemorating the unmanned test flights of AS-203 and AS-202, often improperly known as 'Apollo 2' and 'Apollo 3'.

To everyone's surprise, including the designers, the launch proceeded smoothly amid a deafening din and the Apollo 4 mission ended in great success[5].

NASA was now in the home straight. Everything was ready for the final phase. At least 20 Apollo missions were planned, involving about 60 astronauts. In June 1965, NASA selected a fifth group of astronauts, who christened themselves, somewhat ironically, the 'Original Nineteen', in parody of the original seven Mercury astronauts. There were now 46 active NASA astronauts, but when it came to assigning the crew for the first Apollo flight, scheduled for February 1967, there was definitely a pecking order of seniority involved. The first flight was assigned to Gus Grissom (the second Mercury astronaut and one of the 'Original Seven'), supported by Ed White (the first American spacewalker) and Roger Chaffee, both astronauts from Group 2.

---

[5] No Saturn rocket ever failed at launch: a record still unparalleled in space history.

Figure 5.15: (above) Cover commemorating the launch of Apollo 4. (below) Cover commemorating the 50th anniversary of the successful maiden flight of the "monster" rocket Saturn V, designed by Austrian artist Alfred Gugerell, who also designed the stamp and the special pictorial postmark. (right, top to bottom) To celebrate the event, stamps and special postmarks were issued in Austria, Germany and the USA. (Zazzle custom stamps).

"WE'VE GOT A FIRE IN THE COCKPIT!"

One of the most dramatic accounts of what happened in the cockpit of Apollo 1 on the day of the tragedy can be found in the book *The All-American Boys* by Walter Cunningham (Apollo 7), *"the best of all astronaut books,"* according to the *Los Angeles Times*. The book opens with a dramatic chapter about what remains one of the most tragic moments of the American space program: the fire, on January 27, 1967, that resulted in the deaths of the three Apollo 1 astronauts, Grissom, White and Chaffee.

Cunningham, together with Wally Schirra and Donn Eisele, had been working in close contact with the Apollo 1 crew, as part of the backup crew, until, ten weeks before the tragedy, it had been decided to cancel the Apollo 2 mission, the

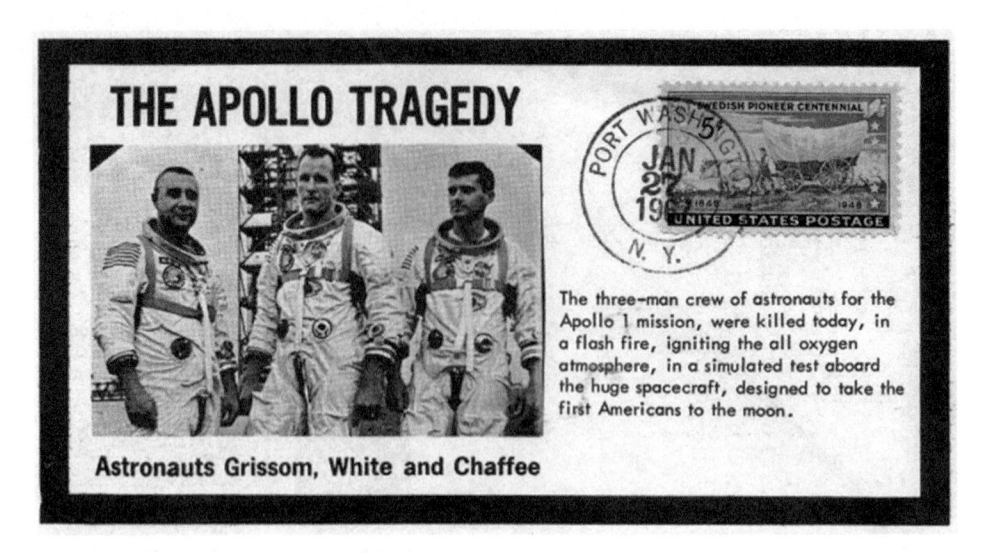

Figure 5.16: A commemoration of the crew of Apollo 1, killed in the fire on Pad 34 on January 27, 1967.

flight to which Cunningham and his colleagues had been assigned. During a routine test on Pad 34 – "*a simple test, a piece of cake,*" Cunningham recalled – the Apollo 1 crew, helpless in the cockpit at only 65 meters above the ground, burned alive in little more than 12 seconds.

It was not a complicated test: the backup crew had run a similar test the night before but crucially the hatch was not closed, so that the spacecraft's atmosphere was not 100 percent oxygen. However, both tests were classified at the time as non-hazardous, because there was no fuel on board.

Apollo 1 was the first tragic accident of the American space program. The seven years of successes accumulated during Mercury and Gemini had brought the whole world, including the Americans, to believe that everything in the space program was matter-of-fact. This accident was a brutal reminder that this was not so and that "*even the daily routine of the astronauts involved risk. The public was reminded that astronauts belonged in that category of men who roll the dice, who – like race car drivers or bullfighters or the cliff divers of Acapulco – put their hides on the line every day.*" [14]

During the following years, this accident would unfortunately be followed by other tragic events that would cost the lives of 17 American astronauts. In a cruel twist of fate, all the fatalities in the American space program happened in this same period of the year. After the Apollo 1 fire of January 27, 1967, the Challenger disaster (STS-51L) led to deaths of the Shuttle's crew of seven on January 28, 1986, while the seven crewmembers of Shuttle Columbia (STS-107) were killed during reentry on February 1, 2003.

Cunningham's book provided portraits of the three Apollo 1 astronauts. Gus Grissom was "*a decisive guy, a team leader and an independent thinker, who nevertheless encouraged input from the rest of the crew.*" Five weeks after Alan Shepard's historic first Mercury flight, Grissom had become the second American in space, aboard *Liberty Bell 7*. He had then flown with John Young aboard Gemini 3, *Molly Brown*, the first manned flight in the Gemini program, before moving on to become commander of the test flight of the Apollo program.

A taciturn, grizzled man, Grissom was scheduled to become the first astronaut to make three spaceflights and to participate in all three American programs. An Air Force cadet at 18, he had flown 100 combat missions in Korea. One of the stories told about him was how, when he first arrived in Korea, he was told that pilots who had not been shot at by a MIG were not allowed a seat on the bus to the hangar. Grissom stood only once. He was shot at on his first mission and qualified for a seat – and a place in the "*brotherhood of the right stuff.*" [15]

"*If you lived by the sword, you could expect to die by the sword: flying is a death-oriented business. You either accept the odds or you stay the hell out.*" Nobody understood this better than Grissom: "*If we die,*" he once said, referring to the U.S. space program, "*we want people to accept it. We are in a risky business and we hope that if anything happens to us it will not delay the program. The conquest of space is worth the risk of life.*"

Ed White was one of the Group 2 astronauts and on June 3, 1965, he became the first American to walk in space. It made him feel "*red, white and blue all over.*" [16] White was a West Point graduate and the son of a retired Air Force General. He was also an athlete who had only just missed qualifying for the 1952 Olympics as a 400-meter hurdler. He was a golden boy; meticulous, tall, clean-cut, and a fierce advocate of all the basic virtues: God, country, mother and faith. On Gemini 4, he took with him a gold crucifix, a St. Christopher medal, and a Star of David.

Roger Chaffee was the rookie of the crew. He was one of the younger astronauts with a short but illustrious navy career behind him. During the "*eyeball-to-eyeball*" 1962 missile crisis, Chaffee had flown most of the photo-reconnaissance missions over Cuba, mapping the missile sites. He was success-orientated and had his eye on the Moon landing.

Within the Apollo spacecraft, the atmosphere was originally of 100 percent pure oxygen, as it had been in the Mercury and Gemini capsules. In our daily lives, we breathe a mixture of nitrogen and oxygen, but maintaining the correct mixture in a capsule during a spaceflight is not easy: if the nitrogen content is excessive, this can inadvertently lead to death by asphyxiation. Moreover, during the depressurization phase, nitrogen bubbles can be formed in the blood and cause embolisms. For this reason, NASA had traditionally chosen an atmosphere of pure oxygen for its spacecraft. Yet, while this was the most inflammable option, the possibility of a fire on the ground had been given precious little consideration.

228 Two tragedies block the race in space

The Apollo 1 cabin did not even have a fire extinguisher and, at the time of the test, there were no appropriate emergency procedures. Firefighters, rescue teams and medical care were all absent, because the test had not been classified as dangerous.

Accidents involving a pure oxygen atmosphere had already happened, however, as in the case of Valentin Bondarenko, although little was known of this in the West[6]. North American Rockwell's engineers, who designed the Apollo capsule, were quite reluctant to have the pure oxygen atmosphere, but at NASA, referring to their experience with Mercury and Gemini, it was estimated that the risk was acceptable.

Another factor in the Apollo 1 tragedy was what had happened to Grissom at the end of his Mercury flight aboard *Liberty Bell 7*, when his capsule had sunk to the bottom of the ocean. At Grissom's own insistence, it was decided to abandon the explosive bolts that enabled quick hatch opening in favor of a complex, inward-opening hatch to avoid the possibility that an accidental door opening could cause fatal decompression in the capsule. The Apollo 1 hatch would have taken several minutes to open and had to be done inwards from the inside. The crew simply did not have the time and, in short, Grissom, White and Chaffee stood no chance.

Following the tragedy, and despite the fear of seeing the Soviet flag suddenly planted on the Moon, the Apollo program took its time to learn from this and, over the next 21 months, made numerous and radical changes both to the Apollo capsule Command Module and the launch procedures. All the remaining astronauts were caught up in accident investigation procedures, but the commission never fully determined the cause of the fire, nor was there total agreement about who had said what, and when.

The investigative work of the commission revealed that all aspects of the Apollo 1 test were inadequate: emergency exits; action plan; first aid kits; wiring; and plumbing. Shortcomings were ascertained in design, production, quality control and test procedures. *"The investigation's conclusions became a virtual indictment*

---

[6] In his book, Alexei Leonov commented: *"We discussed the Apollo 1 fire among ourselves a great deal. From a professional point of view, I viewed the deaths of the three American astronauts as a sacrifice which would later save the lives of others. But I was also very angry at how stubborn the American engineers were in continuing to use a pure oxygen atmosphere in their spacecraft. I couldn't understand why they had not switched to the system we adopted after the death of Valentin Bondarenko: regenerating oxygen during a flight.*

"The Americans must have known of the tragedy that had befallen Bondarenko in a pure oxygen environment. He had been given a big funeral, and the American intelligence services would not have been doing their job properly if they had not informed NASA about what had happened." (Scott-Leonov [2004], p. 192.) Also see the footnote on page 73 in Chapter 2. It is hard to find evidence that Bondarenko was given a big funeral. What is known is that his tombstone merely mentions that he was a pilot, not a cosmonaut.

of everyone connected with Apollo, including those of us who were to fly it," Walt Cunningham recalled. "*We contributed to the disaster by our willingness to tolerate questionable designs, equipment, and testing procedures; by ignoring our own good sense and accepting borderline safety margins; in short, by our eagerness to blow the bolts and get off the ground.*" [17]

A great number of things needed to be improved, but changes meant delays. "*The Program Manager weighed the crew's demands for performance, safety and other operational improvements against the payload of the spacecraft, schedules and, not least, cost. But while all this had alarmed us, it gave no premonition of what was to come. The closer we came to launch date, the more omnipotent we felt. Any astro worth his salt would fly the crate anyway – or die trying. This was no time for the 'right stuff' to waver.*" Moreover, there was a lot of pressure within NASA to launch on time. They feared that the Russians would arrive at the Moon first, but also time was money, in a very real sense, and the annual funding battle with Congress had been getting tougher.

"*The fact that the Investigation Commission found no pilot error left us all with some concern about the hardware,*" Cunningham continued. "*All along, our fear had not been that we might have an accident but that the hairy mishap, when it came, would occur in outer space, leaving no traces, offering no clues – only an eternal silence.*" After the Apollo 1 accident, NASA shifted personnel, clamped down hard on test procedures, and put a foot to the neck of North American Rockwell. At a cost of $500 million, the Command Module was redesigned.

"*The relentless pressure to meet a schedule was gone for the first time since the program's inception. With the federal purse loosened a bit to make the spaceship safe for 'our boys', many of our earlier rejected changes suddenly had a new life. As a conclusion, there was scarcely a system that didn't benefit from the exercise. All told, over 1,300 modifications were made to the Command Module following the Apollo 1 fire. And out of the whole mess, North American was to bring forth one of the greatest machines ever built by man.*" [18]

Unfortunately for the astronaut corps, 1967 would be a tragic year, marked by the passing of two other young astronauts. Ed Givens was killed in a car accident on June 6, and C.C. Williams was killed when a mechanical failure caused his T-38 trainer aircraft to crash on October 5.

SOYUZ 1: AN EMINENTLY PREDICTABLE TRAGEDY

The Apollo 1 fire occurred at a time when tensions between the two competitors in the Space Race were at their highest. The CIA had warned for years that an American victory was not guaranteed: "*Given their ability to concentrate human and material resources on priority objectives, we estimate that with a strong national effort, the Soviets could accomplish a manned lunar landing in the period 1967 to 1968."* [19] Spy satellites had been monitoring the construction of giant launching ramps that the Soviets were building for the gigantic N-1, which was proceeding smoothly without too much trouble. The NIE report[7] in 1967 identified *"a major new launch facility (Area J) at Tyura-Tam that will be able to take vehicles with a first-stage thrust in the 8,000,000–16,000,000-pound range."* [20]

Two similar launch pads were being built at a distance of 500 meters from each other. The launch pad foundation was about 30 meters in diameter and had a cylindrical well in the center with three flame trenches. A 125-meter rotating service tower and two 180-meter lightning conductors were installed near each launch pad, and there were two railway tracks between the launch pads and the vehicle assembly building. [21] A CIA report in March 1967 advised that the Soviets were accelerating their plans to arrive on the Moon before the Americans. The first circumlunar manned flight was scheduled for the 50th anniversary of the 1917 October Revolution.

At the beginning of 1967, the first cosmonauts assigned to the circumlunar project were following specific training on a modified version of a Soyuz known as the L1 or Zond, as well as for lunar landing in another modified Soyuz known as the L3. [22] Vasiliy Mishin's cautious plan called for three circumlunar missions to be carried out with three different two-man crews, one of which would then be chosen to make the first lunar landing. The initial plan was for Alexei Leonov to command the first circumlunar mission, together with Oleg Makarov. They would then expect to be able to accomplish the first Moon landing – ahead of the Americans – in September 1968.

According to Leonov: *"To train for the extreme difficulties of a lunar landing, we undertook exhaustive practice in modified Mi-4 helicopters. The flight plan of a lunar landing mission called for the landing module to separate from the main spacecraft at a very precise point in lunar orbit and then descend toward the surface of the Moon until it reached a height of 110 meters from the surface, where it would hover until a safe landing area could be identified. The cosmonaut in the*

---

[7] National Intelligence Estimates (NIEs) are United States classified documents prepared for policymakers on particular national security issues. They are reviewed and approved for dissemination by the National Intelligence Board (NIB), which comprises the Director of National Intelligence (DNI) and other senior leaders within the intelligence community, and present what intelligence analysts estimate may be the course of future events. They are read by the President, his advisors, and national security staffers, as well as heads and assistant-heads of national agencies including the FBI and the Atomic Energy Commission.

landing module would then assume manual control of its descent. This would involve split-second decisions: he would have no more than three seconds to assess the landing site and enter its coordinates into the on-board computer of the landing module. If no suitable landing site could be identified, the cosmonaut would have to give the command to shift the landing module back into the orbit of the Command Module. For if the landing module touched down on the edge of a crater, for instance, it would become so destabilized that it would never be able to lift away for the journey home." [23]

As the N-1 rocket was not yet available, Vladimir Chelomey's Proton rocket would be used for the launch, appropriately modified and upgraded with the stage D of Korolev's N-1 rocket. The first test mission of the new generation Soyuz 7K-OK spacecraft, launched on November 28, 1966 did not go to plan. After several attempts at retrofire over two days, it was determined that the capsule would reenter in China, so it was given the command to self-destruct. [24] The mission was declassified under the cover name of Cosmos 133, but the U.S. media had correctly interpreted the true nature of the spacecraft, not as an ordinary scientific satellite, but as a precursor for a new manned vehicle. [25]

The second attempt to launch a Soyuz spacecraft came a few weeks later on December 14, but failed on the ground as the main engines shut down. It would end tragically. Orders to flood the pad with water were given immediately and over half-an-hour later, a team of specialists began working at Pad 31, the primary pad for Soyuz launches, to make the spacecraft safe. Suddenly, the escape rocket ignited the third stage of the R-7 which exploded, killing Major Leonid Korostylev and seriously injuring many others. Nikolai Kamanin himself barely escaped death. [26] A few days later, Kamanin would record in his diary: *"The Americans know that we have begun preparations for a new series of manned flights, but they have no idea in what kind of trouble we really are."*

The truth was that, under the challenge of the competing laboratories, the OKB-1 was working in parallel on two new vehicles, the Soyuz 7K-OK for Earth orbit and the Soyuz 7K-L1 for the circumlunar mission. But little progress had been made since the passing of the Chief Designer and the setup of the Soyuz spacecraft was proceeding very slowly. Without the technical genius of Korolev, there was also no one able to work out the problems of pogo vibration in the N-1.

It was decided that a further unmanned mission should be mounted before launching a crew on a Soyuz, but with the explosion in December, Pad 31 had been seriously damaged and was expected to be out of commission for the following six months. Pad 1 was quickly modified to accept the larger Soyuz vehicle, but all the fueling procedures and Soyuz spacecraft checking could only be handled in Area 31 and the vehicle therefore had to be loaded and prepared there before being slowly transferred to Pad 1 by rail, some 30 km away. Finally, Cosmos 140 was launched on February 6, 1967, but the spacecraft suffered attitude control problems and excessive fuel consumption in orbit. A ballistic reentry was planned, but the heat shield burned through, producing a 300 mm hole. The capsule crashed through the ice of the frozen Aral Sea, about 510 km away from its intended landing target.

On March 16, 1967, the first test of the new Soyuz 7K-L1, launched under the cover name of Cosmos 146, succeeded in placing the craft into lunar orbit. No recovery was planned and the test was deemed successful. However, the next mission, Cosmos 154 on April 8, failed to achieve translunar trajectory and remained in low Earth orbit, reentering the atmosphere and disintegrating two days later. The Soyuz program was suffering serious development problems and all the OKB-1 projects continued to be behind schedule. The Kremlin, noting that the American program had been paralyzed by the Apollo 1 fire, impatiently decided that it was time to take advantage of the delays to Apollo that had given the Soviet Union the chance to leapfrog the Americans. [26]

A great deal of political pressure was exerted by General Secretary Leonid Brezhnev and Defense Minister Dmitriy Ustinov to have something spectacular, making headline news around the world, to celebrate the 1967 International Workers' Day (May Day), one of the most important Soviet holidays in the 50th anniversary of the Bolshevik Revolution. Lacking Sergei Korolev's charismatic style and wide experience with the upper echelons of the Soviet bureaucracy, Vasiliy Mishin could not resist the pressure from the leadership. To the consternation of his collaborators, Mishin approved the launch of the new 7K-OK to Earth orbit, despite knowing that the first three unmanned tests had all failed and the spacecraft was still affected by many problems.

After rumors that a new generation spacecraft would be launched shortly had been circulating in Moscow, quoted by Reuters since April 19, Radio Moscow announced on the morning of April 23, 1967, after one successful revolution around the Earth, that Soyuz 1 had lifted off piloted by veteran cosmonaut Vladimir Komarov[8]. What really captured the attention of Western observers was not so much the launch of the new spacecraft, nor that Komarov had returned to space (thus making him the first cosmonaut to fly in space twice), but the name given to the spacecraft: 'Soyuz', meaning 'Union'. It was also remarkable that the spaceship was named Soyuz 1 because it was the first time that a new-generation manned Soviet spacecraft had carried a number designation. Yuri Gagarin's spacecraft had simply been called Vostok and Gherman Titov's was Vostok 2. The same applied regarding Voskhod and Voskhod 2. Many were convinced that the 'Soyuz

---

[8] A very accurate analysis of the flight of Soyuz 1 is provided by Sven Grahn in "*Sven's space place*" (Grahn svengrahn.pp.se]) under "*An analysis of the Soyuz-1 flight*" (accessed in February 2018.)

1' number designation might have some significance and could be a reference to a possible docking in orbit with a second vehicle. They were right. Soyuz 2 was already on the launch pad. [27] Three cosmonauts would form the prime crew for the second Soyuz mission: the veteran Valery Bykovsky (Vostok 5), plus Alexei Yeliseyev and Yevgeny Khrunov. The two spacecraft would then perform the first ever Soviet docking in orbit.

Figure 5.17: As with the Americans, the Soviet manned space program suffered a tragic loss in 1967, with the death of Vladimir Komarov during the return of Soyuz 1 on April 24.

Yeliseyev and Khrunov would also spacewalk from Soyuz 2 to Soyuz 1, in a complicated EVA procedure that had troubled engineers for months. The only previous Soviet EVA experience had been the dramatic spacewalk of Alexei Leonov two years previously. One concern was that the hatch in the 'orbital module' of the Soyuz was much too small – just 66 cm in diameter – for a fully-suited cosmonaut to pass through easily. If successful, the two cosmonauts would return to Earth with Komarov on Soyuz 1, while Bykovsky would return alone in Soyuz 2. Cosmonaut Vladimir Komarov had been selected to fly the Soyuz 1 mission, an honor that every cosmonaut would have wished for, but everyone involved in the project knew that a record 203 engineering issues that required solving had been detected during the Soyuz tests and this put Komarov in a dangerous and unenviable position.

Engineers on the spacecraft's development team knew that Soyuz was not ready for a manned flight, but the politicians, as often happens, were not very sympathetic to such 'technical details' and, above all, had no interest in carrying bad news to Brezhnev, despite the insistence of some officials who understood the seriousness of the problems. An atmosphere of pessimism prevailed at the cosmodrome.

Kamanin wrote in his journal on April 15: "*In all the previous flights we believed in success. Today, there is not such confidence in victory. The cosmonauts are prepared well, and the ships and the instruments have gone through hundreds of tests and verifications, and all seems to have been done for successful flights, but (still) there is no confidence. This can perhaps be explained by the fact that we are flying without Korolev's strength and assurances; we were spoilt by Korolev's optimism.*" [28] May Day was just around the corner.

Gagarin, who was Komarov's backup for this mission, prepared a ten-page memo, addressed to party leaders, detailing the 200+ design issues with the Soyuz, which he gave to Venyamin Russayev, his KGB escort and close friend. Russayev has been quoted as having passed this document to his superior, but it is not entirely clear what happened to that report. [29] The document never reached its intended recipients, with everyone terrified of Brezhnev's reaction to the mission being delayed or scrubbed. All the officers who passed on the letter or who were involved with it in any way were demoted, transferred, sent to Siberia, or fired. Russayev himself would be barred from ever associating with a cosmonaut or anyone affiliated with the space program again.

Allegedly, Gagarin tried to persuade Komarov to swap with him, in the hope that once they realized that Gagarin – a national asset – was in the craft, they would cancel the flight. Equally, Komarov feared that if he refused to fly himself, Gagarin would be forced to take his place, so both men were trying to protect each other and prevent the other from taking on the mission. According to an unverified source, quoted in a 2011 book, Gagarin also tried in vain to disrupt launch procedures sufficiently to force the mission to be cancelled. [30]

Soyuz 1 was launched on April 23, 1967, in the first nocturnal launch in history. For all his efforts, the only result Gagarin obtained was that his image – which, in the eyes of Brezhnev who was initiating the Stalinist restoration, recalled the Khrushchev era too closely – began to fade. With the help of the hostile and ambitious fellow cosmonaut Georgy Beregovoy, Gagarin would quickly be expelled from the space program. A little less than a year later, the first man in space would die in a tragic aircraft accident, one that has been surrounded in mystery and conspiracy theory ever since.

Soyuz 1: An Eminently Predictable Tragedy 235

Soyuz 1 'French' Fakes

Figure 5.18: Sadly, the proliferation of 'French' fake covers also included some that 'commemorated' the Soyuz 1 mission.

The very rich production of 'French' fakes also included several 'commemorative' covers of the Soyuz 1 mission, with the usual unlikely cancellation of Baikonur-Karaganda produced in the standard 'limited edition' of 150 copies.

Upon entry into orbit, the left solar panel of Soyuz 1 failed to deploy, remaining wrapped around the service module and beginning a chain of problems. The first of these was the shortage of electrical power, which meant that the capsule was not receiving sufficient power for its guidance computer. [31] The telemetry antenna associated with the solar panel was also inoperable, preventing omnidirectional coverage. The undeployed panel also obstructed some of the navigation equipment, namely the sun and star sensors required for attitude control, which were essential for stabilizing the spacecraft as well as for firing the engines accurately. The asymmetrical single-panel configuration was itself causing attitude control issues, with unconfirmed reports suggesting that Komarov even tried to knock the side of the ship to try to jar open the recalcitrant panel. [32] The cosmonaut unsuccessfully tried to orient the spacecraft visually, while intermittently losing communication. After five hours of trying, he was still unable to orientate to the sun.

"*As soon as it became apparent the mission was running into trouble,*" recalled Komarov's fellow cosmonaut Alexei Leonov, who was in the control center at the time, "*an emergency committee was convened, under Chertok's command, to try to work out a solution. I also sat in this committee. It was soon obvious that the problem could not be solved. Once this was clear, we recommended that the flight be terminated early and that the second Soyuz spacecraft not be launched.*" [32] The decision was taken to cut short the mission and get Komarov home as quickly as possible, which meant commencing reentry after the 16th orbit. The Soyuz 2 mission was immediately cancelled. The faulty attitude control system prevented Komarov from reentering the atmosphere with the right orientation during the 16th orbit, causing the automatic system to halt retrofire. Things continued to go wrong on the 17th and 18th orbits. Unverified accounts have referred to the reception of voice signals at the listening post near Istanbul in Turkey which suggested that the cosmonaut was aware that the problems he was facing were insurmountable.

"*I had no doubt*, Boris Chertok would recall in his book, "*that Komarov had long since grasped the complexity of the situation. He was not a young fighter pilot, but an experienced engineer and test pilot. More than once, he had risked his life during aircraft tests. Now, his own composure and faultless actions, rather than automatic controls, would determine his return from space.*" [33] Komarov was given new instructions to make a manual retrofire on the next orbit. Reentry was finally successful and the drag parachute was deployed. According to Leonov, "*Once the spacecraft began its reentry, we lost contact with Komarov. This is normal. It is only after the landing capsule's main parachute, which contains an antenna, opens that contact may be reestablished. But we never did regain contact.*" [34]

In fact, the main parachute failed to deploy. The unusual, asymmetrical shape of the spacecraft caused it to drift. Komarov released the reserve chute but it became entangled in the cords of the drogue chute and also failed to open. Rescue forces were deployed to the area that Soyuz 1 was expected to have made its emergency landing, east of the city of Orsk in the southern Ural Mountains, along Russia's border with Kazakhstan. [35] "*Shortly afterwards,*" Leonov explained, "*I*

Soyuz 1: An Eminently Predictable Tragedy 237

joined a team that left from Baikonur for the site. When we arrived, we found little more than a lump of crumpled metal. Komarov's 2-meter-high landing capsule had been reduced to a tangled mess little more than 70 cm high. His spacecraft had landed at a speed of 25 meters per second killing him instantly, and then a fire had started. Lying on the ground near the wreckage were three parachutes. We realized immediately that they had not opened properly." [35] Locals who had witnessed the final stages of Komarov's descent reported seeing the parachutes of the Soyuz simply turning and not filling with air.

Komarov was honored with a state funeral in Moscow and his remains – after being displayed at the Red Banner Hall of the M.V. Frunze Central House of the Soviet Army (TsDSA) – were buried in the Kremlin Wall at Red Square. The first Soviet cosmonaut to fly into space more than once had also become the first human ever to die on a space mission.

Figure 5.19: Vladimir Komarov's remains were buried in the Kremlin wall.

This terrible loss made a huge impression even in America. NASA requested the Soviet government to allow them to send a representative and was ready to send two astronauts, Gordon Cooper and Frank Borman, to attend the funeral in the Soviet Union, but the Russians insisted that this was an *"internal affair."* Komarov was posthumously awarded his second Order of Lenin and also the Order of Hero of the Soviet Union.

238 Two tragedies block the race in space

While Designer Mishin continued to believe that the parachute had been incorrectly packed during preparations, an investigation commission was set up, with seven subcommittees. Their work continued for more than a month, studying all the spacecraft's equipment and remains. [36] The investigation committee, headed by Yevgeny Utkin of the Flight Research Institute of the Aviation Industry, concluded that the Soyuz 1 parachute container had opened at an altitude of 11 km, but had become "*deformed.*" This had squeezed the main canopy and thus prevented it from opening correctly. Subsequent tests revealed that the parachute container was not rigid enough and for future missions, the decision was taken to enlarge and strengthen those containers.

There was also an unofficial, and more likely, version of the cause of the accident, one that suggested gross negligence on behalf of technicians at TsKBEM's manufacturing plant. [37] During preflight preparations, the two Soyuz capsules had to be coated with thermal protection materials and then placed into a high-temperature test chamber to polymerize the synthetic resin. Apparently, technicians tested the vehicles in the chamber with their parachute containers but without the protection[9]. Because of the omission of the covers, the interiors of the parachute containers were coated with a polymerized coating, which created a very rough surface. The consequence was that the interiors of *both* spacecraft's parachute containers had this same rough surface coating, which contributed to the failure of the Soyuz 1 parachute deployment on reentry. Shockingly, if Komarov had not faced all his troubles in orbit and the Soyuz 2 launch had gone ahead as planned, all four cosmonauts would certainly have been killed upon their return.

Echoing the thoughts of Walt Cunningham mentioned at the start of this chapter, Alexei Leonov concluded: "*Before Komarov's death, the Soviet press and public had paid little attention to the extreme risks we took. Spaceflight had seemed easy. All the marches, parades, grand music and medals in honor of the cosmonauts had made the space program seem like an elaborate exercise. Now, people realized that being a cosmonaut did not necessarily lead to fame and public acclaim: it could also lead to death. The public started to appreciate the real dangers involved as, I believe, the American public did after the Apollo 1 fire. More resources were made available to our space program, although they never matched the huge amount of money available to NASA.*

"*It was quite clear that no future flights could go ahead until all systems on the spacecraft had been completely reviewed. A very high-level government decision was taken to carry out a series of tests. That meant that there was no way the original schedule for future spaceflights would be kept.*" [38]

---

[9] In Deputy Chief Designer Chertok's investigation into the matter in the early 1990s, he could not find anyone still living who could remember why the covers had been left off.

References

1. **The All-American Boys**, Walt Cunningham, iBooks, New York, 2003, pp. 297.
2. *A secret uncovered: The Soviet decision to land cosmonauts on the Moon*, Asif Siddiqi, *Spaceflight*, Vol. 46, May 2004, pp. 205–213.
3. Reference 2, p. 207.
4. *The N1-L3 Programme*, Daniel A. Lebedev, *Spaceflight*, Vol. 34, September 1992, p. 288.
5. *Designer Mishin speaks on early Soviet space programmes and the manned lunar project*, Vasiliy Mishin interview, *Spaceflight*, Vol. 32, March 1990, p. 104.
6. **Rockets and People – Vol. 3: Hot Days of the Cold War**, Boris Chertok, NASA SP-4110, Washington D.C., 2009, pp. 588–9.
7. **The Russian Space Bluff – The inside story of the Soviet drive to the Moon**, Leonid Vladimirov, Dial Press, London, 1973, p. 50.
8. Reference 6, pp. 528–536.
9. *Soviet Scientists Hail Apollo Flight*, New York Times, December 27, 1968, p. 21.
10. **Unmanned Space Project Management – Surveyor and Lunar Orbiter**, Erasmus H. Kloman, NASA SP-4901, Washington D.C., 1972, p. 10.
11. *Surveyor 1, America's First Lunar Landing*, Paul D. Spudis, in www.airspacemag.com, June 2, 2016 (accessed in February 2018).
12. **Spaceflight Revolution: NASA Langley Research Center from Sputnik to Apollo**, James R. Hansen, NASA SP-4308, Washington D.C., 1995, pp. 312–3
13. Reference 1, pp. 271–2.
14. Reference 1, p. 6.
15. Reference 1, p. 10.
16. Reference 1, p. 11.
17. Reference 1, p. 13.
18. Reference 1, pp. 17–19.
19. CIA top-secret report 1961, accessed February 2018 at www.cia.gov/library/readingroom/docs/DOC_0000242846.pdf
20. *Declassified American Government documents show a broader and in-depth interest in Soviet space activities*, Peter Pesavento, *Journal of the British Interplanetary Society (JBIS)*, Vol. 56, 2003, p. 176.
21. Reference 4, p. 289.
22. **Two Sides of the Moon: Our Story of the Cold War Space Race**, David Scott and Alexei Leonov, St. Martin's Press, New York, 2004, pp. 187–8.
23. Reference 22, p. 188.
24. **Soyuz: A Universal Spacecraft**, Rex Hall and David J. Shayler, Springer-Praxis Books, 2003, p. 124.
25. Reference 24, p. 127.
26. Reference 24, p. 124–5; see also *Encyclopedia Astronautica, "Soyuz failure in detail"*, Mark Wade, www.astronautix.com, (accessed in February 2018).
27. **The First Soviet Cosmonaut Team: Their Lives and Legacies**, Colin Burgess and Rex Hall, Springer-Praxis Books, 2009, p. 263.
28. *Mourning Star*, Asif Siddiqi, in *Quest*, Vol. 2, Winter 1994, p. 7.
29. **Escaping the Bonds of Earth: The Fifties and Sixties**, Ben Evans, Springer-Praxis books, 2009.
30. **Starman: The Truth Behind the Legend of Yuri Gagarin**, Jamie Doran and Piers Bizony, Walker Publishing, 2011.
31. Reference 22, p. 196.

32. **Challenge to Apollo: The Soviet Union and the Space Race, 1945–1974**, Asif Siddiqi, NASA SP-4408, Washington D.C., 2000, p. 582.
33. Reference 6, p. 638.
34. Reference 22, p. 197.
35. Reference 32, pp. 585–586.
36. Reference 6, p. 647.
37. Reference 32, p. 589.
38. Reference 22, p. 198.

6

The final leap

A NEW BEGINNING AT NASA

After the tragic Apollo 1 fire, NASA once again conducted an internal reorganization and began a complex redesign of the Apollo Command and Service Module (CSM). North American Rockwell (NAR) had won the contract to design the Apollo spacecraft, ousting McDonnell, the former producer of Mercury and Gemini, without having previous experience of manned flight in space. NAR, too, would have to undertake some strategic changes, and a new project leader, John P. Healey, was assigned. Under his direction, over 1,300 changes would be implemented on the 'Block I' Apollo. Walt Cunningham, a former member of the Apollo 1 backup crew and by now assigned to the Apollo 7 mission, actively participated in the design of the new module. He defined the old spacecraft 012 (as the first Block I was known) as *"a piece of junk,"* put together in a hurry, *"pressed by the Russians breathing down our necks."*

The Command Module was entirely redesigned. Among the changes was a new one-piece hatch that swung outwards and could be opened in ten seconds. To extinguish any fire that might develop, an emergency venting system was added that could reduce cabin pressure in seconds. From now on, while the spacecraft was on the pad and during launch, a mixture of 60 percent oxygen and 40 percent nitrogen would replace the 100 percent oxygen atmosphere formerly used. A rapid re-pressurization system, an emergency oxygen breathing system, and electric power system changes were all introduced. From brand new crew couches to an additional urine dump nozzle, there was scarcely a system that did not benefit

242 The final leap

Figure 6.1: The commemorative Apollo 7 'Orbit' Cover pays tribute to the three astronauts killed in the Apollo 1 fire on January 27, 1967. The cover is signed by John P. Healey, the new North American Rockwell Apollo Project Manager.

from the rebuild. "*I am convinced,*" Cunningham has commented, "*that we would never have reached the Moon in only five missions had we not gone through this rebuilding process – the inevitable result of the fire on Pad 34.*" [1]

FIVE GIANT STEPS

It was now time to aim straight for the goal. The Apollo program was moving ahead and in April 1967, Deke Slayton called a summit meeting in his office, convening the group of 18 selected astronauts to present the "*strategic plan*" of five giant steps.

- Step 1: Apollo 7 – engineering test of the CSM in Earth orbit;
- Step 2: Apollo 8 – flight test of the CSM and Lunar Module (LM) in Earth orbit;
- Step 3: Apollo 9 – lunar mission simulation in Earth orbit;
- Step 4: Apollo 10 – a full dress-rehearsal of the lunar mission around the Moon;
- Step 5: Apollo 11 – first lunar landing attempt.

In parallel, at the Marshall Space Flight Center (MSFC) in Huntsville, Alabama, where Wernher Von Braun was developing the Saturn V, they were working feverishly to prepare the first 'all-up test', following the new directive instigated by George Mueller.

THE RUSSIANS ARE COMING!

There was a clear feeling that the Soviets had recovered the advantage that the Americans had established with the Gemini program, and were again in the lead. The veil of secrecy behind the Iron Curtain was impenetrable, but there were signs that something was happening. The propaganda machine was in full swing and the Soviets – it was believed – never did anything by chance.

Figure 6.2: Soviet propaganda postcard issued in 1967.

Philatelic items issued at the beginning of 1967 featured spacewalks and rockets in Earth orbit. Halfway through the year, a philatelic leaflet appeared praising the 'Great October Endeavors', followed by a set of stamps celebrating 'cosmic navigations' featuring evocative sketches and wording such as "*In circumlunar orbit*" and "*On the Moon*" which worried the Americans.

It appeared that despite the tragic loss of Vladimir Komarov on Soyuz 1, the Soviets had never stopped their activities. A lunar orbiter, Luna E-6LS, was launched on a Molnya booster on May 17, 1967. This was a modified version of the Luna E-6LF lunar imaging orbiter, with an additional payload to help test the communications and tracking systems for a crewed Moon mission. However, the probe never left Earth orbit and would subsequently be designated Cosmos 159. A full-specification Soyuz 7K-L1 Moon orbiter spacecraft (not a prototype) was then launched on September 27, 1967, but its booster failed shortly after liftoff

244   **The final leap**

Figure 6.3: The souvenir sheet issued in 1967 for the 50th anniversary of the Russian Revolution was entitled "Слава великим свершениям октября!" (*"Glory to the Great October Endeavors!"*)

Figure 6.4: This stamp, issued on March 30, 1967 to mark Cosmonautics Day, depicts a spacecraft flying around the Moon.

and fell to Earth about 65 kilometers (40 miles) away. The escape tower separated from the booster, taking the upper module with it. Max Faget, who had invented the escape tower in 1958, would later joke about never having been paid royalties for it by the Soviets! [2]

Figure 6.5: On October 20, 1967, a new set of stamps entitled *Space Exploration* was issued. The four-Kopek stamp (left) was titled "На селеноцентрической орбите" (*"On a selenocentric orbit"*), while the six-Kopek stamp (right) was called На Луне (*"On the Moon"*).

A Soyuz 7K-OK, under the cover name Cosmos 186, was launched on October 27, followed three days later by a second, Cosmos 188. The two unmanned spacecraft – dedicated to the 50th anniversary of the October Revolution – performed the first ever automated docking in orbit, finalizing what should have been done by the unfortunate Soyuz 1 and 2. [3] However, the docking adapters did not work completely, due to mechanical negligence during assembly at the engineering facility (TP) which prevented complete retraction and mating of the electrical connectors of the two vehicles. [3] The modules were mechanically, but not electrically docked, and the maneuver cost more fuel than planned. Cosmos 186 returned to a soft landing in a predetermined part of the USSR on October 31.

Unfortunately for Cosmos 188, an incorrect orbital attitude meant that it reentered at an excessively steep angle and its emergency self-destruct system was activated as it was heading towards a landing north of the Mongolian border. However, another space first had been established with these two flights, in a display of advanced robotics that proved the feasibility of Soyuz-style docking procedures and removed some of the fears over lunar orbit rendezvous when such procedures would have to be carried out some 400 million kilometers away. More unmanned tests were still needed, however, and the crewed mission was once again delayed by a few months.

Figure 6.6: Stamps depicting the joint flights of Cosmos 186 and Cosmos 188.

Figure 6.7: Commemorative covers issued by the Philatelic Club of Tartu to celebrate the launch of Cosmos 186 and Cosmos 188.

Tests of the Soyuz 7K-L1 lunar module also intensified. After two more failed launches, the fifth model of the 7K-L1 finally launched successfully on March 2, 1968, under the generic name of Zond 4 (Зонд, in Russian means 'probe'), which was sent out to the equivalent distance of the Moon. There was no attempt made to circle the Moon itself, however, as the probe was sent in exactly the opposite direction so that its orbit would be minimally distorted by the Moon's gravitational field. This was actually a test of spacecraft systems and procedures, in particular aimed at testing the critical Earth atmosphere reentry maneuver through the narrow entrance corridor. Unfortunately, the capsule returned at the wrong inclination, heading at high speed towards West Africa, so the self-destruct command was given. [4]

Figure 6.8: Cover commemorating the launch of Zond 4, serviced in Tartu by the local Philatelic Club.

On July 15, another launch of the 7K-L1 failed on the pad, due to the explosion of a fourth-stage tank, and three technicians were killed. Unfortunately, accidents of this kind were not infrequent in the period 1967-69. Haste is always a poor advisor.

THE MYSTERIOUS DEATH OF YURI GAGARIN

A few weeks after the failure of Zond 4, a more serious accident shook the spirits of the Soviet people and of the whole world: the death of Yuri Gagarin during a mysterious aircraft crash, for reasons never completely explained. After a solemn

248   **The final leap**

state funeral, Gagarin's ashes were buried at the foot of the Kremlin Wall. The Cosmonaut Training Center at Star City would be renamed after him. Five different Commissions of Inquiry, systematically thrown off by the KGB, failed to clarify the details of the incident.

Figure 6.9: (left) Yuri Gagarin's headstone in the Kremlin Wall Necropolis. (right) A 'French' fake cover commemorating the death of Gagarin, with the usual run of 150 items.

Within days of the crash, a government committee of investigation was established under the leadership of the Minister of Defense. Gagarin's best friend Alexei Leonov wrote later: *"Gherman Titov and I were assigned to it [the investigation] as representatives of the cosmonaut corps. Different possible causes of the accident were closely examined. One theory was that Yuri's plane had gone into an uncontrolled spin after maneuvering to avoid a flock of birds. Another was that it had collided with a hot air balloon... After a while rumors started to circulate. One claimed that Yuri had been drinking before he flew. Another speculated that he and Seregin had been taking potshots at wild deer from their plane, causing it to spiral out of control. Yet another claimed that Yuri was not dead at all, but had been thrown into prison after tossing a cognac in Brezhnev's face. Another had it that he was languishing in a mental asylum... We tried very hard to have the investigation into the crash re-opened. I wanted to conduct my own independent inquiries. We spoke at numerous scientific symposiums stating that we did not believe the reasons for the crash had been looked into thoroughly enough. In the beginning Titov stood by us, but with time even he distanced himself from the controversy."* [5]

It would be 25 years before all the documents concerning the crash were declassified and Leonov could gain access to them: *"When I studied them carefully, I found a document I had written at the time, describing the one-and-a-half to two-second interval between the two booms I had heard. The document had been altered: in handwriting that was not my own, the interval between the two booms had been changed to twenty seconds."*

The documents also revealed that, in their last transmission, the crew had reported they were flying at a height of 4,200 meters (about 14,000 feet). Gagarin said that they had finished their maneuver and were returning to base. The crash site was at almost exactly the position from which that last transmission was made, indicating that they had gone down almost immediately. Leonov explained: *"At the time of the accident, it was known that a new supersonic Sukhoi SU-15 jet was in the same area as Yuri's MiG. Three people who lived near to the crash site confirmed seeing such a plane shortly before the accident. According to the flight schedule of that day, the Sukhoi was prohibited from flying lower than 10,000 meters. I believe now, and believed at the time, that the accident happened when the pilot of the jet violated the rules and dipped below the cloud cover for orientation. I believe that, without realizing it because of the terrible weather conditions, he passed within 10 or 20 meters of Yuri and Seregin's plane while breaking the sound barrier. The air turbulence created overturned their jet and sent it into the fatal flat spin.*

"To complicate matters, Yuri and Seregin's MiG had been fitted with external, expendable, 260-litre fuel tanks, the purpose of which was to allow a plane to fly much further in combat. The tanks were designed to be dropped before entering a combat zone, where complicated maneuvering was called for, because they severely compromised the plane's aerodynamic performance. Yuri and Seregin were not expected to perform such [maneuvers] that day, but it was clear from the way their plane had chopped through the treetops that they had tried to recover from the spin and it seemed they were short of doing so by a matter of just one-and-a-half to two seconds.

"The investigating committee would never have admitted at the time that that is what had happened because it would have meant admitting that flight controllers were not adequately monitoring the airspace close to sensitive military installations. I believe ordinary people were unable to accept the real explanation because the technical details of Yuri's plane being intercepted by an SU-15 jet were too complicated for most to understand. But now, nobody repeats any nonsense about Yuri being drunk, irresponsible or mad." [6]

NASA SPEEDS UP AND MODIFIES ITS PROGRAMS

Following the successful docking of Cosmos 212 with Cosmos 213 in April 1968, the Soviet Cosmos 238 flight was being viewed as a dry run for returning cosmonauts to orbit for the first time since Soyuz 1. In America, the CIA advised that the Soviets had overcome the problems of fine-tuning the N-1 launcher and were planning to test their circumlunar flight profile with an automatic probe in October that year. According to the intelligence service, a 7K-L1 manned mission would follow in December 1968, commanded by Alexei Leonov and piloted by Oleg Makarov.

250 The final leap

The news shocked NASA's top managers, who quickly revised the schedule of missions already planned. At Grumman, unexpected difficulties were being encountered with the Lunar Module (LM), in part because no one had ever designed a spacecraft able to soft-land on the Moon with men onboard before, and the vehicle was seriously behind schedule. In order not to drop the pace of progress while waiting for the LM to be ready, the decision was taken to revise the 'five giant steps' program. On August 12, 1968 – at the suggestion of George Low to Robert Gilruth, and after consulting with Deke Slayton and Chris Kraft – it was decided to modify the goal of Apollo 8, which was originally planned to test the LM in Earth orbit. Instead, the mission would fly around the Moon in December, without the LM.

The revised plan saw the LM slip back one mission.

- Step 2: Apollo 8 – test of insertion into lunar orbit (without LM) to try the LOR concept;
- Step 3: Apollo 9 – test of the LM in Earth orbit.

This would give NASA experience of handling a lunar flight that would help to cut the schedule for the first Moon landing. It would also provide lunar orbital path data for follow-on flights. Most importantly, it would allow the U.S. to beat the Soviets to the Moon with a crew. A manned circumlunar expedition was not a full Moon landing, but it would be a major trump card. A Moon landing mission could then be simply represented as a natural follow up. [7] The change of plans also meant changes for the astronauts. The original crew for Apollo 8, commanded by LM specialist Jim McDivitt, was switched to Apollo 9 along with the Lunar Module, while Frank Borman's Apollo 9 crew, who had already been training for a high orbit mission, were switched to Apollo 8 – and a much different orbit to the one they had expected. The Saturn V, even without the LM onboard, was programmed to make a flight to the Moon, thus beating the Soviets to the draw.

When the news was announced that the Americans would send a virtually untested rocket to the Moon, Vasiliy Mishin was skeptical and did not take it seriously. A manned circumlunar mission would be a high-risk adventure. The Americans had not accomplished any unmanned lunar fly-bys to demonstrate that their systems would function correctly and of the only two Saturn V flight tests to date, *"the second was a failure."*

ZOND 5: A NEW SOVIET RECORD

The Soviet N-1 rocket still had numerous problems at this point, but Mishin did not give up. In an extremely desperate attempt to salvage national pride, the Soviets took a gamble with Zond 5. Launched on September 14, 1968, Zond 5 was formally a space probe, but was in fact a Soyuz 7K-L1 designed for a manned mission and modified for an automatic flight around the Moon. It would become the first spacecraft to circle the Moon and return to land on Earth. The crew was replaced by a biological payload of two Russian tortoises, wine flies, mealworms, plants, seeds, bacteria and other living matter. In the pilot's seat, there was a 5-foot 7-inch (170 cm) mannequin, weighing 155 pounds (70 kg) and filled with

Figure 6.10: Cover commemorating mission Zond 5.

radiation detectors. No information was provided on the mission, with TASS only announcing that a probe had been launched into open space.

On September 18, the spacecraft flew around the Moon, at its closest at a distance of 1,950 km. High-quality photographs of the Earth were taken at a distance of 90,000 km. On its return to Earth, a problem in the attitude control sensor saw the spacecraft reach 20G during its ballistic reentry, coming down in a backup splashdown area in the Indian Ocean. However, this did not appear to affect any of the biological specimens onboard, all of which were alive and in good shape when the descent module was finally opened four days after landing. It was announced that the tortoises had lost about ten percent of their body weight but remained active and showed no loss of appetite.

252 The final leap

The United States had been tracking Zond 5 for its entire flight, collecting intelligence information. Photographs taken of the descent module bobbing in the ocean by USS McMorris raised concerns at NASA that the Soviets were planning a manned circumlunar flight soon and, eventually, a lunar landing. The recording of choirs of the Red Army that had been used to test the communications system added to the suspicion that this had been a test for a manned flight. However, Zond 5 had reentered the atmosphere in a trajectory steep enough to heat the craft to levels beyond human tolerance. But the Soviets had gained a new 'first', with Zond 5 becoming the second spacecraft to circumnavigate the Moon and the first to return safely to Earth. Furthermore, the number of Soviet launches in the last year (twice that of the Americans) was confirmation that Mishin was trying to regain the upper hand.

APOLLO 7: THE FLIGHT OF THE PHOENIX

At NASA, who were aiming for victory in this *"Great Olympiad in the sky,"* they were feeling the pressure of the Soviets breathing down their necks. Now, 21 months after the Apollo 1 tragedy, human flights were resuming as scheduled.

On October 11, 1968, the Apollo 7 mission began. This was the first American capsule to be launched with three men onboard – Wally Schirra, Donn Eisele and Walt Cunningham. The first goal of the mission was a technology test of the new 'Block II' CSM: *"It was the first model of a new generation of space vehicles,"* Cunningham wrote, *"and it was being checked out by a new contractor whose only previous experience resulted in a spacecraft being incinerated on the pad thirty days before launch."* [8] Apollo 7 was required to demonstrate the capability of the Apollo craft to perform rendezvous maneuvers and the ability of the crew to perform on a long-term mission.

But after the tragedies of Apollo 1 and Soyuz 1, Apollo 7 had also, above all, to restore everyone's faith in the space program. This was the first time that the Saturn rocket – the most powerful in the world – had been used for a human mission, but as there was no intention of using the LM on this flight (because it was not yet completed), Apollo 7 flew on the 'light' version, the Saturn IB, rather than on the Saturn V which would be used for all subsequent missions. In his book, Cunningham's evocative description almost gave the impression of being on a public stage:

"The rocket rises from the pad so slowly, so ponderously at first, that you could be imagining it even moved at all. It is an agonizingly slow ten seconds before Apollo 7 clears the tower. One million three hundred thousand pounds is balancing on an arrow of flame. Painstakingly, it climbs, trailing a fireball as vivid as the colors of hell. At the spectator bleachers, two-and-one-half miles

Apollo 7: The Flight of the Phoenix 253

Figure 6.11: (above) A cover commemorating the launch of Apollo 7 (from the collection of astronaut Walt Cunningham). (below): An emblem of Apollo 7, flown in space with Walt Cunningham.

away, the earth actually trembles. The vibrations, the noise, the shock; they roll over you in waves. And, from out of the second fire on Pad 34 rises our modern-day Phoenix." [9]

The mission returned to Earth after spending 11 days in space; more than the time needed for the trip to the Moon and back. Apollo 7 demonstrated the suitability of the spacecraft and rocket as well as the accuracy of all the related procedures from launch protocol to the flight direction. In particular, the mission fully tested the Service Propulsion System (SPS), the engine – featured in the mission patch – that would take Apollo on to lunar orbit and then reposition it in Earth orbit. The engine was successfully ignited eight times.

The first step towards the lunar landing goal had been accomplished and paved the way for the next mission, Apollo 8, which received the go-ahead. The Americans had regained their confidence in being able to maintain the commitment made by President Kennedy seven years earlier.

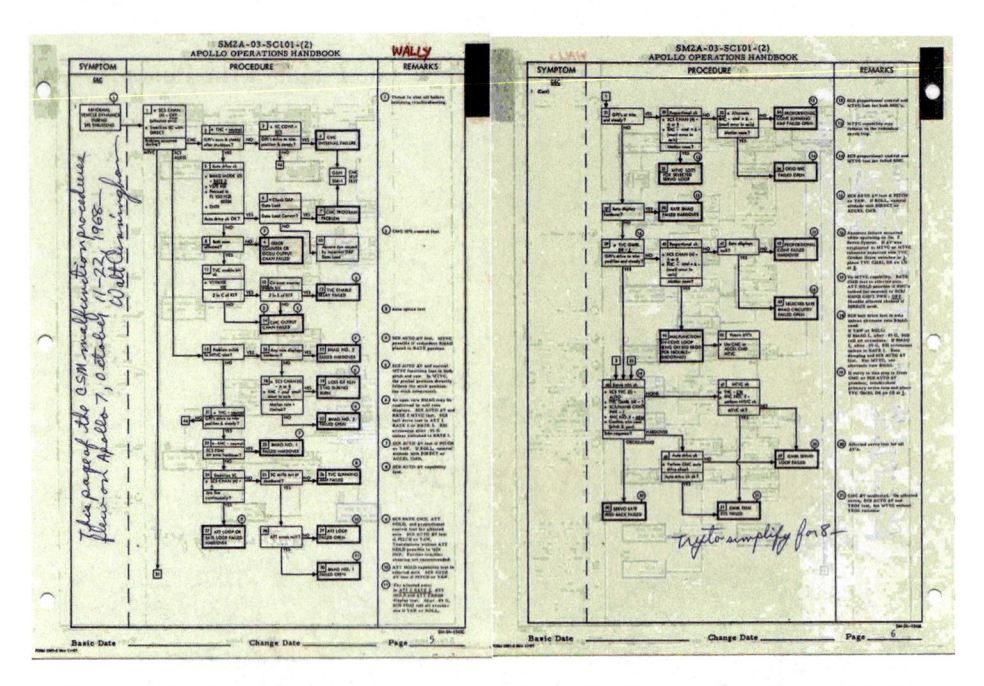

Figure 6.12: The malfunction procedures for the Apollo Command and Service Module (CSM) were tested for the first time during the Apollo 7 mission. This image shows an extract from the manual that was actually flown and used aboard Apollo 7 during the mission. (From the collection of Walt Cunningham.)

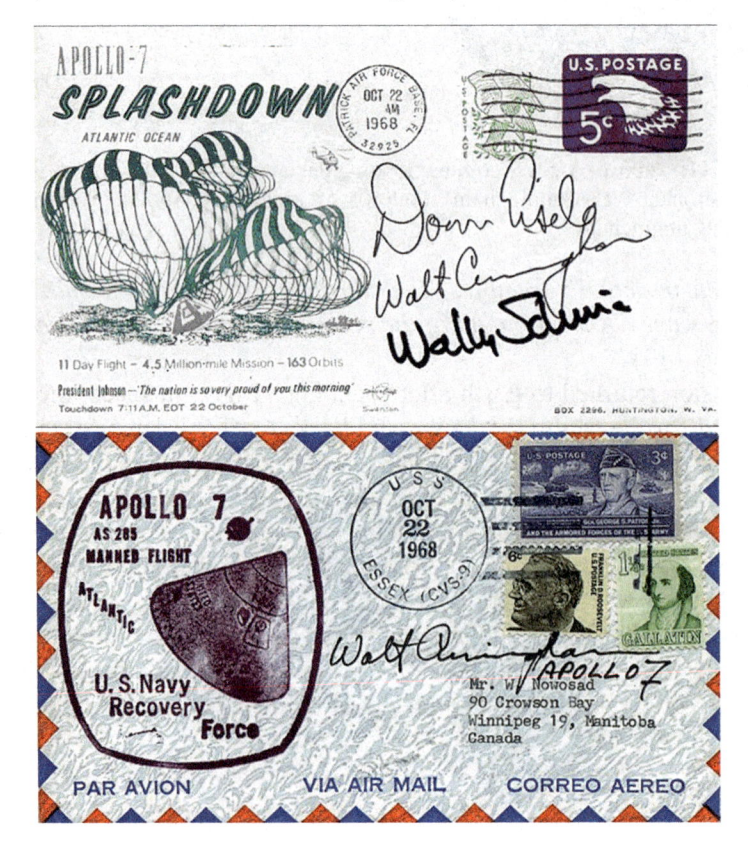

Figure 6.13: (above) A cover commemorating the splashdown of Apollo 7 (from the collection of astronaut Walt Cunningham). (below): A cover commemorating the Apollo 7 recovery by the aircraft carrier USS Essex.

The 'Prisoner' Recovery Covers

With the relaunch of the Apollo missions, public interest, and that of collectors in particular, began to grow quickly towards the race to get to the Moon first. It provided a big opportunity for counterfeiters, whose own numbers during this period grew in parallel.

Many forged philatelic covers suddenly appeared on the market, including the two shown in Figures 6.14 and 6.15: the NASA-like Essay Cachets and the 'faded' KSC Type I Die Machine Cancellations of the Kennedy Space Center on October 11, 1968. [10]

But the most unique fake covers of this period were the 'Prisoner's Covers' (which were sometimes postcards) with the fictitious USS Essex cancel that originated in a prison printshop. Two of them (an envelope and a postcard) are shown in Figure 6.16. They bear the postmark of the aircraft carrier USS Essex that recovered the Apollo 7 spacecraft in the Atlantic Ocean. They were not an attempt to counterfeit, but were creations of pure fantasy that had nothing to do with the original postmark. A simple comparison with the original USS Essex cancellation in Figure 6.13 is enough to realize this. These, too, were widely distributed.

Figure 6.14: Apollo 7 genuine NASA Essay Cachet (left) and Apollo 7 NASA-like Essay Cachet (right).

(continued)

(continued)

Figure 6.15: A genuine KSC Die Machine cancellation (above) and a counterfeit version (below).

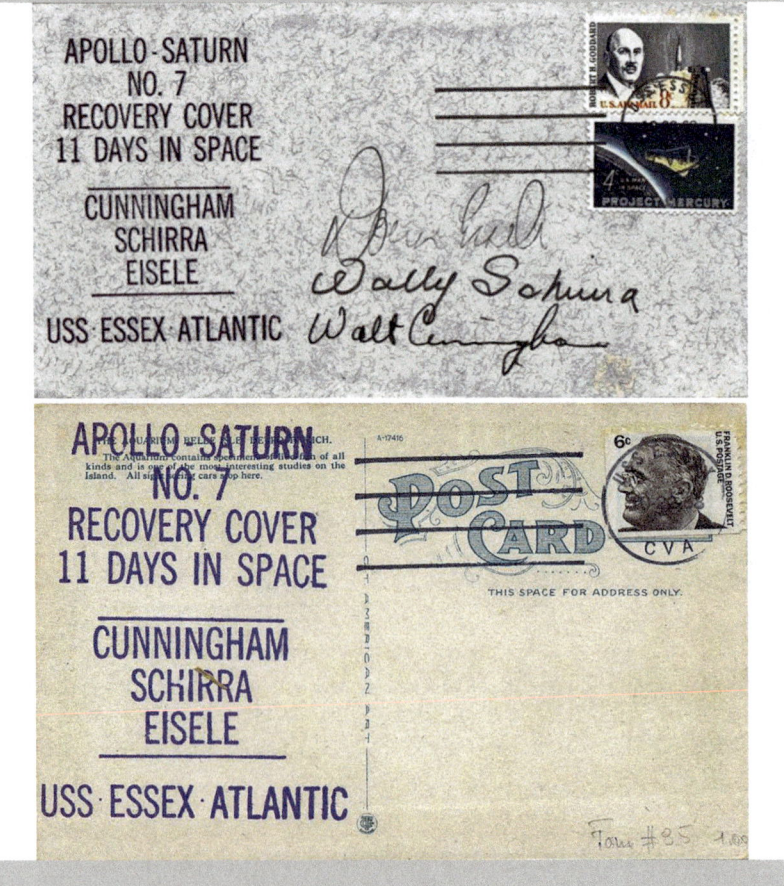

Figure 6.16: 'Prisoner Cover' envelope and postcard with the fictitious USS Essex cancel.

SOYUZ 3: THE SOVIET REPLY

The Soviets did not take long to respond. On October 25, 1968, two days after the return of Apollo 7, the USSR launched an automated spacecraft without announcing its purpose. The following day, Mishin sent a new Soyuz into orbit for a journey that would last four days. It was the first crewed mission since the tragedy of Soyuz 1 eighteen months earlier. The Soviet political leadership was particularly anxious to resume space missions after the long gap, especially following NASA's well-publicized launch of Apollo 7. Chief Designer Mishin succeeded in setting the 'return to flight' Soyuz mission in time for the 51st anniversary of the Great October Revolution.

It would eventually be revealed that different solutions had been evaluated behind the scenes, with some of them on again, then off again. They had wanted to launch two spacecraft and try a docking in Earth orbit, but technical and training problems still remained. Crew-rating the Soyuz spacecraft would be critically important for the future of the Soviet space program, but for the short-term the focus remained on the Moon, and in particular on the L-1 circumlunar program. After a long debate about the goals to be assigned to the mission, it was given the name Soyuz 3. It was decided to launch a single cosmonaut to dock with an unmanned Soyuz 2 spacecraft. The commander of the mission was Georgy Timofeyevich Beregovoy, a 47-year-old wartime combat veteran and test pilot, who had been uneasily inserted into the cosmonaut corps thanks to political pressure. He had already come into conflict with the group on more than one occasion.

Figure 6.17: Commemorative cover (with enlarged depiction of the stamp on the right) for the Soyuz 3 mission of Georgy Beregovoy.

On October 23, the day after Apollo 7 splashed down, the State Commission for Soyuz met at the Baikonur Cosmodrome to discuss preparations for the two launches. Nikolai Kamanin presented Beregovoy as the primary candidate, with Vladimir Shatalov and Boris Volynov as his backups. There seem to have been some serious doubts about Beregovoy's qualifications for the flight. He had failed

258 **The final leap**

his prelaunch examination, receiving a score of '2' ('bad') out of a possible '5' ('excellent'), but instead of flying his backup Shatalov, Air Force officials organized a second examination, in which Beregovoy managed to rate a '4' ('good'). [11] All three men – Beregovoy, Shatalov and Volynov – had trained for the Voskhod 3 flight in 1966, the cancellation of which had been one of Mishin's first actions after his official appointment as Chief Designer.

Beregovoy tried to replicate the maneuver that Neil Armstrong and Dave Scott had performed on Gemini 8 in March 1966, but his approach with Soyuz 2 failed. He did not even succeed in bringing his Soyuz craft close to the unmanned capsule. [12] His repeated errors not only failed to achieve the docking, but used up so much propellant that mission control had to interrupt the operations, refusing to allow him to perform a repeat approach again when they realized that there was only enough fuel remaining for the maneuvers needed to return to Earth[1]. Inevitably, the Soviets announced that a docking hadn't been planned at all, and the truth was not revealed for decades.

The commander's performance was judged to be disappointing and Beregovoy was never assigned to any subsequent missions. However, thanks to his powerful political support, he did manage to carve out a career and would later be appointed Director of the Yuri Gagarin Cosmonaut Training Center at Star City.

ZOND 6: OFFICIALLY "A COMPLETE SUCCESS"

On November 10, 1968 (as preannounced by a CIA alert), Zond 6 was successfully launched towards the Moon. Zond 6 was a cover name for an unmanned version of Soyuz 7K-L1. Onboard the probe was a scientific payload that included unidentified animals (probably tortoises again), cosmic ray sensors, micrometeoroid detectors and photography equipment, namely an AFA-BAM camera with 400mm lens, shooting 13 x 18 cm frames of isopanchromatic film. A batch of 111 frames were taken at a distance of 9,290–6,843 km and another batch of 58 from 2,660–2,430 km. If the mission was successful, there would still be a chance to launch another manned mission in December.

The flight profile was similar to that of the previous mission, but with an improved 'skip reentry return trajectory' to reduce the acceleration to 7–8G. The target landing point was only 16 km from the pad which had launched the mission towards the Moon.

---

[1] See Chertok [2011], p. 477-482. *"Tempers flared* [on the ground] *but we had no time for squabbling."* Boris Chertok, who was serving at Baikonur during the mission, provided a detailed story, concluding *"I have dwelled on the story of Beregovoy's flight in such detail because it was very instructive."*

Figure 6.18: Zond 6 commemorative cover prepared by the Tartu Philatelic Club, who were very active in this period.

The probe reentered at the wrong angle and bounced back into space. It skipped across the atmosphere like a stone skimming across water and during its second reentry a gasket failed, leading to cabin depressurization and killing all the animal test subjects onboard. The parachutes also deployed too early, ripping the main canopy. The probe crashed and was destroyed on impact with the ground near the designated landing area. [13] The film canister was flattened and broke open, but 52 photographs were recovered, albeit with some degree of laceration and fogging. A few were published, including a picture of the Earth and Moon (similar to those that would be taken a month later by Apollo 8)[2], and the world was told that the mission had been "*a complete success*." Once again, the planned crewed flight around the Moon would have to be postponed and Mishin's only hope of beating the Americans now was a failure or delay in the Apollo 8 flight planned for December.

THE FINAL SPRINT

On November 11, 1968, the Americans publicly announced the decision to fly Apollo 8 to the Moon. The last stage of the head-to-head race to put the first man on the Moon had begun. The curtain of secrecy around Soviet plans remained impenetrable, and NASA was unaware of the many problems that had arisen on

---

[2] A good selection of these pictures can be found at the Don. P. Mitchell site mentallandscape. com under 'Soviet Moon Images' (accessed in March 2018).

260 **The final leap**

Zond 5, or the more serious issues with Zond 6. They did not know that the Russians were now in no position to fly to the Moon and tensions remained high at NASA during the final preparations for the Apollo 8 mission. They had pieced together the rumors that were circulating and their experiences with the behavior of the Soviets and had begun to fear the worst. The next favorable launch window at Baikonur for a circumlunar shot would open on December 6, two weeks before the window for Apollo 8. The Americans assumed that the trajectories for upcoming Soviet missions would follow those of Zond 5 and 6.

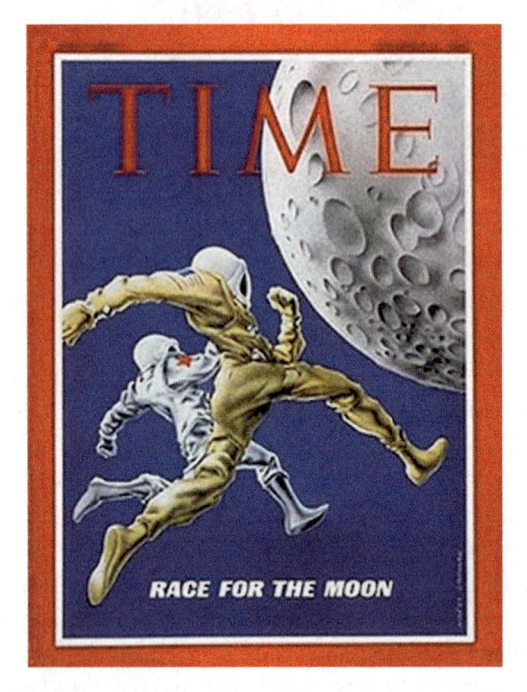

Figure 6.19: The cover of *Time* magazine on December 6, 1968. © *Time* Magazine.

The cover of *Time* Magazine on December 6, 1968 (see Figure 6.19) featured an American and a Russian in spacesuits elbowing each other in the "*race for the Moon*", aptly summarizing the atmosphere at NASA and the mood of the Western world. Years later, Alexei Leonov would tell Tom Stafford that he and Oleg Makarov were prepared to take the risk and ride a Zond for seven days to the Moon and back: "*We could have beaten the Apollo 8 crew, but Mishin was a blockhead!*" [14]

The Apollo 8 crew – Frank Borman, Jim Lovell and Bill Anders – were placed into quarantine in preparation for the launch. After the experience of the Apollo 7 flight – with two of the three crew unwell – NASA had learned to isolate its astronauts for a few days before a flight. This sort of 'house arrest' ensured that the flight crew did not catch a bug and allowed them to focus on their flight preparations without interruption. Michael Collins had originally been part of the crew, but had had to undergo a shoulder surgery and was replaced by Anders, almost at the last moment.

Figure 6.20: A cover commemorating the launch of Apollo 8. (From the collection of astronaut Walt Cunningham.)

On the morning of December 21, 1968, it was finally time for the long-awaited launch. This would be the first time that the Saturn V was used for a human flight and it was also the inaugural launch from the new Pad 39A. The launch was perfect, but shortly after leaving Earth orbit on their way to the Moon, Frank Borman began to suffer from motion sickness, with nausea, vomiting and diarrhea. With the lack of gravity in the spacecraft, his illness got a little messy. When one gets sick in space, it is three-dimensional. Lovell and Anders were worried about Borman, but they were also worried about living in a pretty disgusting spacecraft for seven days as well. Fortunately, Borman's discomfort did not last long.

On Christmas Eve morning (December 24), Apollo 8 reached lunar orbit, passing behind the Moon for nearly an hour and losing all radio contact with Earth. The crew were the first human beings to observe the far side of the Moon directly with their own eyes. At the precise moment they were due to reappear on the eastern rim – if they had successfully entered lunar orbit – the Capcom at Mission Control tried to reach them: "*Apollo 8, Houston... Apollo 8, Houston, over.*" For seconds that seemed like minutes there was silence, until Jim Lovell's voice confirmed that they were in lunar orbit: "*Go ahead Houston, Apollo 8.*"

Lovell then provided Mission Control with man's first impression of the lunar surface: "*The Moon is essentially gray – no color – looks like plaster of Paris – sort of gray sand...*" Bill Anders then added: "*Looks like a sand pile my kids have been playing in for a long time – it's all beat up – no definition, just a lot of bumps and holes.*" [15] Over the next 20 hours, the Apollo spacecraft would pass around the Moon ten times.

Apollo 8 was indelibly established in the minds of millions of TV viewers around the world when, later on Christmas Eve, Frank Borman began a shared

262 The final leap

reading with his crew of the first chapter of the Book of Genesis: *"In the beginning, God created the heaven and the earth..."* Even today, much of the public thinks of Apollo 8 as the Christmas Mission and as the first flight in the Apollo series. The Bible reading was universally well received, with the exception of one Madalyn Murray O'Hair, a militant atheist from Austin, Texas, who filed a lawsuit against NASA and the government for violations of the First Amendment with such a reading on a government-sponsored mission. The case was dismissed. [16]

The Apollo 8 crew took several photographs of possible landing sites for future missions. The most suspenseful moment of the mission came on the tenth and final trip behind the Moon. Shortly before reestablishing contact, the Apollo 8 crew had to burn the service module engine for two-and-a-half minutes if they wanted to return home – without the support of Mission Control. The maneuver was executed perfectly. The mission ended happily on December 27, 1968, with a splashdown in the Pacific Ocean, only 2.5 kilometers from the intended target point.

Although it had been decided upon at the last minute, the Apollo 8 mission was one of the least problematic of the entire program and helped to lift the spirits of Americans at the end of what had been a notorious year, remembered for the revival of the Vietnam War, the assassinations of Martin Luther King and Robert Kennedy, and the violent student protests. America had reached the Moon first and *Time* magazine devoted its cover to Borman, Anders and Lovell, naming them *"Men of the Year"*.

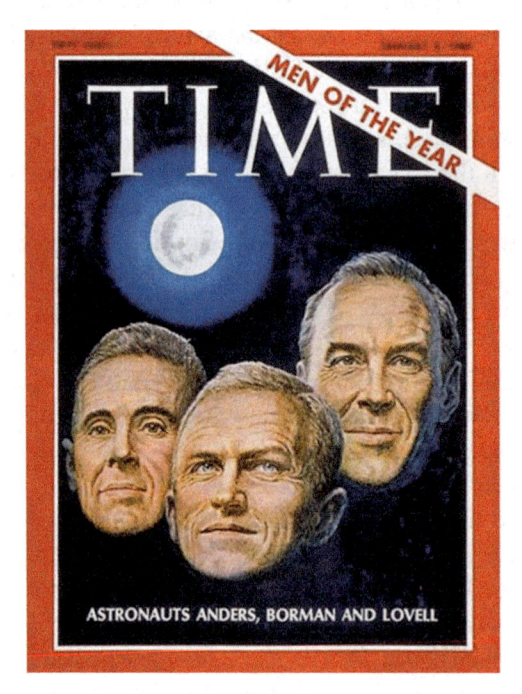

Figure 6.21: Anders, Borman and Lovell, Men of the Year, on the cover of *Time* Magazine, January 3, 1969. © *Time* Magazine. Cover credit Hector Garrido.

The Apollo 8 Stamp

Apollo 8 was the first human mission to leave Earth orbit, the first to be captured by the gravitational field of another celestial body and the first to return from another celestial body to planet Earth. The American Post Office decided that the event had to be celebrated by issuing a stamp.

Figure 6.22: "Earthrise" taken by Bill Anders on the lunar horizon. Courtesy NASA.

The assignment was given to Leonard Buckley of the Bureau of Engraving and Printing. Buckley prepared a sketch using the NASA photograph of "Earthrise" taken by Bill Anders on the lunar horizon on December 24, 1968, and quoted the first verse of the Bible, "*In the beginning...*" to remember the reading of the first chapter of the Book of Genesis that the three astronauts had given on Christmas Eve on live TV. The stamp had a face value of six cents, the price for shipping normal first-class envelopes, and was printed in six colors with the revolutionary 'Giori Press', created by the Italian designer Gualtiero Giori, who had designed the new American dollar, the ruble, the mark and the main European currencies.

There were 187,165,000 copies of the stamp issued, in sheets of fifty. First day ceremonies were held at the Johnson Space Center in Houston, Texas, on May 5, 1969, where two different postmarks were used.

(continued)

264 **The final leap**

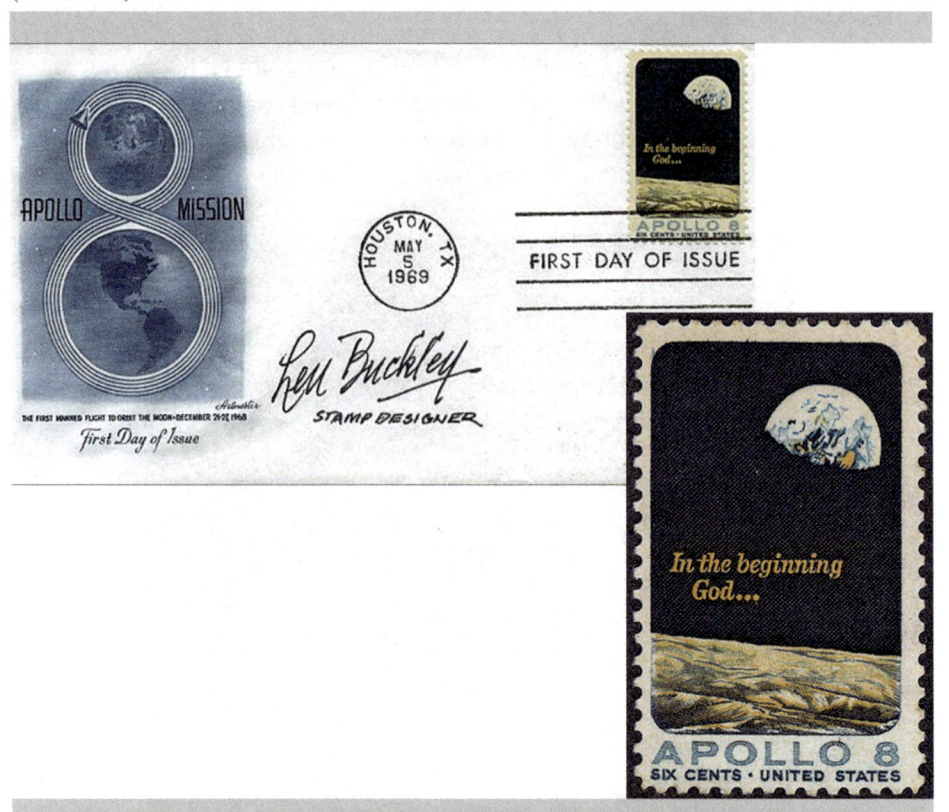

(continued)

Figure 6.23: The Apollo 8 stamp designed by Leonard Buckley can be seen on this signed cover.

Figure 6.24: Two different postmarks were used in Houston for the First Day Covers.

THE SOVIETS INTENSIFY TESTING AND THE SUCCESS OF SOYUZ 4 AND 5

The Soviets now needed to recover lost time and staked all on landing on the Moon ahead of the Americans. They began docking tests in Earth orbit once more. On January 4, 1969, Soyuz 4 lifted off, piloted by rookie cosmonaut Vladimir Aleksandrovich Shatalov. The launch had been planned for the previous day but, for the first time in the history of the Soviet space program, it had been delayed due to adverse weather conditions.

The day after Shatalov's launch, Soyuz 5 lifted off, this time carrying three cosmonauts on their first missions: Boris Valentinovich Volynov, Aleksei Stanislavovich Yeliseyev and Yevgeny Vasilevich Khrunov. The goal of the mission was finally to test the main phases and most critical techniques of the Soviet Moon landing program: including the transfer of crewmembers between two manned spacecraft with the aim of preceding the pre-announced Apollo 9 mission. Even the Soviet lunar program would require the transfer of a cosmonaut from the command module to the lunar module. These maneuvers were originally intended for the Soyuz 1 and 2 missions in 1967. Shatalov maneuvered his spacecraft to rendezvous with Soyuz 5 and the two spacecraft linked up and interconnected their electrical and mechanical couplings. This was a new record; for the first time, two manned spacecraft had docked in space. TASS was quick to broadcast: "*Today was born the first ever Space Station.*"

However, there was no direct way for the cosmonauts to transfer internally from one spacecraft to the other and after docking, Yeliseyev and Khrunov began their preparations for the EVA required to reach Soyuz 4. The preparation phase was broadcast live by Soviet TV. During the 35th orbit, the cosmonauts began their egress from the spacecraft, on only the second ever Soviet EVA. A problem occurred when Khrunov got snagged on wires while exiting and this distracted Yeliseyev enough to make him forget to switch on the camera. Only a few pictures of this historical event were video-recorded by the external camera and no TV images exist. Shatalov lowered the pressure in his spacecraft to allow his cosmonaut comrades to enter Soyuz 4.

The two EVA cosmonauts took with them some letters for Shatalov, as well as copies of the *Izvestia* and *Pravda* newspapers issued on the day of the Soyuz 4 commander's launch. The two spacecraft remained docked for 4 hours and 35 minutes before undocking and reentering the atmosphere separately. This created another new record: the first time that a crew had returned to Earth aboard a spacecraft other than the one they had launched in.

During the Soyuz 5 reentry, with Volynov returning alone, another tragedy was only just avoided. The retrofire module failed to separate completely, even though the explosive bolts had fired. A similar problem had previously occurred during Vostok and Voskhod missions, as well as during the Mercury mission of John

Figure 6.25: (above) The commemorative cover, serviced by the Tartu Club on the day of docking, emphasizes the docking mechanisms (below): Official Kniga cover signed (after their return) by the four cosmonauts who met in space. The red ink used for the postmark underlines the solemnity of the event

Glenn, but the Soyuz Service Module was far bigger and heavier. Once the Soyuz reached the atmosphere, Volynov lost control of the spacecraft, with the two modules assuming the most aerodynamically stable position by themselves: the heavy descent module and heat shield at the rear and the light metal wall at the front. This was the worst possible configuration because that was where the spacecraft's skin was thinnest. The gaskets sealing the hatch began to burn, filling the craft with dangerous fumes and smoke. Wearing no spacesuit, Volynov realized that he only had seconds to live, his body strained against the restraining straps rather than back in his seat as expected. Luckily, the struts between the descent and service modules burned through completely, separating the two and causing the descent module to swing around to the correct orientation. With its reentry speed exceeding 9G, however, Soyuz 5's main parachute deployed irregularly and the fuel for the control thrusters, that were supposed to stabilize the module, was exhausted.

Inside the capsule, Volynov fainted due to the toxic smoke. Soyuz 5 landed in the snowy Ural Mountains, thousands of miles from the planned landing site. Even though it came down in a snowbank, the capsule landed heavily, with the shock throwing Volynov across the cabin and breaking several of his front teeth. But he survived. The temperature outside was minus 38 degrees Celsius and Volynov realized that it would take hours for the rescue team to locate him. Spending many hours inside Soyuz 5 in sub-zero temperatures would mean certain death, so he clambered outside. Many hours later, helicopters spotted the downed spacecraft and landed nearby. Finding the capsule's hatch open, with no one inside and no trace of the cosmonaut, the rescue team followed his footprints and the bloody spots he had spit in the snow. They caught up with him a few kilometers away, in the hut of some peasants who were keeping him warm. [17] No news of this near-disaster was ever printed in the Soviet press at the time and it would remain secret until 1997. Volynov would not fly again until Soyuz 21, seven years later.

To celebrate the first completely successful piloted space mission in the post-Korolev era, a parade was organized at the Kremlin, including several cosmonauts and Premier Leonid Brezhnev. A young Soviet army lieutenant shot at Brezhnev, but fired wildly and missed him, instead hitting the car in which some of the cosmonauts, including Beregovoy, Leonov, Andrian Nikolayev and Valentina Tereshkova, were sitting. No one was injured, but the ceremony was abruptly cancelled.

"Earth-Space-Space-Earth" The First Special Cancel of the Baikonur Cosmodrome

In preparation for the Soyuz 4/5 docking mission – with cosmonauts transferring in outer space from one spaceship to another – the Soviet Ministry for Communication prepared a special postmark with a changeable date, to be used at the Baikonur Cosmodrome at launch and during the mission between January 14 and 18. The postmark did not bear the name of the launch site, since Baikonur was still strictly a secret site at that time.

As reported by Julius Cacka, covers exist that were postmarked on January 13, the day on which the launch of Soyuz 4 was originally planned. [18] It is believed that they were sent by employees of Korolev's Design Bureau who actually brought the special postmark to the Cosmodrome on the previous day. They had the opportunity to use it and in fact sent covers on January 13, 1969 (supposedly about 18 items) cancelled with the special postmark using violet ink. All these covers were sent as 'normal' letters because, for security reasons at that time (i.e., for maintaining the secrecy of the site), it was severely forbidden to ship covers simultaneously

(continued)

(continued)

cancelled with both the special postmark and the usual 'Leninsky' cancellation. This is why no genuine registered letter with this special postmark exists. During the mission, the postmark was used at the launch site on four official dates: January 14, 15, 17 and 18, 1969.

Figure 6.26: A rare cover with the four official dates in purple ink. (From the collection of Viacheslav Klochko, Russia.)

The postmark was only officially supposed to be used with black ink, but there are also covers officially cancelled in purple. According to Klochko, genuine postmarks also exist in red, which were against the rules of the time. [19] After January 18, the cancelling device was sent from Baikonur back to Moscow, where all the items for collectors were produced. Before returning it, the postmark was intentionally defaced with a horizontal scrape on the missile head, about 2.8 mm from the top. It is not completely clear who decided to scratch it, but the scratch makes it easier, and more reliable, to distinguish the genuine Baikonur cancels from those subsequently produced in Moscow and Star City.

Since covers with the genuine special postmark from Baikonur were highly sought after by collectors, counterfeiters inevitably flooded the philatelic market with many fakes. An example can be seen in Figure 6.28, where the fake differs from the original mainly by its size and also because its layout is less accurate.

The Soviets Intensify Testing and the Success of Soyuz 4 and 5 269

(continued)

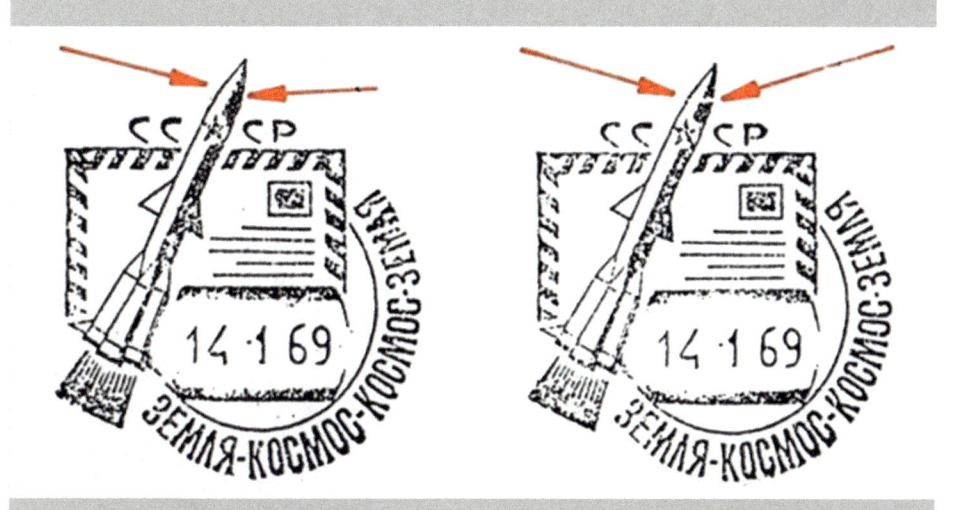

Figure 6.27: Imprint of Baikonur (left); Imprint of Moscow (right)

Figure 6.28: Imprint of Moscow (left); Fake (right)

The First Soviet Space Mail

For this mission, the Soviet Ministry for Communication prepared 10,000 stationery items with a four-Kopek imprinted stamp. The cachet, designed by the artist Yuri Levinovsky, featured a rocket and an envelope with the description "Earth-Cosmos-Cosmos-Earth".

On the mission, Khrunov was in charge of delivering to Shatalov:

- Copies of the *Izvestia* and *Pravda* newspapers issued on the day of Shatalov's launch;

(continued)

(continued)

- An envelope postmarked on January 14, 1969, officially addressed to Commander Shatalov by General Kamanin, director of the Cosmonaut Training Center at Star City;
- A letter from Shatalov's wife, enclosed in an official envelope and marked as 'Mail of Pilot-Cosmonauts of the USSR'. The envelope did not bear stamps or postmarks.

Shatalov signed both of the covers and added the handwritten notation "*Onboard Soyuz 4 15-1-1969.*" He then recorded the cover using the onboard TV camera. This was the first Soviet mail to fly in space.

Figure 6.29: The flown Soyuz 4/Soyuz 5 envelope. (From the collection of Renzo Monateri, Italy.)

The four-Kopek stationery was sent to Shatalov by Kamanin, Director of Star City's Cosmonaut Training Center. One ten-Kopek stamp featuring Berezovoy was added in order to cover the 'space mail' tariff, but was left uncancelled. The handwritten notation in brown (partially unreadable because of the cachet) reads: "*Onboard spacecraft Soyuz 4*" and is signed by Shatalov. The envelope reads:

Addressee: Outer Space – to the Commander of the craft Soyuz 4 Shatalov, Vladimir Alexandrovich

Sender: Earth, Launching Site – Kamanin

At the bottom, the vendor (Bolaffi) has added a statement in Italian: "*La primera lettera dalla Terra allo Spazio*" (The first letter from Earth to Space). The cover was signed by Alberto Bolaffi and expert Enzo Diena to certify the item's authenticity.

The Soviets Intensify Testing and the Success of Soyuz 4 and 5

(continued)

> The Kamanin envelope contained four pages addressed to Shatalov:
> - The first by General Nikolai Petrovich Kamanin himself (responsible for the training and tutoring of the cosmonauts, and nicknamed "*the cosmonauts' mother hen*") – not shown here;
> - A second message by the members of the State Commission (Figure 6.30);
> - A third page by the testing team of the base (Figure 6.31);
> - A fourth by the Cosmodrome military unit that had supervised the launch (Figure 6.32).

> **Message from the State Commission**
>
> Translation:

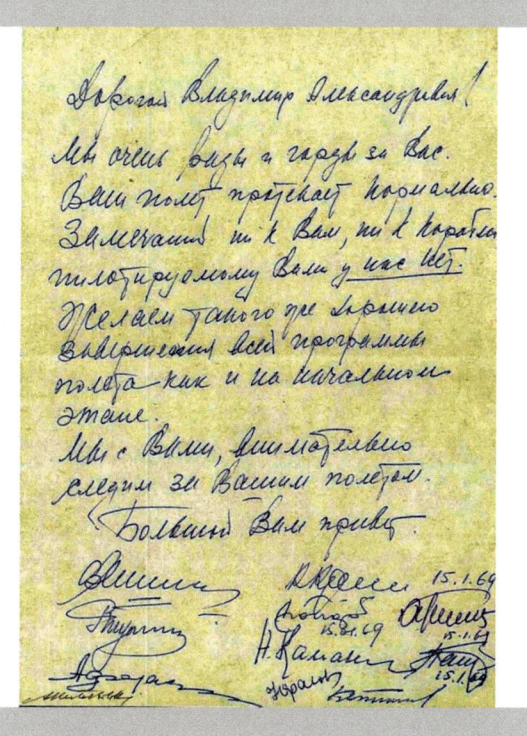

Figure 6.30: Message from the state commission to Vladimir Shatalov (From the collection of Renzo Monateti, Italy.).

(continued)

The final leap

(continued)

> Dear Vladimir Alexandrovich!
> We are very glad and proud of you.
> We trust your flight goes on ok.
> No comment to you or to the craft you are piloting.
> We wish that the same success of the initial phase will continue during the whole flight.
> With our best regards
> 15.1.69
> Kimirov (Commission President)
> Afanasyev (Minister of Metal and Mechanics Industry)
> Mishin (the Chief Designer, successor to Korolev)
> Kamanin
> (others)

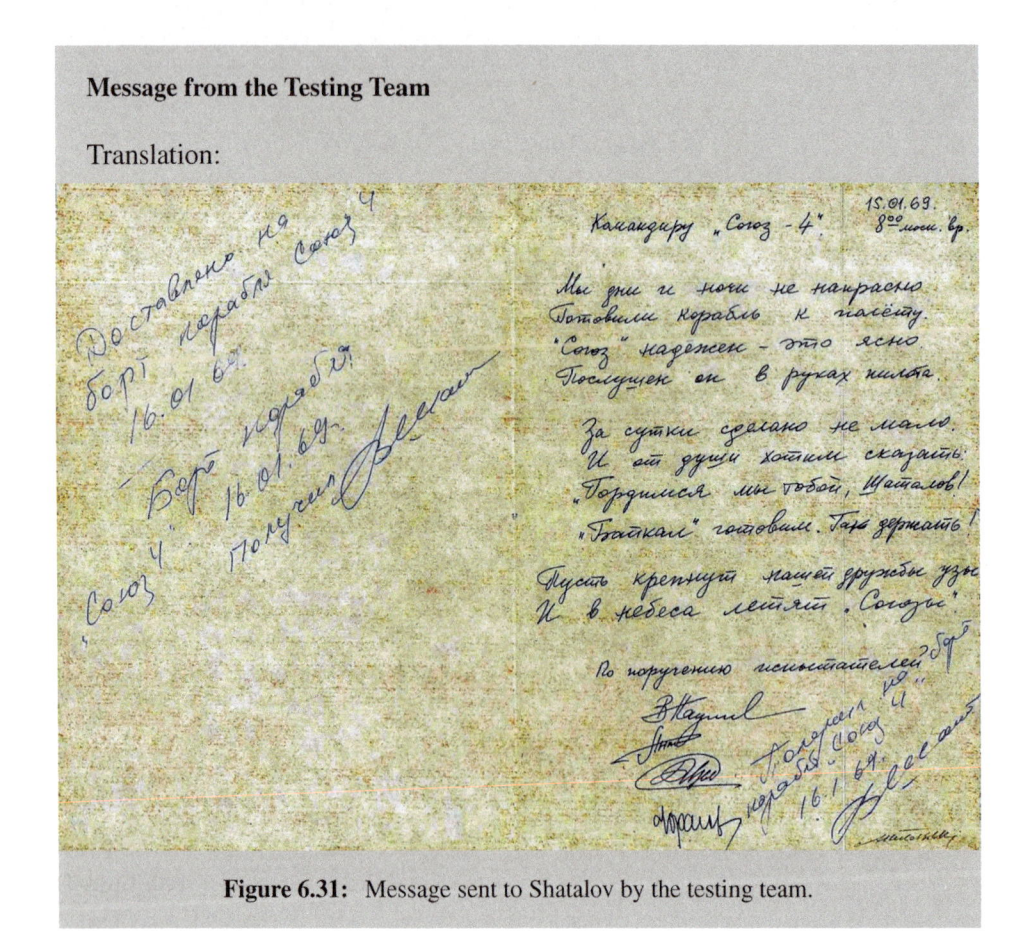

Message from the Testing Team

Translation:

Figure 6.31: Message sent to Shatalov by the testing team.

(continued)

On the left (oblique):	On the right:
Delivered onboard Spacecraft Soyuz 4 16/1/69	to the Commander of Soyuz 4, 15/1/69 8:00, Moscow time.
Onboard spacecraft Soyuz 4 16/1/69 V. Shatalov	Night and day we prepared the craft for the flight Soyuz is therefore safe – this is for sure. It is obedient in the hands of the pilot. What had been done in these days is not a little.
	And with all our heart we want to say: "We are proud of you, Shatalov!" Let's prepare the Baikal…! Let's our friendly tie make stronger And let many Soyuz fly in the sky! On behalf of the Testing Team V. Naumov, Yurasov (and others) Handwritten in blue (oblique): Received onboard spacecraft Soyuz 4 16/1/69 V. Shatalov.

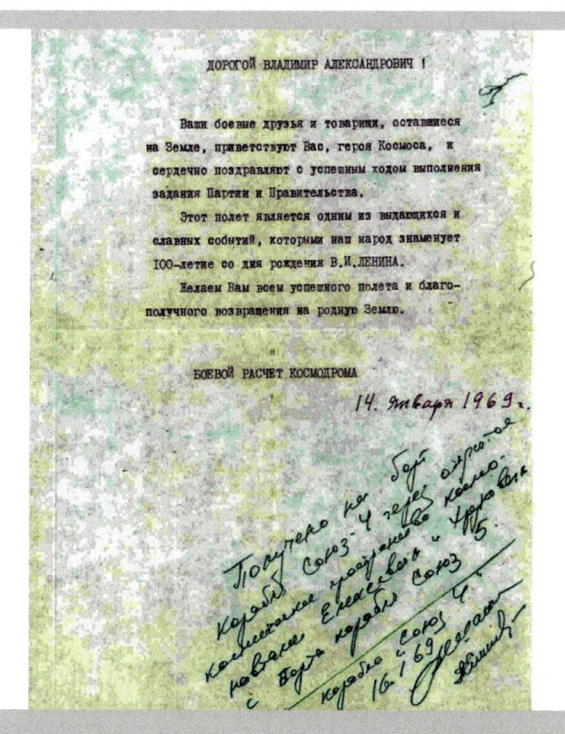

Figure 6.32: Message from the military unit of the Cosmodrome. (From the collection of Renzo Monateri, Italy).

(continued)

274 The final leap

(continued)

Message from the Military Unit of the base

Translation:

Dear Vladimir Alexandrovich!

Your friends and buddies remaining on Earth greet you, Hero of space and with all our heart congratulate you for your commitment in performing the task the Party and the Government have entrusted you.
This flight is one of exceptional significance and glory for our people celebrating the 100th anniversary of the birth of V.I. LENIN.
We wish you all a successful mission and a good flight back to your home country.

Cosmodrome's Military Unit
January 14, 1969
(handwritten, oblique)
Received onboard spacecraft Soyuz 4
Through Outer Space by cosmonauts Yeliseyev and Khrunov from craft Soyuz 5.

Spacecraft Soyuz 4
V. Shatalov, Yeliseyev

The letter from Shatalov's wife Muza

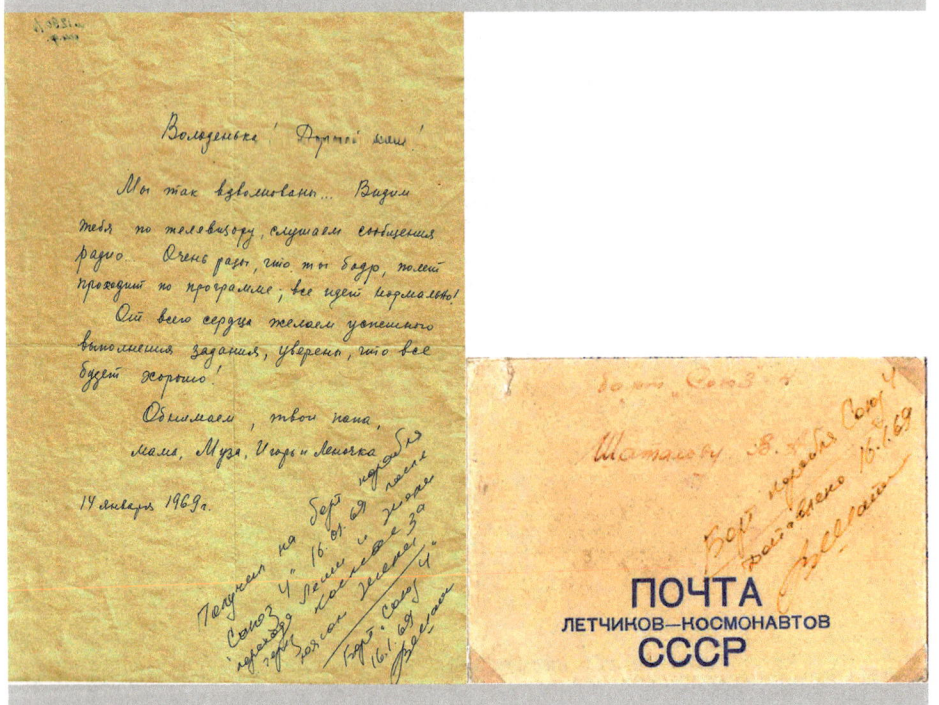

Figure 6.33: Letter to Shatalov from his wife. (From the collection of Walter Hopferwieser, Austria.)

(continued)

Translation:
Volodenka[3]! Our loved one!
 We are so excited… To see you in television, hear the news on the radio… We are very happy you are well. The flight is as scheduled. Everything is OK.
 With all our hearts we wish you successful fulfillment of your duties. We are sure everything will go well!
 We embrace you
 Your Papa, Mama, Muza, Igor and Leonichka.
 The letter was enclosed in an official envelope, with no stamps or postmarks, marked as 'Mail of Pilot-Cosmonauts of the USSR'. Both the cover and the letter bear Shatalov's oblique annotation acknowledging that they were delivered aboard Soyuz 4.

---

[3] Familiar name for Vladimir.

A NEW ATTEMPT AT A CIRCUMLUNAR TRIP

After the impressive achievement of Apollo 8, a Soviet manned mission around the Moon would have been politically meaningless, but the Soviet establishment wanted to wrest the initiative back and pressed to accelerate preparations for a manned lunar landing mission. This was despite the protests of technicians who were not yet ready: the lunar module had never been tested in flight. [20]

Senior space officials convened at Tyura-Tam, amid the cold, snowy weather, for a meeting which was presided over by the Minister of General Machine Building, Sergey A. Afanasyev, to discuss how to neutralize the success of Apollo 8 and, more generally, to talk about the bigger picture of the direction of their entire piloted space effort. Many senior government officials were beginning to shift their thinking towards automation and after this meeting, the Soviet circumlunar project was half-heartedly continued in its automated variant. [21] Plans for piloted missions were postponed, while the remaining 7K-L1 spacecraft were prepared for use only in robotic mode. [22]

Aware of the vague deadlines and lack of prospects, even Mstislav Keldysh, the President of the USSR Academy of Sciences, was in favor of accelerating the Ye-8-5 project (Lunokhod): *"We can show that our way of studying the Moon is through automatic spacecraft. We have no intention of foolishly risking human life for the sake of political sensation."* The decision was made to give this astonishingly hypocritical explanation to the mass media. [23]

276 The final leap

Meetings followed regularly over the next few weeks, at the end of which serious proposals were made to use Mars to neutralize the success of Apollo. For the first time, a three-step Mars exploration program was discussed:

- Mars '73 – a robotic vehicle to Mars (on the N-1) for sample return
- Mars '75 – a piloted satellite to Mars (on the improved N-1 F-V3)
- Mars '77 – a piloted landing on Mars using an N-1 with nuclear rocket engines.

With the recent American success of Apollo 8 in mind, however, there was little support for completely abandoning lunar projects and new attempts at reaching the Moon would follow. The first of these, the unmanned Soyuz 7K-L1 lunar orbiter launched on January 20, 1969, failed to reach Earth orbit due to a malfunction in the Proton launcher. The reentry module was recovered in Mongolia. For the next attempt, Mishin decided to launch Lunokhod, an eight-wheeled remote-controlled lunar rover which was ready to go as a backup for L3 manned lunar expeditions. Lunokhod's original primary mission was to survey sites for manned lunar landings and bases. After locating the best landing area, Lunokhod would then have provided a radio homing beacon for the precise landing of the LK manned spacecraft that would follow. In the event of a rescue mission, the single cosmonaut on the lunar surface would be able to walk between the primary and backup LK lunar landers using the extra life support supplies stored aboard the Lunokhod. [24]

Lunokhod 201, or 'Ye-8 No. 201' was launched on February 19, 1969. Unfortunately, the first stage of the Proton that should have carried the rover to orbit disintegrated within a few seconds and the rover was lost. The rest of the world would not learn of the rocket's valuable payload for several years. Two days later, on February 21, the N-1 rocket was launched, in the secret hope that something would go wrong with the American Apollo program giving Mishin time to get to the Moon first. The unlucky N-1 had already been installed on launch pad number 1 on May 7, 1968, but launch preparations had been halted when cracks were discovered in the structure of the first stage – probably caused during installation of the payload. The booster was returned to the assembly building, where the cracks were repaired. [25] It was returned to the launch pad in November.

After launch preparations lasting 28 days, the N-1 was ready to launch. Following years of delays, its big day had finally arrived. Its payload included an improved Soyuz 7K-L1S craft and a dummy model of the LK Moon lander, as well as cameras that would take close up images of the sites chosen for the Moon landings. [26] The first stage of this monster rocket booster was the most powerful single stage of any rocket ever built, including the Saturn V. The rocket was 344 feet (105 meters) tall with its L3 payload, and almost 56 feet (17 meters) in diameter, weighing over six million pounds (2,750 tons). The N1-L3 consisted of five stages in total, with the first three (N1) used for insertion into a low Earth parking orbit and the remaining two (L3) for the translunar injection and lunar orbit insertion.

The monster rocket came to life and began to rise into the sky. Engines No. 12 and 24 shut down because of an error in the KORD control system, but the flight was able to continue because the remaining engines compensated for the failures. At T+25 seconds, the engines were throttled down until the vehicle safely passed through the period of maximum dynamic pressure. At T+66 seconds, a flame was observed developing at the base of the first stage. The oxidizer pipeline of one engine had broken, spilling liquid oxygen. The KORD control system was unable to shut the engine down quickly enough and a fire broke out, causing the surrounding engines and turbopumps to overheat and explode. The 'Launch Escape System' quickly separated the capsule from its launcher and the booster blew apart *"in a spectacular detonation that flings debris for thirty miles,"* falling from the sky onto the steppes. [27]

"It isn't alarm and it isn't dismay," Deputy Chief Designer Boris Chertok, who was responsible for the flight control system, later wrote in his book. *"It's more a certain complex mixture of intense inner pain and a feeling of absolute powerlessness that you experience while watching a crashing rocket approaching the ground. Dying before your eyes is a creation with which you have become so intertwined over a period of several years that it sometimes seemed that this inanimate 'article' had a soul."* [28]

The commission of inquiry would attribute the cause of the disaster to the heat and vibrations caused by the 30 engines of the first stage. Mishin would later claim that due to the lack of a first-stage ground test facility, he had to improve the launch vehicle's design on the basis of results from actual flights: *"The Americans were able to test an entire assembled engine module on their test stands and then install it on the launch vehicle. But we tested in pieces and did not even dare to think of firing all 30 motors in the first stage as a full assembly. The individual pieces were then assembled without guarantees that they were properly run in."* [29] Shortly afterwards, those around Vasiliy Mishin began to notice that he was drinking more than he had been.

APOLLO 9: FLIGHT TEST OF THE LUNAR MODULE

By doing their job methodically, the Americans were successfully implementing their plan. Two more test flights were scheduled before the actual landing and they were now moving on to the third 'giant step'. On March 3, 1969, with the Lunar Module (LM) finally ready, they launched Apollo 9 into Earth orbit. The crew of Jim McDivitt, Dave Scott and Russell 'Rusty' Schweickart were all Air Force people. The main purpose of the ten-day mission was to evaluate both the LM and rendezvous operations with the Command and Service Module (CSM). To simplify communications between the two craft, two distinctive callsigns would be used for the first time: one for the CSM (*Gumdrop*) and one for the LM (*Spider*),

278 **The final leap**

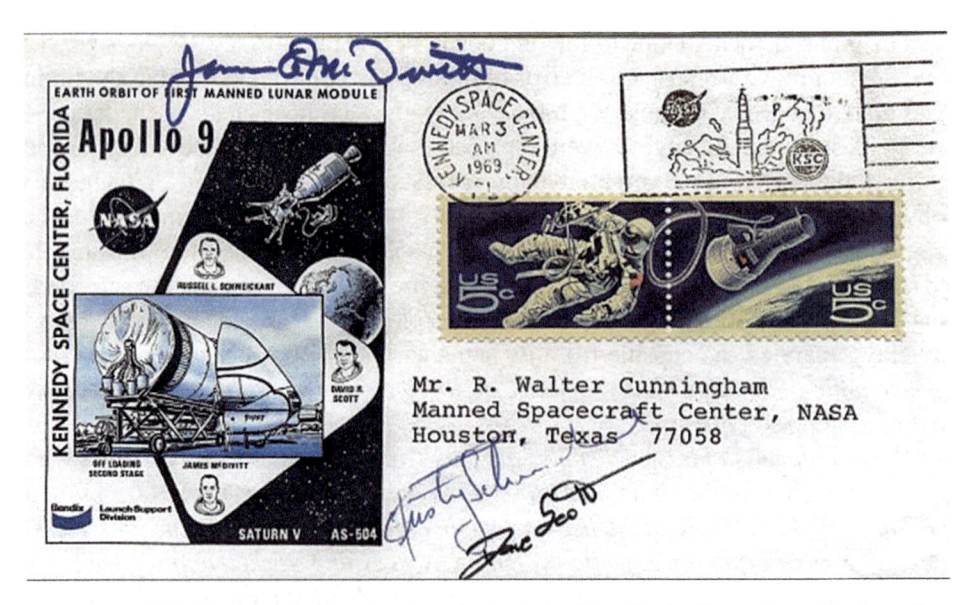

Figure 6.34: A cover commemorating the launch of Apollo 9. (From the collection of astronaut Walt Cunningham.)

thus resuming the tradition that had been discontinued after the Gemini 3 mission of allowing the astronauts to choose a name for their spacecraft.

Once they reached Earth orbit, and after checking out the CSM and LM systems, the Apollo 9 crew carried out all the maneuvers that would be required for the Apollo 11 landing to take place. The lunar module was perfectly extracted from the top of the Saturn V only three hours after the launch. On the third day of the flight, Schweickart and McDivitt passed through the connecting tunnel from the CSM to the LM, the first time ever that an astronaut had passed directly from one spacecraft to another. A live television broadcast was transmitted from inside the lunar module.

On the fifth day of the mission, the lunar module undocked from the CSM and separated to a distance of 180 kilometers (111 miles). After six hours and 22 minutes, *Spider* and *Gumdrop* redocked, with McDivitt and Schweickart returning to join Dave Scott in the CSM. The propulsion engines of the LM were then re-ignited and burned until their fuel was exhausted. This maneuver would keep *Spider* in Earth orbit until 1981, when it returned to its destruction in the atmosphere. The Apollo 9 capsule splashed down without problems on March 13 in the Atlantic Ocean, where it was recovered by the aircraft carrier USS Guadalcanal.

At NASA, there was complete satisfaction with the success of the mission, so much so that a proposal was put forward to let the next mission, Apollo 10, make the Moon landing. However, *Snoopy*, the Apollo 10 LM, was a prototype that was too heavy to be used for a landing and, rather than waiting for a month to replace

Apollo 10: Dress Rehearsal of the Moon Landing

it with a full-specification LM, it was decided to stick to the program as it had been defined.

APOLLO 10: DRESS REHEARSAL OF THE MOON LANDING

With Apollo 10, NASA undertook its second circumlunar mission, this time carrying the LM to practice for an actual landing. The mission was launched on May 18, 1969, carrying the crew of Tom Stafford, Gene Cernan and John Young, and would last for eight days. With the tradition of naming their spacecraft restored for Apollo 9, this time the crew chose the names of *Charlie Brown* and *Snoopy*, respectively, for their capsule and the lunar module, after the *Peanuts* comic strip characters created by Charles M. Schulz. The origin of these call-signs appears to have been from a nickname given to John Young by his crewmates, but their 'frivolity' caused some concern within the NASA hierarchy, who subsequently passed down a rule that future Apollo spacecraft names would have to be more dignified.

During the launch phase, both the first and second stages suffered from 'pogo' oscillations that seriously worried the crew. But the worst occurrence happened during the trans-lunar injection, with the third stage vibrating violently and causing Stafford to keep his hand on the abort handle that would dump the spacecraft into a high elliptical orbit and then return them safely back towards Earth. After a few moments of tension, however, the vibrations finally ceased and the journey to

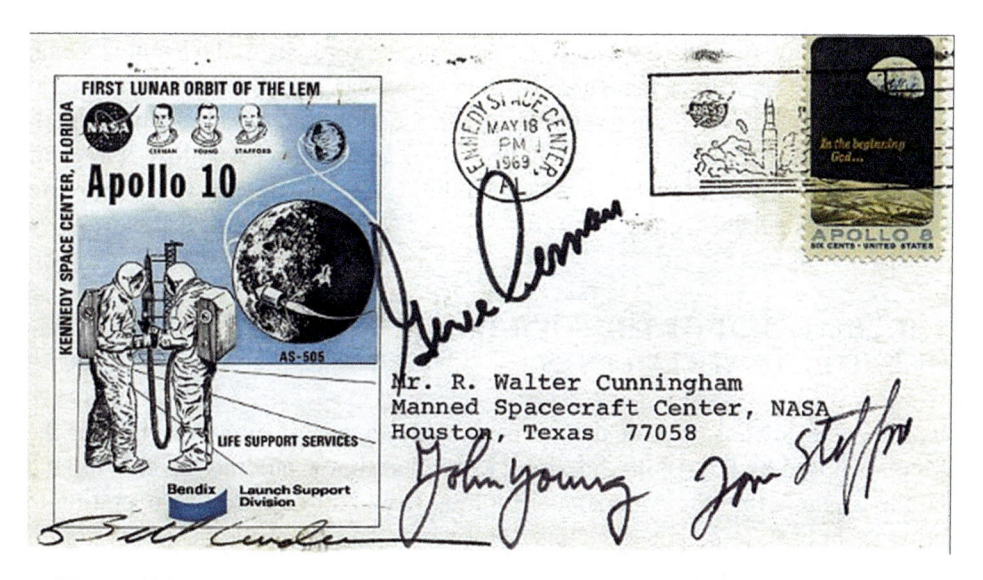

Figure 6.35: A cover commemorating the launch of Apollo 10. (From the collection of astronaut Walt Cunningham.)

280 The final leap

the Moon could continue. The problem would promptly be identified and fixed before the next mission.

At about 110 kilometers (68 miles) from the Moon, the lunar module undocked from the command module and began the descent maneuver, approaching the lunar surface to within 15,600 meters (8.4 nautical miles), the minimum height after which it would no longer be possible to reverse direction. The crew took a number of photos of the planned landing site for Apollo 11. Having completed their close approach, Stafford and Cernan prepared to jettison the LM descent stage and fire the ascent stage engine to rejoin Young in the CM. They had probably trained for this maneuver repeatedly on the simulators dozens of times, but now, on the one occasion when it really mattered, something went wrong, as Cernan recalled: *"As we went through the checklist, I reached over with my left hand and switched navigation control from Pings to Ags. We were so busy and familiar with the systems that we hardly ever looked at the switches we manipulated. A moment later, Tom (Stafford) reached out with his right hand and instinctively touched the same switch, knowing that it needed to be changed from one setting to another, and moved it back to where it had been a second before."* [30] The error caused the LM to start searching for the CM prematurely by trying to orientate itself in three axes to bring its radar into position to lock onto Young's capsule. After 15 seconds of pure pandemonium and whirling oscillations, Stafford activated the manual controls and stopped the carousel. With the LM microphone still open, Cernan uttered an expletive that was retransmitted around the world!

During the return from the Moon, on May 26, 1969, Apollo 10 set the record for the highest speed attained by a manned vehicle, according to the 2002 Guinness World Records: 39,987 km/h (24,791 mph or 11.08 km/s), a record the mission still holds today. Apollo 10 splashed down without problems in the Pacific Ocean, where the spacecraft was recovered by the aircraft carrier USS Princeton. Apollo 10 had performed every step required to get to the Moon except for the lunar landing itself. The success of the mission, just seven months after Apollo 7, was clear proof that NASA could make the Moon landing. Now, the agency would move to the fifth, and decisive, 'giant step'.

THE FINAL SHOT OF THE SOVIETS: A FLURRY OF UNFORTUNATE FAILURES

By now, the Soviets, who had seen their achievement of circumnavigating the Moon eclipsed by the Apollo missions, knew that they would not be able to land on the Moon before the Americans. But they would still attempt an operation to bring them back to center stage and 1969 would be one of the busiest years for lunar-related space launches in the history of the Soviet space program. [31]

On June 16, 1969, the Soviets made a second attempt to launch Luna Ye-8-5, a probe designed to land on the Moon and return to Earth. Once again, a

The Final Shot of the Soviets: A Flurry of Unfortunate Failures

malfunction of the Proton, this time in its third stage, caused the probe to fall back into the atmosphere. This was better than a previous attempt on April 15, when the Proton had exploded on the launch pad. However, the secrecy surrounding the Soviet program kept these setbacks from the eyes of the outside world. Mishin had also not given up on another attempt to launch the monster N-1 rocket booster on a mission to fly around the Moon.

A rehearsal meeting had been held in Tyura-Tam a few weeks before by the technical management – without the top brass – with the aim of *"closing the issues"* from the previous failed launch of the N-1 and moving on with the new launch. Boris Chertok, who was responsible for the flight control system, had explained to his superiors that it was impossible to predict the behavior of the KORD system if the cable networks were damaged by fire, as had happened with the first rocket. Viktor Litvinov, the head of the department at the General Machine-Building Ministry, who was chairing the 'rehearsal', suggested that Chertok *"had better not mention this"* during the full State Commission meeting planned for the following day. [32]

Vladimir Barmin, in charge of the launch infrastructure for the N-1, prophetically asked whether anybody could guarantee that the accident with the first rocket would not reoccur 50 seconds earlier in the launch sequence, when the rocket was still on the pad. In order to preserve the expensive launch facility, Barmin proposed blocking the emergency engine cutoff for 15–20 seconds, so that even an uncontrollable rocket could be moved a safe distance away, into the desert. After a heated debate, Barmin promised not to raise this issue at the official meeting either. In turn, Chertok and his colleagues promised to study Barmin's proposal, although they felt that there would be insufficient time to make such a change for the upcoming launch. [32]

Two weeks before the Apollo 11 mission, everything was ready at the Baikonur Cosmodrome for the second attempt with the N-1. On July 3, 1969, the rocket slowly lifted off the launch pad, but nine seconds later, at 50 meters off the ground, something went wrong. Suddenly, several of the thrust indicators dropped to zero and all of the first stage engines shut down, apart from one – No. 18. The giant N-1 continued rising up to 200 meters, then seemed to freeze in mid-air before falling back to the launch pad. With most of its propellant still on board, a violent explosion followed, comparable to that of a small nuclear bomb, with a red mushroom cloud silently rising over the launch pad and illuminating the steppe under the night sky for dozens of kilometers. The inhabitants of the town of Leninsk, 35 kilometers away, observed a bright glow and shuddered at the possible consequences: they had family and friends at the launch site.

At the launch pad, ramp No. 2 was completely destroyed and No. 1 was rendered unusable. Windows were broken within a range of 40 kilometers from the pad and debris would be found some ten kilometers away. The damages amounted to millions of rubles but fortunately the evacuation measures proved to be

282 The final leap

effective, and all reports from the various sites mentioned *"no fatalities."* Amazingly, it would later be estimated that as much as 85 percent of the propellant onboard the rocket had not detonated, reducing the force of the blast from a potential 400 tons to just 4.5–5 tons. Regardless, this would still be the largest explosion of any rocket in history.

Under pressure to find a culprit for the initial explosion, the propulsion engineers at the Kuznetsov design bureau insisted that some foreign object must have entered the pump. Any suggestion that the pump could have exploded by itself would have been politically unacceptable, as it would stall the entire Soviet lunar program. It was impossible to prove or disprove such a scenario and from then on, for the lack of better explanations, a 'foreign object' became the favorite excuse for engine failures. [33]

The CIA discovered the disaster of 'Complex J' (as the experts at the National Photographic Interpretation Center, or NPIC, had labeled the massive launch pad facility where the Soviets launched their N-1 rockets) a month later, while examining the imagery taken by the CORONA and GAMBIT reconnaissance satellites. Years later, the ruins of the launch pad would still be visible to the LANDSAT satellite.

The last hope for the Soviets now rested with the Ye-8-5 program, which had already suffered several setbacks in recent months. Finally, however, the Ye-8-5 probe was launched on July 13, 1969. The launch – publicly designated as Luna 15 – was successful and the Proton rocket headed towards the Moon. The flight took place roughly in parallel with the Apollo 11 Moon landing, which was followed on live TV all over the world. The mystery Soviet probe reached lunar orbit

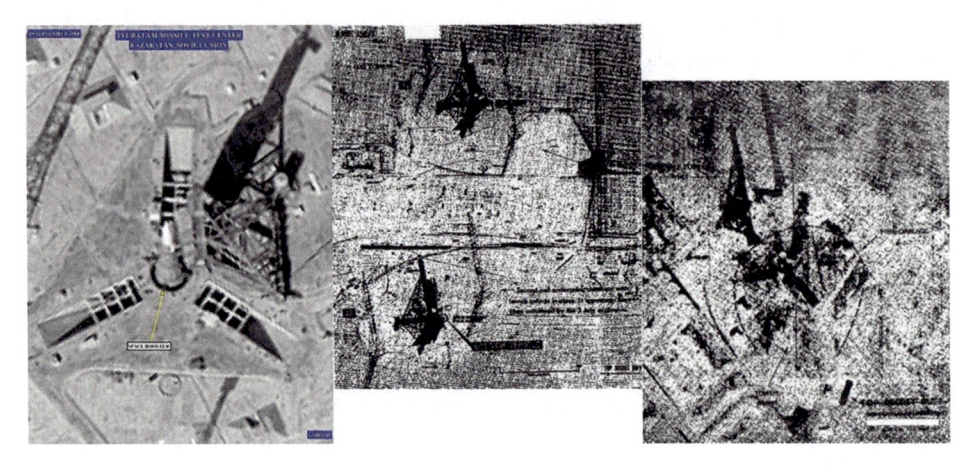

Figure 6.36: (left): N-1 rocket imaged by KH-8 Gambit on September 19, 1968. (center): Declassified image taken by KH-8 Gambit on June 11, 1969. (right): Image of the facility damage in the aftermath of the rocket explosion. Taken by CORONA on August 3, 1969 (The latter two pictures are from the declassified 'Central Intelligence Bulletin' (known as a CIB Report) dated August 15, 1969. (See Vick [2011].)). (Courtesy NOR and NASA.)

The Final Shot of the Soviets: A Flurry of Unfortunate Failures 283

during the landing of Armstrong and Aldrin, raising some concerns within NASA that the Soviets might be trying to sabotage the Apollo mission.

Colonel Frank Borman, the Commander of Apollo 8 who had just returned after a nine-day tour of the USSR (the first American astronaut ever to visit the country), suggested using the famed hotline established between Moscow and Washington after the Cuban Missile Crisis in 1962. Borman officially sent a message to Mstislav Keldysh, the head of the USSR Academy of Science, whom he had met a few days before. Keldysh promptly replied that *"the orbit of probe Luna 15 does not intersect the trajectory of the Apollo 11 spacecraft as announced in the flight program."* As usual, however, there was no information given about the goal of Luna 15.

The probe actually attained lunar parking orbit and began its descent, but instead of landing on the Moon as expected, it crashed into the lunar surface on July 21, the day after Apollo 11's successful landing. The probe's descent was tracked by British technicians in the radio telescope facility at Jodrell Bank in the

Figure 6.37: A cover commemorating the launch of Luna 15, serviced by the Philatelic Club of Tartu. (From the collection of Jim Reichman, USA.)

UK. The Soviets' last hope of beating the Americans had failed, and in the West, the media struggled to understand the purpose of the mission. All would become clear over a year later when Luna 16 successfully repeated the operation, landing on the Moon and returning to Earth with 105 grams of lunar rock, and again a few months later when Luna 17 landed on the Moon carrying Lunokhod 1, the first remote-controlled roving robot to land on another celestial body.

In a face-saving exercise, the Soviet media began to claim that the Luna probes were ten times cheaper than Apollo and far less risky than a manned mission. Denying that the USSR had ever been in a 'Moon Race' at all, they claimed that the Kremlin had never intended to squander money that could be used for the

284   The final leap

well-being of the proletariat on a business that could be carried out with much less risk and much greater efficiency using significantly less expensive automatic probes.

Only on August 18, 1989, during Mikhail Gorbachev's Glasnost, would the Soviet Union officially acknowledge – after years of denial by silence and misinformation – that they had indeed been racing the United States in the 1960s to be

Figure 6.38: The propaganda poster "I am Walking on the Moon" by Veniamin M. Briskin and Valentin Viktorov (1970) made reference to the popular Soviet movie "*I am walking down the streets of Moscow*" produced by Mosfilm studios in 1964.

the first to send men to the Moon. After the initial release of information by the Soviet newspaper *Izvestia*, an increasing number of photographs and blueprints of Soviet hardware started to become available to Western analysts and space observers. An impressive selection of rare images from a laboratory within the Moscow Aviation Institute revealed the never-flown Soviet lunar lander and several spacecraft that were used for training. [34] Resolution No. 655-268 of the Soviet Central Committee of the Communist Party, declassified in June 2003, provided historians with a unique view into the Soviet lunar program. The Soviets had considered a manned lunar landing as the most important path to compete with the American space program, as philatelists already knew well (see pages 243-245). [35]

APOLLO 11: AMERICAN FOOTPRINTS ON THE LUNAR SURFACE

Finally, the launch window for the American mission to land on the Moon opened and they were 'go' to make history. Apollo 11 left Cape Canaveral at 09:32 on July 16, 1969, watched by half-a-million viewers who had gathered on the beaches around the launch site. The journey to the Moon lasted three

Figure 6.39: 1999 stamp issued to commemorate the 30th anniversary of the Apollo 11 Moon landing and the first human steps on another world.

days and everything proceeded normally until the "*last mile*," when the onboard computer was overloaded due – it would be discovered later – to an error by the crew. The "*12-0-1 Alarm*" that flashed up indicated that the computer could no longer calculate the altitude and descent speed, creating tension in the Mission Control room with everyone holding their breath. A young technician, Steve Bales, took the courageous decision that would make his mark in history and earn him the Freedom Medal: he reset the onboard computer to clear its memory. Bales would later receive his medal from President Richard Nixon himself at the White House, where he had been presented along with the crew of Apollo 11.

The landing site, in the southern part of the Sea of Tranquility, had been chosen because it was considered fairly smooth according to the surveys carried out by Ranger and Surveyor, as well as by the maps drawn by the Lunar Orbiter. In fact, the astronauts soon realized that the site was much rockier than shown by the

286   **The final leap**

Figure 6.40 (above): One of the insurance covers postmarked on the day of launch of Apollo 11 (a full description is provided on pages 291–297. (below): Commemorative cover postmarked on the USS Hornet recovery ship on the day of the Apollo 11 splashdown, with a rare autograph of John Hirasaki, the NASA mechanical engineer who volunteered to be isolated with the astronauts inside the Mobile Quarantine Facility (MQF) onboard the ship and during the quarantine period for necessary maintenance.

photographs. Armstrong took manual control of the lunar module *Eagle*, which landed at 20:17:40 with only 17 seconds of fuel left. The rest is recorded history.

Before returning to *Eagle*, Armstrong and Aldrin left a medal on the Moon that had been given to them by the family of Yuri Gagarin, as a tribute to the world's first man in space. They also left a medal that had belonged to Vladimir Komarov.

THE RACE IS OVER

The promise made by President John F. Kennedy had finally been fulfilled and the Americans claimed victory in the 'Moon Race'. Neil Armstrong and Buzz Aldrin's excursion on the Moon was witnessed by some 500 million television viewers back on Earth. The only country that did not carry the live broadcast of the first ever moonwalk was the USSR. Coverage of the event in Soviet newspapers was also sparse. *"The Russian people had many problems in day-to-day life [so] they were not too concerned about the first man on the Moon,"* former Soviet Premier Nikita Khrushchev's son Sergei would grudgingly comment some 40 years later when interviewed by scientificamerican.com. [36]

Figure 6.41: One of the rare Soviet Apollo 11 commemorative covers postmarked on July 21, 1969 at the Post Office of Panevėžys in Lithuania. The rubber stamp on the left of the cover reads (in Russian and English) *"The First Man on the Moon"* (From the collection of Renzo Monateri, Italy).

It is important to remember that the Soviet lunar program was still secret at the time, so for the general public of the USSR there was no feeling of being in a race with the Americans, nor of losing it. Well-informed collectors, however, managed to rush to the Post Office in Panevėžys, Lithuania – when it opened the following morning – to have a few commemorative covers postmarked on that historic day.

288 The final leap

Alexei Leonov, the cosmonaut that the Soviets had designated to be their 'first man on the Moon', commented that after the success of Apollo 10, "*it was clear that we were in no position to carry out such a mission. The problems we had had with the N-1 rocket meant we would not be able to attempt a lunar landing that year, or even the following one. So, it was with mixed emotions that I stood watching events unfold on our television monitors that July morning.*

"*When Apollo 11 had soared away from Cape Kennedy, I had kept my fingers crossed. I wanted man to succeed in making it to the Moon. If it couldn't be me, let it be this crew, I thought, with what we in Russia call 'white envy' – envy mixed with admiration. On the morning of 20 July 1969 everyone forgot, for a few moments, that we were all citizens of different countries on Earth. That moment really united the human race. Even in the military center where I stood, where military men were observing the achievements of our rival superpower, there was loud applause... That achievement filled me with pride for all humanity. At a gathering of cosmonauts a few days later, we drank a toast to the safe return of the crew of Apollo 11.*" [37]

Leonov went on to mention that some in the USSR were not so pleased with what Apollo 11 had achieved. "*Mishin was very upset and began looking for excuses for our failure to achieve what the Americans had done. He attributed their success to the fact that so much more money had been made available to NASA than to our own space program. Some estimates have since concluded that the Apollo program cost $24 billion in 1969 terms, compared with an equivalent in rubles of $10 billion spent on our own lunar missions.*" [37]

But the lack of funding was not the cause of the failure of the Soviet space program. As has been mentioned, the program had suffered from more systemic problems since its inception[4]. The insufficient resources available were, if anything, one of the consequences of these problems[5].

Boris Chertok, Deputy Chief Designer under Sergei Korolev, complained: "*We were not capable – not at the highest levels of political leadership, not at the ministerial level, and especially not within our rocket space engineering community – of concentrating our efforts on a single mission of 'crucial national importance': a lunar landing expedition. Having realized that it was impossible to catch up with the U.S. in the execution of a piloted circumlunar flight and an expedition to the lunar surface, we continued to expend our resources on a number of disparate goals: on an unpiloted circumlunar flight using L1 vehicles; on the automatic delivery of lunar soil; on accelerating the piloted flight program using 7K-OK*

---

[4] See 'Rivalry and Inefficiency in the Soviet Space Program' in Chapter 1, and 'The Soviet Lunar Program' in Chapter 5.

[5] Some well-informed observers have mentioned smaller amounts. According to Siddiqi, the total cost of the Soviet manned lunar program was about $4.7 billion (Siddiqi [2004] pp. 211–2).

model Soyuz vehicles; and on designing new, more advanced vehicles – modifications of the Soyuz." [38]

Poor government – and the personal rivalries and dispersion of resources that led to duplication of effort and parallel projects – can also be seen as a consequence of more fundamental systemic problems at the very highest level. What was missing was a real interest in the space program among the Soviet leadership and, therefore, a coherent strategy of space exploration. While the Americans, with their program formally managed by civilians, felt morally obligated to the strong mandate they were given from the goal set by President Kennedy in 1961, in the USSR, governed by the military, there was no central NASA-like space organization and no strong guiding idea driving a global effort. "*A Soviet long-term space research program*" simply did not exist.

Korolev's successes in launching the first satellites and the first humans into space were linked to assignments for military purposes. The military were interested in launching combat rockets into near-Earth space and not in mastering far-space. Politicians, on the other hand, were exploiting the old Stalinist principle of 'divide and rule' to keep the situation under control. Competing schools of rocket and spacecraft designers, as well as personal rivalries, were encouraged, in order to limit the power of each and prevent individual scientists from becoming a powerful and potentially dangerous opposition to the political leaders.

For all these reasons, it is not surprising that, in the end, the Soviets failed to grasp the significance of President Kennedy's challenge, misjudged American intentions and only began marshalling their own resources when it was far too late, thus losing crucial years of development. The lack of cooperation between the leading personalities, such as Korolev/Mishin versus Chelomey/Glushko, was in the end just an additional, marginal complication.

With hindsight, the right question should perhaps be: Did the Soviets really expect to win?

References

1. **The All-American Boys**, Walt Cunningham, iBooks, New York, 2003, pp. 17–18.
2. *Race to the Moon 1957-1975*, Greg Goebel, vc.airvectors.net, Ch 21.2, *Soyuz Emerges / The N-1 Moon Booster.* (Accessed in February 2018).
3. **Rockets and People – Vol. 3: Hot Days of the Cold War**, Boris Chertok, NASA SP-4110, Washington D.C., 2009, p. 156.
4. **Soviet and Russian Lunar Exploration**, Brian Harvey, Praxis Publishing, Chichester, UK, 2007, p. 139.
5. **Two Sides of the Moon: Our Story of the Cold War Space Race**, David Scott and Alexei Leonov, St. Martin's Press, New York, 2004, p. 221.
6. Reference 5, pp. 221–3.
7. *Race to the Moon 1957-1975*, Greg Goebel, vc.airvectors.net, Ch 22.2, *NASA plans a circumlunar trip – Soviet progress.* (Accessed in March 2018).
8. Reference 1, p. 128.

9. Reference 1, p. 148.
10. **Study of Suspect Space Covers**, 2nd Edition, Paul C. Bulver, Reuben A. Ramkissoon and Lester E. Winick, ATA Space Unit, Dallas, TX, 2001, pp. 3.9–3.12
11. **Challenge to Apollo: The Soviet Union and the Space Race, 1945-1974**, Asif Siddiqi, NASA SP-4408, Washington D.C., 2000, pp. 657–8.
12. **The Russian Space Bluff – The inside story of the Soviet drive to the Moon**, Leonid Vladimirov, Dial Press, London, 1973, p. 155.
13. Reference 4, pp. 188–90.
14. **Escaping the Bonds of Earth: The Fifties and Sixties**, Ben Evans, Springer-Praxis Books, 2009, p. 462.
15. Reference 1, p. 248.
16. Reference 1, p. 249.
17. *Soyuz 5's Flaming Return*, James Oberg, *Flight Journal*, Vol. 7, No. 3, June 2002, pp.55–60.
18. *"Typy a Padělky Ruských Razítek Tématu Kosmos"* (*Russian space postmarks and fakes* - in Czech), Julius Cacka, Prague 2006, (2nd Edition), pp. 17–21.
19. **Outer Space Mail of the USSR and Russia**, Viacheslav Klochko, Zvezdnyi Gorodok, Moscow, 2009, pp. 4–5.
20. Reference 11, p. 674.
21. **Rockets and People – Vol. 4: The Moon Race**, Boris Chertok, NASA SP-4110, Washington D.C., 2011, pp. 190–4; also, Reference 11, p. 677.
22. Reference 11, p. 674–6.
23. Reference 21, p. 185.
24. *Encyclopedia Astronautica, "Luna Ye-8"*, Mark Wade, www.astronautix.com
25. *The Ni-L3 Programme*, Daniel A. Lebedev, *Spaceflight*, Vol. 34, September 1992, pp. 289–90.
26. Reference 21, p. 197.
27. **The Last Man on the Moon: Astronaut Eugene Cernan and America's Race in Space**, Eugene Cernan and Donald A. Davis, St. Martin's Griffin, New York, 1999, p. 188.
28. Reference 21, p. 204.
29. Reference 28, p. 217.
30. *Designer Mishin speaks on early Soviet space programmes and the manned lunar project*, Vasiliy Mishin interview, *Spaceflight*, Vol. 32, March 1990, p. 106.
31. Reference 11, p. 678.
32. Reference 21, p. 217.
33. Reference 21, pp. 230–1.
34. *Inside the Soviets' Secret Failed Moon Program*, Att Hardigree, www.wired.com (accessed in March 2018).
35. *A secret uncovered: The Soviet decision to land cosmonauts on the Moon*, Asif Siddiqi, *Spaceflight*, Vol. 46, May 2004, pp. 205–213.
36. *The Moon Landing through Soviet Eyes: Q&A with Sergei Khrushchev, son of former premier Nikita Khrushchev*, Saswato R. Das, www.scientificamerican.com, July 16, 2009 (accessed in February 2018).
37. Reference 4, pp. 247–249.
38. Reference 21, p. 211.

7

First Lunar Landing: The Philatelic Side of Apollo 11

INSURANCE COVERS: FAR MORE THAN PLAIN COLLECTORS' COVERS

One of the philatelic 'innovations' introduced for the Apollo 11 mission was the concept of 'Insurance Covers', the pre-signed envelopes that the astronauts left with trusted individuals as a form of life insurance, which family members would be able to sell to collectors if something went wrong on the mission and the astronauts died. These are therefore unflown covers that, in the context of collectables, have a special meaning[1].

The triumphant trip to the Moon was self-evidently a very risky journey into the unknown. All those involved knew that the risks were high. The most dangerous part of the trip was not landing the lunar module on the Moon, but launching it back from the lunar surface up to the mother ship. If that failed, there would be no way to rescue Neil Armstrong and Buzz Aldrin, and while NASA wanted to remain optimistic publicly, they were also prepared for the worst.

In 1999, the *Los Angeles Times* reporter Jim Mann, who was conducting unrelated research on White House documents concerning U.S. affairs with China in the 1960s, unexpectedly discovered a somber memo titled "*In the Event of Moon Disaster*" that the presidential speechwriter William Safire had addressed to President Richard Nixon's White House Chief of Staff, Harry Robins Haldeman. This was apparently at the suggestion of astronaut Frank Borman, who was at the White House as a NASA liaison during Apollo 11.

---

[1] A first version of this chapter was published as "*Apollo 11 Insurance covers*", in *Orbit* no. 92 (January 2012), pp. 22–25, and as "*The First Man on the Moon: the greatest philatelic success ever*", in *Orbit* no. 93 (March 2012), pp. 22–25.

© Springer Nature Switzerland AG 2018
U. Cavallaro, *The Race to the Moon Chronicled in Stamps, Postcards, and Postmarks*, Springer Praxis Books,
https://doi.org/10.1007/978-3-319-92153-2_7

292 First Lunar Landing: The Philatelic Side of Apollo 11

> To : H. R. Haldeman
> From: Bill Safire
> July 18, 1969.
>
> IN EVENT OF MOON DISASTER:
>
> Fate has ordained that the men who went to the moon to explore in peace will stay on the moon to rest in peace.
>
> These brave men, Neil Armstrong and Edwin Aldrin, know that there is no hope for their recovery. But they also know that the...

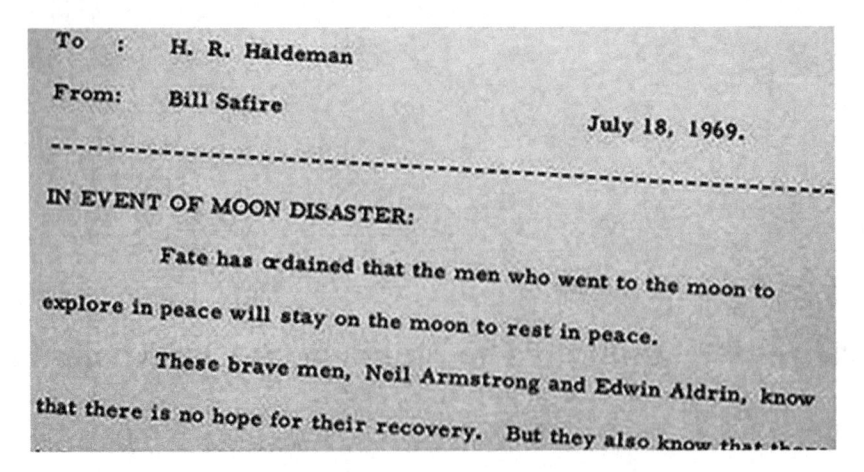

Figure 7.1: The speech that President Nixon would have delivered had there been a tragedy on the Moon during Apollo 11.

The following is the transcript of the undelivered speech that Nixon would have delivered in the event of a tragedy with the first lunar landing:

Fate has ordained that the men who went to the moon to explore in peace will stay on the moon to rest in peace.

These brave men, Neil Armstrong and Edwin Aldrin, know that there is no hope for their recovery. But they also know that there is hope for mankind in their sacrifice.

These two men are laying down their lives in mankind's most noble goal: the search for truth and understanding.

They will be mourned by their families and friends; they will be mourned by their nation; they will be mourned by the people of the world; they will be mourned by a Mother Earth that dared send two of her sons into the unknown.

In their exploration, they stirred the people of the world to feel as one; in their sacrifice, they bind more tightly the brotherhood of man.

In ancient days, men looked at stars and saw their heroes in the constellations. In modern times, we do much the same, but our heroes are epic men of flesh and blood.

Others will follow, and surely find their way home. Man's search will not be denied. But these men were the first, and they will remain the foremost in our hearts.

For every human being who looks up at the moon in the nights to come will know that there is some corner of another world that is forever mankind.

The memo, which remained secret for decades, also contained a contingency plan and suggested a protocol that the Administration might follow for what would happen both before and after the president made that speech. In the event of such

Insurance Covers: Far More Than Plain Collectors' Covers 293

a tragedy on Apollo 11, Nixon would have had to contact Armstrong and Aldrin's widows to express his condolences before addressing the nation with the prepared speech. After that address, NASA would have closed down communications with the lunar module and a clergyman would have come to give their last rites and commend their souls to "*the deepest of the deep*" in a public ritual akin to a burial at sea. As noted in Roger Brun's 2001 book *Almost History*, the closing words of the president's speech echoed those of British poet Rupert Brooke's *First World War* poem, a salute to the fallen whose bodies would remain on foreign soil.

> PRIOR TO THE PRESIDENT'S STATEMENT:
>
> The President should telephone each of the widows-to-be.
>
> AFTER THE PRESIDENT'S STATEMENT, AT THE POINT WHEN NASA ENDS COMMUNICATIONS WITH THE MEN:
>
> A clergyman should adopt the same procedure as a burial at sea, commending their souls to "the deepest of the deep," concluding with the Lord's Prayer.

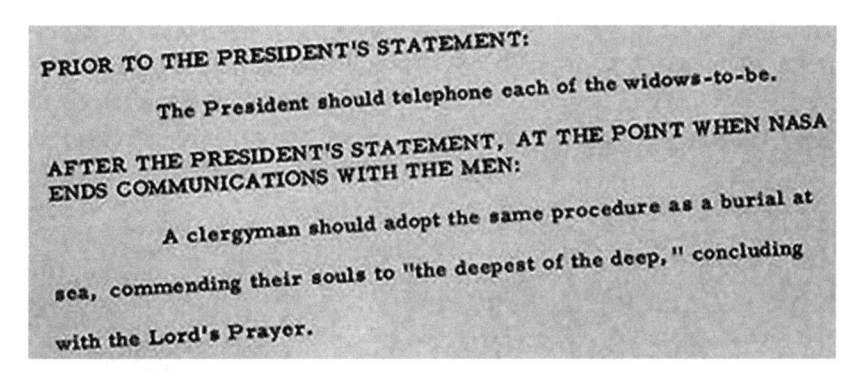

Figure 7.2: The protocol to be followed in the event of the loss of Armstrong and Aldrin on the Moon.

Because of the high risks involved, there was no insurance company willing to provide life insurance for a man about to go to the Moon. While in quarantine prior to launch, however, the crew were approached by a representative of the Houston Manned Spacecraft Center Stamp Club (MSCSC), who suggested the idea of signing a quantity of Apollo 11 covers with the MSCSC cachet and then leaving them behind with their families as a creative way to provide them with funds, since the covers could be sold to collectors if the mission failed with fatalities. The crew agreed and bought a number of MSCSC covers. Some sources have reported the number to be 1,500, although no one knows for certain the exact quantity involved.

At that time, philately was a sort of 'national pastime' and space was attracting the attention of people, and particularly collectors, from around the world, not just in the United States. The expectation for easy earning through space covers (how times have changed!) – carried to excess – led to the 'scandal' of the Apollo 15 covers two years later (see Appendix A: Apollo 15: "The problem we brought back from the Moon").

During the pre-flight quarantine, it seems that the Apollo 11 crew decided to include more than just the Club's own printed cachet envelope, and asked colleagues to find other popular covers available on the market. Due to the short time available before the launch, they locally sourced only two different types of covers and therefore three types of Insurance Covers seem to exist: 1) MSCSC covers; 2) Dow Unicover covers; and 3) Mission patch covers, printed at KSC at the request of the NASA Exchange Council a few weeks prior to the mission. A note of caution must be sounded here, because no official record was kept and some have pointed out that the crews' autographs differ in style on the many covers.

All the covers were affixed with the new Scott 1371 first-class postage stamp, the Apollo 8 *Earthrise* image with the text from the Book of Genesis that was first issued in Houston in May 1969. The latter two types of covers were cancelled at the Kennedy Space Center central post office on the day of the launch using the pictorial NASA postmark 'dark NASA logo'. This adhered to the rules of astrophilately, which prescribe that postal documents (covers and cards) must be cancelled at the post office nearest to the place of any special event, and on the exact date it occurs.

Figure 7.3: Two examples of 'Mission Patch Covers' (Type III) cancelled at the KSC post office on July 16, 1969, the day Apollo 11 launched.

Insurance Covers: Far More Than Plain Collectors' Covers 295

As there were a huge number of items for the crew to sign, the covers (Figures 7.3; 7.4) are likely to have been autographed on different days (or over weeks, in between other duties and training) with different pens and inks. The signatures are quite similar to the ones on the MSCSC covers (Figure 7.5),

Figure 7.4: Two examples of 'Dow Unicover' covers (Type II). Of all the insurance covers, the Dow Unicover ones are the rarest.

Tom Stafford then took care of the MSCSC covers and, after the launch, carried them in his T-38 from KSC in Florida to Houston, home of Mission Control. There, they were machine cancelled on the day of the Moon landing – again according to astrophilately rules – at the post office nearest to the site where the Mission Control Center was providing technical support to the astronauts that had landed on the Moon.

After this operation, all the covers were divided equally into three groups and some 500 covers were delivered to each astronaut's family.

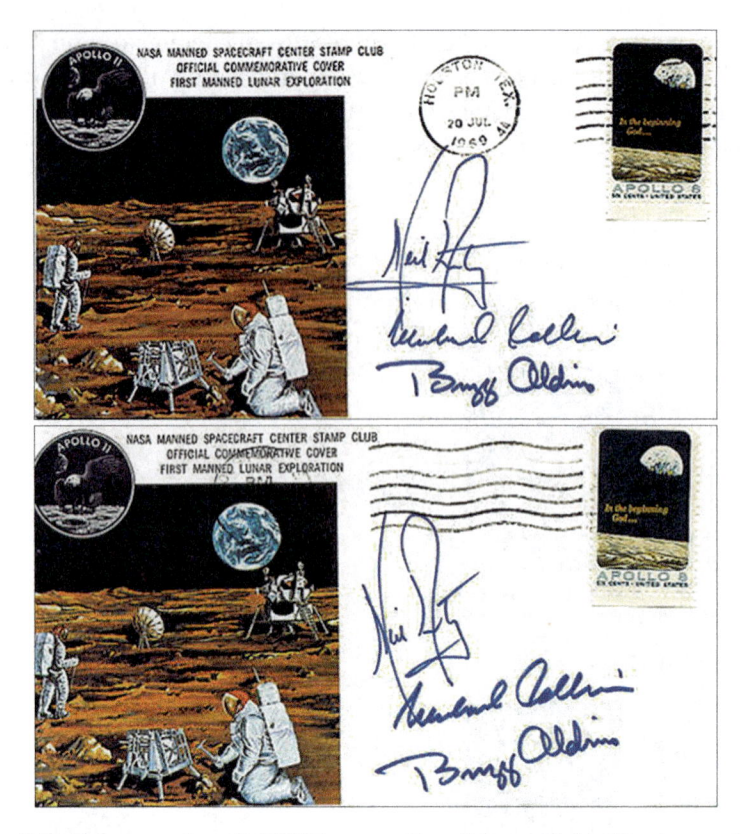

296   First Lunar Landing: The Philatelic Side of Apollo 11

Figure 7.5: Two examples of MSCSC covers (Type I), cancelled in Houston on July 20, 1969, the day of the Moon landing. As far as we know, 1,286 genuine MSCSC Insurance Covers should exist. Unfortunately, the postmark is generally quite indecipherable.

Insurance Covers are much sought after by collectors for several reasons, Firstly, the signatures are both authentic and coeval with the covers, as they were certainly signed during the pre-flight quarantine (unlike covers signed in later years at different locations). Secondly, Insurance Covers transcend generic autograph items because of the role they played and as dramatic reminders of the risks that the astronauts had faced[2].

The tradition of Insurance Covers continued until Apollo 16. It is interesting to note that crew-signed Insurance Covers were also prepared for Apollo 13 and were signed days before the launch by the original crew. The last-minute replacement of Ken Mattingly by Jack Swigert meant that Swigert did not sign the Apollo

---

[2] Such covers reached the market soon after the flight, with the exception of those held by the family of Neil Armstrong, who never sold them. The family did donate one of his covers to benefit the Astronaut Scholarship Foundation in 2014. It was one of the Type III 'Mission Patch' covers and can be seen at www.collectspace.com under "neil-armstrong-auction-scholarship".

13 Insurance Covers despite flying in Mattingly's place. He signed a few of them after the mission, but by that time they had lost their 'insurance' purpose.

The Apollo 17 astronauts did not prepare proper Insurance Covers. According to Gene Cernan, an undefined number of Apollo 17 covers (probably in the hundreds) were prepared and signed by himself and Ron Evans a couple of weeks before the flight, *"with no specific intent in mind."* This was probably because Apollo 17 occurred at the time of the fallout from the Apollo 15 flown postal cover 'scandal' and covers suddenly became taboo. Jack Schmitt chose not to sign them. The covers were therefore left in the crew quarters along with their personal effects, without having them postmarked at the launch. The covers *"indeed could have been disposed of in any way our families so desired were we not to return home,"* Cernan said, and although these covers were never considered to be Insurance Covers, they could have been assumed to be just that if the crew had failed to return. Cernan, who described himself as sentimental, began postmarking the covers on specific milestone anniversaries of the flight *"to be held as personal memorabilia"* and only after the 20th anniversary did he offer several dozen to Jack Schmitt, asking him to sign some of them as well to complete the crew autographs. Some of these have been offered on the market in recent years.

THE FIRST LUNAR POST OFFICE: THE MOON LETTER

During pre-flight preparations for Apollo 11, the U.S. Post Office Department delivered an engraved master die to NASA – created in secret – which would later be used to make the printing plate for the 'First Man on the Moon' stamp. The master die had to be carried to the Moon onboard the lunar module *Eagle*, together with an envelope franked with a non-perforated die proof of the stamp, which had not yet been issued (it would be officially issued in September).

The 'Moon Letter' (see Figure 7.6) had to be cancelled at the lunar landing site, in the 'Lunar Post Office' temporarily in operation at the Sea of Tranquility. For this event, Neil Armstrong had been appointed Post Master and was equipped with a special stamp pad, made for Apollo 11 by the Baumgarten Company of Washington D.C., a manufacturer of rubber stamps for postal use since 1888.

When the cancelling device was delivered, NASA officials told the company that it was too heavy. On the Moon mission, and especially for the *Eagle's* lift-off from the lunar surface, every ounce of weight that could be shed was vital, so a Baumgarten worker drilled a series of holes in the wooden handle and mount. *"It looked like a piece of Swiss cheese when we were finished,"* recalled James A. Baturin, the firm's president. [1] Apart from the 'Moon Landing' cancel, the kit delivered to Armstrong included an ink pad and a handwritten memo by T.E. Jenkins, who was responsible for the operation, with the recommendation *"Envelope enclosed to be cancelled on landing on Moon."* This was corrected to read *"anytime during mission."*

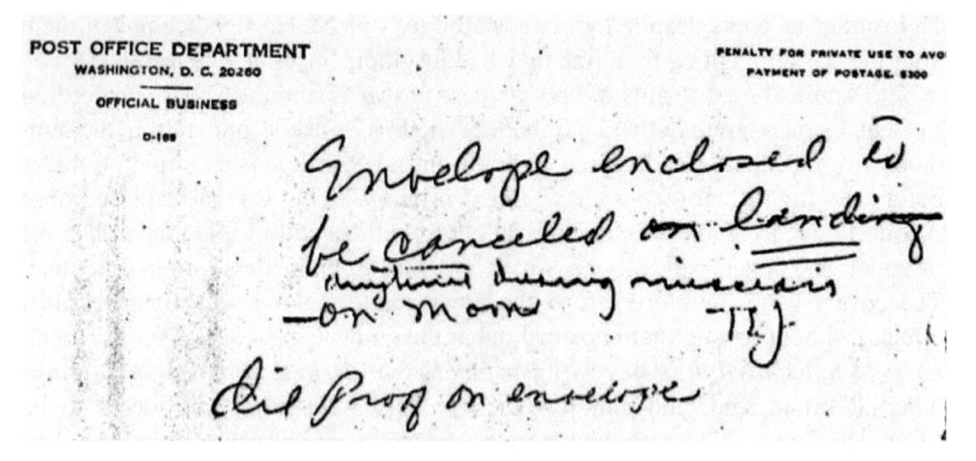

298 First Lunar Landing: The Philatelic Side of Apollo 11

Figure 7.6: Handwritten memo by T.E. Jenkins (responsible for the operation in the U.S. Post Office Department) with the recommendation to cancel the 'Moon Letter' at the Apollo 11 landing site.

For the 'Moon Landing' cancel, the help of Matthew Radnofsky was enlisted. He was a NASA engineer who had been deeply involved in the development of beta cloth, the flameproof material developed to protect the astronauts as a result of the Apollo 1 pad fire in 1967. Radnofsky was also a well-known collector and, at that time, he was serving as the President of the NASA MSCSC in Houston. Radnofsky simulated cancelling in space in his laboratory, where he tested the 'Moon Landing' postmarking device with four proof covers (numbered 1 to 4). Then he prepared an additional 150 specimen reference covers, numbered 1 to 150, which also bore a Webster 'Aug 11, 1969' machine cancel on the reverse. According to collector Walter Hopferweiser, "*At least one unnumbered item 'USA Outer Space Passport' was also produced. It was sold in October 2007 at the Regency-Superior auction.*" [2]

Specimens additionally bore the handstamp "Delayed in Quarantine at Lunar Receiving Laboratory, M.S.C. – Houston, Texas" on the reverse. This additional stamp was used on 214 covers carried around the Moon by Armstrong, Aldrin and Collins.

Most of the 150 specimens were MSCSC covers (prepared by the Stamp Club) but there were some blank covers. All of them were marked "*Specimen for Philatelic Reference (# of 150).*" To avoid any misunderstanding, both the proof and specimen covers were marked with a rubber hand-stamp which clearly read: "*The marking and inscriptions on this cover are examples of the usage of the postmarks and cancellations applied to mail which was carried aboard the flight of Apollo Eleven. This is not a flown cover.*"

Figure 7.7: Proof #4 of 4. Front and reverse of the cover. (From the collection of Walter Hopferwieser, Austria.)

Figure 7.8: Two specimen covers. #19 (above) is a MSCSC cover, while #46 (below, from the collection of Renato Rega, Italy) is a plain cover with no cachet.

300   First Lunar Landing: The Philatelic Side of Apollo 11

Figure 7.9: The reverse of specimen #46 bearing the "Delayed in Quarantine…" handstamp. (From the collection of Renato Rega, Italy.)

The Post Office Department wanted to have this great event recorded through the live TV transmission, with the comments of the astronauts (as would be done later by Dave Scott during the Apollo 15 mission), but NASA refused to add any further tasks to the astronauts' schedule, which was already too busy with scientific tasks and other duties. In the event, that was just as well, because the astronauts forgot, or did not have time, to cancel the 'Moon Letter' on the Moon. It was actually 'back-dated' by the crew on their way back to Earth on July 22, 1969, after leaving the Moon and docking with the *Columbia* CSM.

In his book *Carrying the Fire*, Apollo 11's Command Module Pilot, Mike Collins, described the short 'postmarking ceremony': *"We also have a stamp kit, including a first day cover commemorating the issuance of a new 10¢ stamp showing an astronaut at the foot of the LM ladder about to sample the lunar surface. With the envelope is an ink pad and a cancellation stamp which says, 'Moon Landing, Jul 20, 1969, USA*[3]*.' Never mind that it is July 22. This is the first chance we have had to get to it. We try the cancellation out first, inking it and printing it in our flight plan three times until we get the hang of it, and then we apply it gingerly to the one and only envelope, which we understand the postmaster general will put on tour."* [3]

After the crew's recovery, this 'Moon Letter' and other items returned from the Moon were placed in the decontamination area of the Mobile Quarantine Facility (MQF) carried aboard the recovery ship USS Hornet. The die was especially

---

[3] A small draft of one of these cancels was reproduced in the book by Mike Collins.

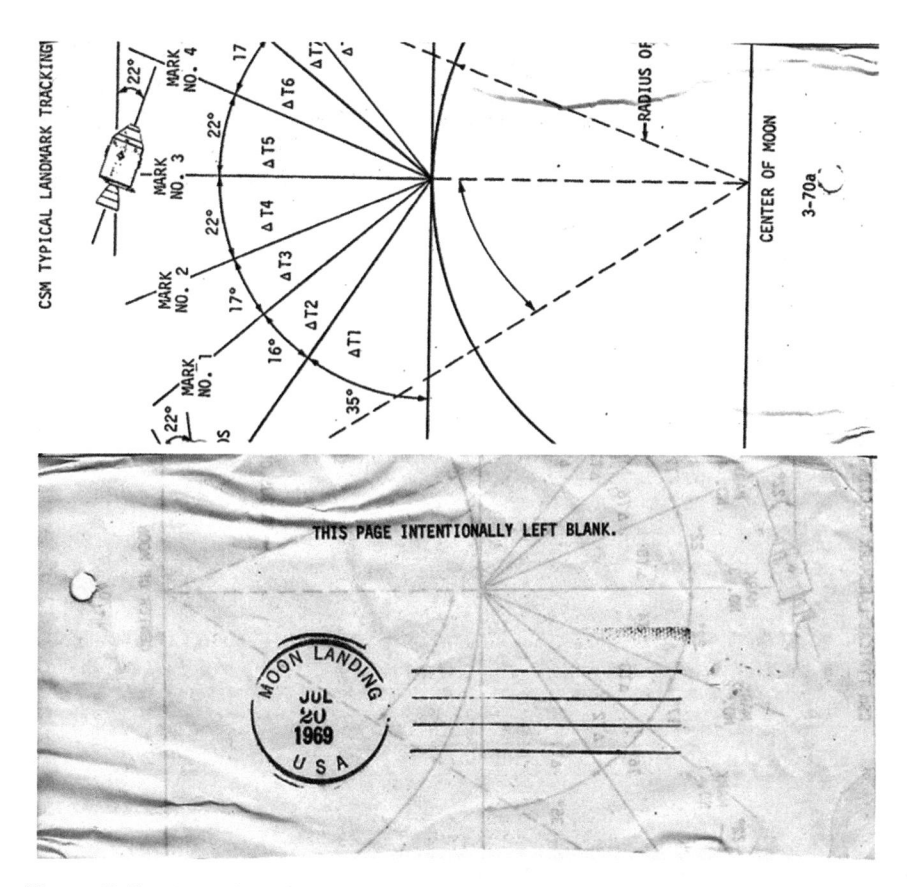

Figure 7.10: One of the three trial cancellations on the back of the flight plan, mentioned by astronaut Mike Collins in his book. (from the collection of Walter Hopferwieser, Austria.)

processed for accelerated decontamination before the prescribed quarantine period in the Lunar Receiving Laboratory had elapsed, and was taken on a special flight from the Space Center in Houston to Washington, where it was delivered back to the Post Office Department. On July 31, Postmaster General Winton M. Blount provided press photographers with a quick look at the die before it was sent to the Bureau of Engraving and Printing where the process of preparing the plates for stamp production began promptly. The 'Moon Letter' was put on show after the prescribed 18 days of quarantine, with the Post Office noting with pride that the letter had traveled more than half-a-million miles, the longest distance any piece of mail had ever gone. The letter is now on display at the Postal Museum in Washington D.C.

302   First Lunar Landing: The Philatelic Side of Apollo 11

Figure 7.11: This image shows the one and only Apollo 11 'Moon Letter' cancelled during the return from the Moon. It is franked with a non-perforated die proof of the C76 stamp designed by Paul Calle. This document is the property of the U.S. Government and may be considered – together with the Apollo 15 'Moon Letter' – one of the rarest postal items existing worldwide. (Picture taken by David Ball, USA.)

THE APOLLO 11 FLOWN COVERS

During their pre-flight quarantine period, as previously mentioned, the Apollo 11 crew were presented with their Insurance Covers by the MSCSC, while at around that same time, the U.S. Post Office Department delivered the secret engraved master die and envelope to be cancelled on the Moon. It is difficult to say for certain whether these initiatives influenced the crew's decision to carry any postal covers to the Moon, but shortly before launch, they decided to take 214 covers aboard Apollo 11 and fly them during the mission.

Having obtained the required permission to do so from Deke Slayton, the legendary Chief of the Astronaut Office, the crew placed these covers in their Personal Preference Kit (PPK) and stored them aboard the CSM *Columbia*, where they remained for the eight days of the mission and thus never made it down to the lunar surface. After the recovery, these covers were stamped *"Delayed in Quarantine"* in red ink (see Figure 7.12) and were held for 18 days in the Lunar Receiving Laboratory to guard against any potential pathogens that may have existed on the Moon. It was not until after the Apollo 14 mission that biologists concluded that there was no threat of contagion and the quarantine process was dropped.

DELAYED IN QUARANTINE AT LUNAR RECEIVING LABORATORY M. S. C. - HOUSTON, TEXAS

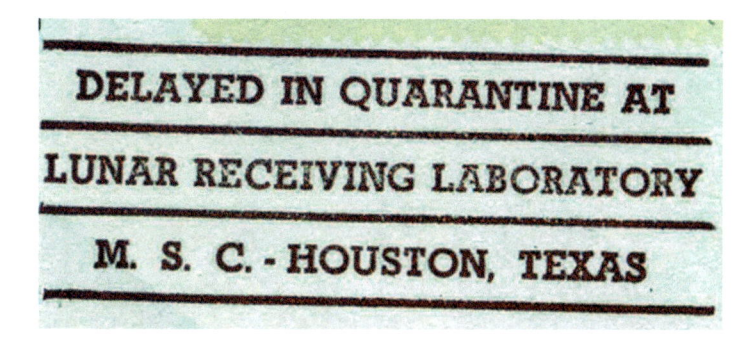

Figure 7.12: All the covers carried aboard Apollo 11 were held in quarantine alongside the crew.

All the flown covers were signed by Neil Armstrong, Buzz Aldrin and Mike Collins while they were in quarantine and were then shared among the crew. On the afternoon of August 10, 1969, at the end of their quarantine, the astronauts entrusted Matthew Radnofsky with the package containing the flown covers and asked him to have them postmarked at the nearest post office. Aldrin wrote on the package the number of envelopes held by each crewmember: 47 by Armstrong, 63 by Collins, and 104 by Aldrin (see Figure 7.13). [4] Radnofsky took the package to the Webster post office, south of Houston, for the covers to be cancelled, and then took them back to Aldrin.

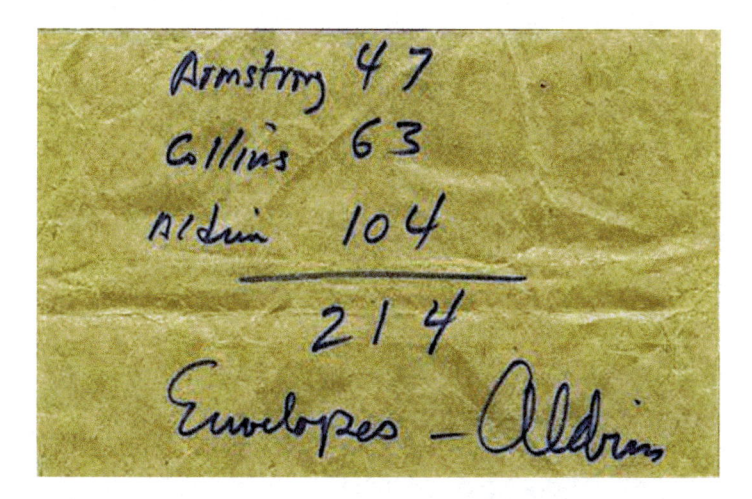

Figure 7.13: The annotation on the outside of the package denoting each astronaut's allocation of flown covers from Apollo 11.

Aldrin used a felt-tip sharpie pen to add a handwritten notation to his 104 covers, in capital letters: "*Carried on the Moon on Apollo 11.*" He also individually numbered each envelope sequentially, using two different annotations: for numbers 1 to 54, each number was preceded by "*EEA-*", while those from 55 to 104 were preceded by "*A-*". (see Figure 7.14).

304   First Lunar Landing: The Philatelic Side of Apollo 11

Figure 7.14:   The handwritten notation Aldrin added to his covers.

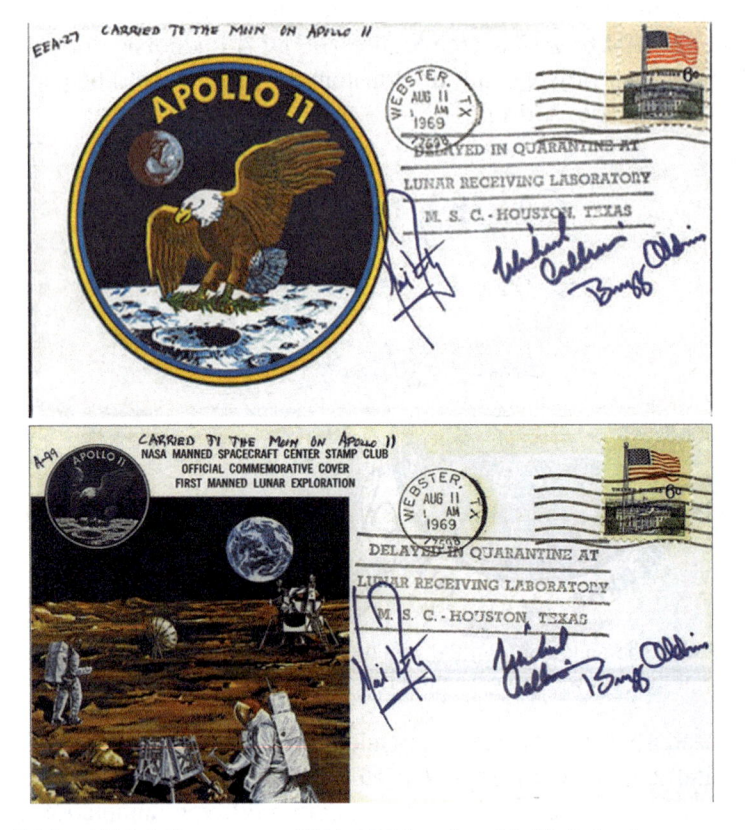

Figure 7.15:   Aldrin's flown covers EEA-27 (above) and A-99 (below).

Mike Collins used a ballpoint pen to add the words *"Carried on the Moon aboard Apollo 11"* or *"Carried aboard Apollo 11 on the moon"* on each of his 63 covers and added the sequential numbers C-1 to C-63 on the Earth image featured in the cachet (see Figure 7.16).

Figure 7.16: Flown covers C-2 and C-54 from the collection of Mike Collins. The serial number is scarcely visible as Collins used a blue ballpoint pen on a blue background. (Picture courtesy of Heritage Auctions, www.HA.com.)

Little was known about the 47 covers held by the Armstrong family (other than that they were numbered consecutively in the upper left corner with NA-1 to NA-47), until two of them were offered on the market in November 2018 through Heritage Auctions. (see Figure 7.17) [5]

Needless to say, all the Apollo flown covers (not just those flown aboard Apollo 11) were postmarked on the Earth. With two exceptions. There are two covers in existence with a postmark applied during the Moon missions: the Apollo 11 'Moon Letter' (which, as mentioned, was carried to the surface of the Moon but was cancelled aboard the CSM *Columbia* during the return to Earth) and the Apollo 15

Figure 7.17: Two flown covers, NA-18 and NA-28, from the collection of Neil Armstrong were publicly offered for the first time through Heritage Auctions in November 2018. (Courtesy of Heritage Auctions, www.HA.com.)

cover officially cancelled by Dave Scott at Hadley Rille. Neither of these are in private possession as they are both owned by the U.S. Government. Three more items were postmarked in outer space – the Apollo 11 trial cancellations mentioned previously – but they were not envelopes. Private collectors, however, may have specimens of the three postmarks prepared for the Moon (one for Apollo 11 and two for Apollo 15), which are occasionally seen at specialized auctions.

'FIRST MAN ON THE MOON': THE GREATEST PHILATELIC SUCCESS EVER

The Apollo 11 'First Man on the Moon' stamp designed by Paul Calle[4] is probably the most famous American stamp, both at home and overseas. It was a unique stamp for many reasons. Firstly, it was one of the last to be issued by the old U.S. Post Office Department before it was replaced by the U.S. Postal Service, an independent government agency. Secondly, it was the largest stamp ever issued by the United States up to that time, some 50 percent larger than conventional U.S. commemorative stamps. As Calle recalled, the post office wanted "*a truly spectacular commemorative issue*" and this 'jumbo' format was the result. [6] But the main reason for its uniqueness is because "*the master die, from which all subsequent plates were made and stamps printed... was carried to the surface of the Moon by the Apollo 11 crew, and the 'Moon Letter' with its die proof, was cancelled by the astronauts on their way back to Earth after the landing on the Moon.*" [6]

After obtaining the approval of President Nixon, the production of the stamp design and the die were carried out in secrecy by officials of the Post Office Department and the Bureau of Engraving and Printing, as had been done with the Project Mercury commemorative of 1962. As Calle recalled: "*The assignment came as an outgrowth of a series of assignments executed for the NASA Fine Art Program. Proceeding in complete secrecy, the Postmaster General, Mr. Winton M. Blount, advised Stevan Dohanos, chairman of the Postmaster's General Citizens Stamp Advisory Committee, of the plan to issue a commemorative stamp, and I was recommended for the assignment on the basis of my previously assigned Twin Space stamps of 1967. The fact that I was also working for a NASA project in connection with Apollo 11 mission afforded me easy access to material and key personnel at NASA who could help with technical problems.*" [6]

As he had previously done with the Gemini twin stamps, Calle started by drawing a series of pencil sketches, attempting different solutions (see Figure 7.18). "*My initial rough thinking sketches explored the concept of a design incorporating the Moon, Earth and the lunar landing module, and a Peace Dove representing the mission objectives and the concept of 'We came in Peace for all Mankind'.*" [7] The more realistic representation of the astronaut on the Moon was the preferred option and the idea of the Peace Dove was soon left aside. The idea then developed into a series of portrait format sketches representing the Lunar Module (see Figure 7.19). Initially, the image included the entire LM but this was progressively reduced to the ladder from which Armstrong would descend. At the suggestion of

---

[4] Known since the early 1950s as an illustrator of science fiction stories. Paul Calle showed a keen interest in space for almost 50 years. He was selected as one of the first eight artists in the NASA Arts Program in 1962, which was established with the purpose of recording space exploration for posterity through the eyes of artists.

Figure 7.18: The initial sketches explored the concept of a design incorporating the Moon, Earth and the lunar landing module, as well as a Peace Dove. (Courtesy Chris Calle.)

the Citizens Committee, the focus was switched to the astronaut, leaving the technology in the background.

"*In the evolution of the design, it quickly became obvious that the first step on the Moon was the most dramatic moment, and with that final sketch* [Figure 7.19, bottom right], *we knew we had our design!*" Eventually, it was decided to return to a landscape format (see Figure 7.20) and the familiar and famous color painting was finalized.

The 'First Man on the Moon' stamp artistically recreated the moment that Neil Armstrong placed his foot onto the lunar surface for the first time (see Figure 7.21). The Earth, rising above the horizon over the astronaut's shoulder – taken from a photograph shot during the Apollo 8 mission and furnished by NASA – was added to the stamp under artistic license (the Earth could not have been seen in that configuration, at that landing site, at that time) to commemorate the home of the Moon's human visitors. Initially, it was intended to release a 6¢ stamp, covering the basic postal rate required at that time for 'first class', i.e., for shipping a normal letter within the U.S. The decision to issue it as a 10¢ airmail stamp was made at the last minute.

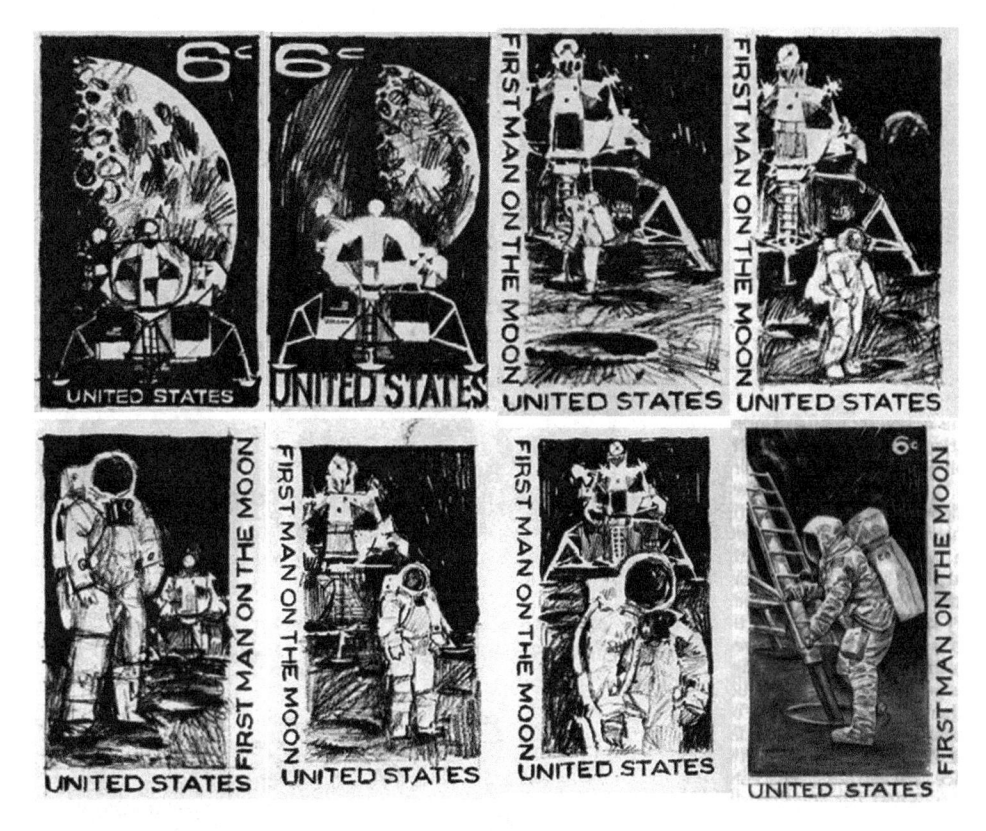

'First Man on the Moon': The Greatest Philatelic Success Ever 309

Figure 7.19: The idea then developed into a set of vertical images representing the Lunar Module. (Courtesy Chris Calle.)

As U.S. Federal law forbids the use of an image of a living person on U.S. postage, the Post Office was careful to describe the subject of the stamp simply as "*a spaceman*" in its press releases. The individual on the stamp is completely hidden by a spacesuit, the idea being that the picture was symbolic, not literal: it was not a tribute to a *person*, but rather to the *concept* of the 'First Man on the Moon'[5]. The painting was delivered to the Bureau of Engraving and Printing for preparation of the master die. The modeler was Robert J. Jones and the engraving was done by Edward R. Felver (vignette) and Albert Saavedra (lettering). [8]

To maintain the project's secrecy, there was no documentation involved. Rather than using messengers to carry materials between the Post Office Department and the Bureau, official staff workers served as couriers. Those who did not need to

---

[5] Neil Armstrong appears on more commemorative postage stamps than any other human in history, save for Yuri Gagarin. According to the American collector and itemizer Peter Hoffman, there had been 536 Armstrong stamps (and 605 Gagarin stamps) issued up to March 2018. (From personal communications with the author.)

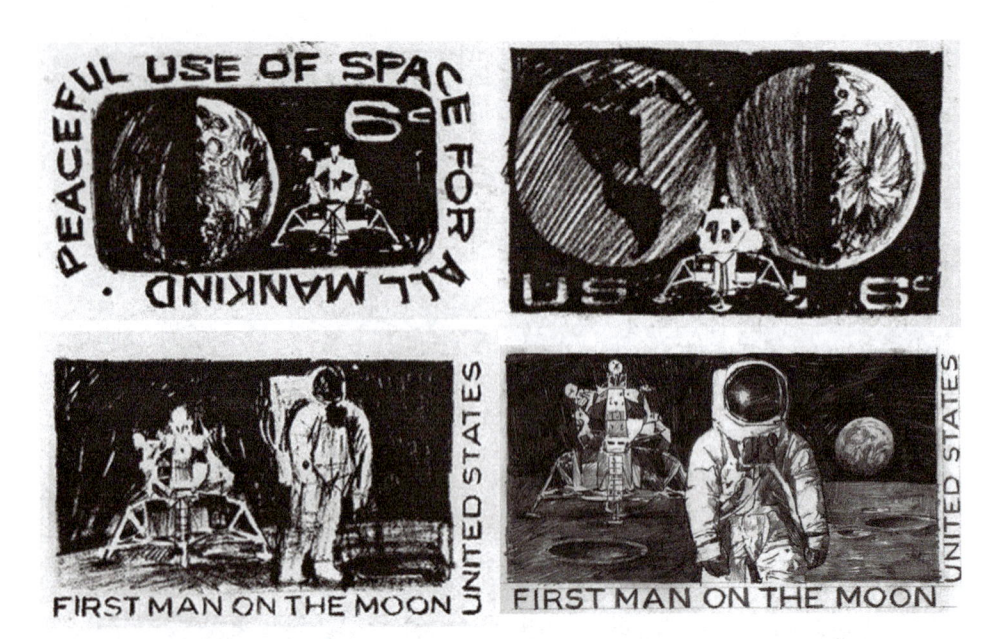

310   First Lunar Landing: The Philatelic Side of Apollo 11

Figure 7.20: The format eventually chosen reverted back to a landscape orientation. (Courtesy Chris Calle.)

Figure 7.21: Closeup of the artwork of the 'First Man on the Moon' stamp. (Courtesy Chris Calle.)

know about the stamp were kept out of the loop. The plan was officially disclosed only a week before the scheduled launch of the mission, with the announcement: "*Apollo 11 will mark America's first mail run to the Moon.*" [9] The master die was then sent to Cape Kennedy and started its long trip to the Moon, together with

Figure 7.22: This cover, one of a kind, has been signed by Paul Calle, the designer of the 'First Man on the Moon' stamp, as well as by the employees of the U.S. Bureau of Engraving and Printing who secretly worked on the project: the modeler Robert J. Jones, the letter engraver Albert Saavedra and the picture engraver Edward R. Felver. The 'Fleetwood' cover is known in two versions.

Figure 7.23: This version of the 'Fleetwood' cover, which is the rarest, features a different version of Neil Armstrong's historic *"One Small Step..."* speech. By scrutinizing the recording of Armstrong's words, it was definitively decided that the correct version was *"One small step for Man,"* as shown in the cover in the previous image (Figure 7.22).

the 'Moon Letter'. Once the die had been returned to the Bureau of Engraving and Printing, the process of producing the stamps began, using a combination of offset photolithography and recess engraving.

The colors – yellow and light blue, and then red and dark blue – were applied in two passes through the two-color Harris offset press. A single pass through a Giori press then added black for the picture, blue for the bottom inscription 'First Man on the Moon', and red for the vertical 'United States' inscription on the right. Due to its size, the stamp was produced in sheets of 32 instead of the usual 50, and a total of 152,264,000 stamps were printed and distributed. [6] The stamp was issued in Washington on September 9, 1969, in conjunction with the National Postal Forum attended by several business executives and Post Office officials, as well as the three Apollo 11 astronauts (see Figure 7.24). The special postmark included the September 9 Washington D.C. date stamp and a replica of the July 20 'Moon Landing USA' date stamp that the astronauts had applied to the 'Moon Letter'.

Anticipating a great interest among First Day Cover (FDC) collectors, the U.S. Post Office Department ordered 25 postmarking devices in advance to cope with demand. But the success of this stamp exceeded every expectation and

Figure 7.24 (left to right): Apollo 11 astronauts Michael Collins, Neil A. Armstrong and Edwin E. 'Buzz' Aldrin, with General Postmaster Winton M. Blount, at the unveiling ceremony of the new stamp honoring the mission. © NASA 69HC1119

'First Man on the Moon': The Greatest Philatelic Success Ever 313

triggered an unprecedentedly high demand. Prior to the official ceremony, the U.S Post Office Department had issued a press release on August 25, which declared that the 10¢ stamp was inspiring great interest worldwide. During the three weeks after the issue, the Post Office had received 500,000 requests for FDCs, demand which continued to grow by 60,000–80,000 *per day*. Such requests came in particularly large numbers from Australia, Great Britain, France and Belgium, but altogether they arrived from over 100 countries. Eventually, 8,743,070 FDCs were postmarked: an "*astronomical*" figure (in the words of the Post Office). By comparison, even the Elvis Presley commemorative stamp of January 8, 1993, one of the most phenomenally popular stamps ever printed, received only 4,451,718 requests. The First Day Cover processing crew was quickly expanded from 40 to 100 employees, but even so, it took five months to complete the task.

Due to the new technology being used, the Bureau of Engraving and Printing had some difficulties with registration on the stamps, and numerous copies have been found with noticeable shifts in the offset colors. These were not all identified and destroyed in time and some of them escaped detection and reached the market. One such variety of the stamp with a major error was first found in El Paso, Texas, in October 1969. One month after the issue, a sheet arrived with some specimens missing the offset red color, which was used in the flag stripes on the astronaut's shoulder patch, and the series of light red dots over the yellow portions of the lunar module and the astronaut's face plate (see Figure 7.25). Because of

Figure 7.25: The 'Unknown Astronaut' variant of the stamp, missing the offset red color.

this extraordinary technical error, no American flag appears on the shoulder of the astronaut, and this variety was quickly dubbed 'The Unknown Astronaut' as a symbol for mankind conquering space without flags.

References

1. *Inside Story of the First Men on the Moon Stamps*, George Amick, www.unicover.com
2. Communication from Walter Hopferwieser in April 2018; also, *Matthew Radnofsky: il quarto astronauta dell'Apollo 11 (Matthew Radnofsky: the fourth Apollo 11 astronaut)*, Alberto Bolaffi, *Il Collezionista* # 6, 2009 pp. 8–17 [in Italian].
3. **Carrying the Fire**, Michael Collins, Farrar, Strauss and Giroux, New York, 2009, p. 426.
4. *Mail from the Moon*, http://postalheritage.wordpress.com/2010/04/30/mail-from-the-moon/
5. One of these covers was publicly shown for the first time, on request of the author, on www.collectspace.com/ubb/Forum20/HTML/001018.html in May 2018
6. **The Pencil**, Paul Calle, published by North Light Publishers, Westport, Conn, distributed by Watson-Guptill Publications, New York, 1974, p. 121.
7. **Celebrating Apollo 11 – The Artwork of Paul Calle**, Chris Calle, AeroGraphics Inc., Bradenton, Florida, 2009, p. 22.
8. *Astrophile* (journal of Space Unit – American Topical Association) #309, 2009, p. 134.
9. Reference 8, p. 131.

Appendix A
Apollo 15: "The Problem We Brought Back From the Moon"

Postal Covers Carried on Apollo 15[1]

Among the best known collectables from the Apollo Era are the covers flown onboard the Apollo 15 mission in 1971, mainly because of what the mission's Lunar Module Pilot, Jim Irwin, called *"the problem we brought back from the Moon."* [1] The crew of Apollo 15 carried out one of the most complete scientific explorations of the Moon and accomplished several firsts, including the first lunar roving vehicle that was operated on the Moon to extend the range of exploration. Some 81 kilograms (180 pounds) of lunar surface samples were returned for analysis, and a battery of very productive lunar surface and orbital experiments were conducted, including the first EVA in deep space. [2] Yet the Apollo 15 crew are best remembered for carrying envelopes to the Moon, and the mission is remembered for the *"great postal caper."* [3]

As noted in Chapter 7, Apollo 15 was not the first mission to carry covers. Dozens were carried on each flight from Apollo 11 onwards (see Table 1 for the complete list) and, as Apollo 15 Commander Dave Scott recalled in his book, the whole business had probably been building since Mercury, through Gemini and into Apollo. [4]

People had a fascination with objects that had been carried into space, and that became more and more popular – and valuable – as the programs progressed. Right from the start of the Mercury program, each astronaut had been allowed to carry a certain number of personal items onboard, with NASA's permission, in

---

[1] A first version of this material was issued as *Apollo 15 Cover Scandal* in *Orbit* No. 87 (October 2010), pp. 25–29 and in AD*ASTRA (quarterly Journal of the ASITAF - Italy), No. 6, June 2010, pp.2-7 (in Italian and English).

Table 1. Covers flown during the Apollo missions*

Apollo 11	214
Apollo 12	87[+]
Apollo 13	50
Apollo 14	55
Apollo 15	552
Apollo 16	28
Total	**986**

[+]Flown on Apollo 15

*not including the three covers flown respectively on Apollo 11 and Apollo 15 at the request of the USPS

their Personal Preference Kits (PPK). All such items had to be listed and approved prior to launch by Deke Slayton, the Chief of the Astronaut Office, and were intended for private use or as personal gifts after the flight. They could not be used for commercial purposes or personal gain. Astronaut PPKs typically included badges, jewelry, coins, medals, flags, stamps, postal covers, currency, printed materials, and similar, easily packed, lightweight mementos. As the flights became more significant, the number and type of items carried increased.

Aside from personal mementos, each crew also carried medallions. [5] The number of silver medallions had also grown steadily with each mission. In his book, astronaut Walt Cunningham referred to a rumor around the Astronaut Office that Apollo 14 had carried onboard a personal package weighing 19 kilograms (42 pounds). [6] The Franklin Mint had even advertised the proposed sale before the flight. After the mission, the deal was never completed and all went quiet. Nothing about it was published in the media[2], but some members of the U.S. Congress were unhappy with the situation. [5]

Given this background, it was perhaps inevitable that the *"lapses of judgment"* by the Apollo 15 crew would be a step too far. Dave Scott carried a total of 641 postal covers onboard the mission (including the two official covers requested by the U.S. Postal Office), but only 243 of these had been listed and authorized before the flight. The remaining 398 were not and he secretly carried these aboard in a pocket of his spacesuit[3]. Had they been listed as being in Scott's PPK, they would probably have been routinely approved for inclusion on the mission. [7]

---

[2] According to Apollo 15 Command Module Pilot Al Worden, because one of the astronauts involved was Alan Shepard, the first American in space, NASA preferred to gloss over this issue.
[3] NASA News Release 72-189, September 15, 1972, p. 3. In his book (p. 330), Scott provided a slightly different version: "*Usually the list was certified by Deke. But before our flight, for some reason, he never asked us personally for each of our lists, as it was customary, nor signed off on the list personally. He said the flight-crew support team had already logged everything. Whereas we had purchased the covers ourselves, the Astronaut Office at the Cape had prepared the covers for the flight and had had them stamped and franked on the day of launch. Somehow, however, the support team had missed them when they prepared the PPK flight manifest.*" A variant of this ver-

The 243 listed and authorized covers included:

I. 2 official U.S. Postal Service covers, one of which – the 'Official Cover' – was publicly cancelled on the Moon by Scott at the request of the U.S. Postal Service (Figure A.3).
II. 1 'Wright Brothers' commemorative cover, dated 1928 and autographed by Orville Wright, which was carried by Al Worden for a friend (Figure A.4).
III. 1 cover bearing a 'First Man on the Moon' stamp and a Bliss Centennial 3¢ stamp, carried by Jim Irwin for Barbara Baker (Figure A.5).
IV. 8 'Shamrock Covers' carried by Irwin[4]. The covers were provided by the collector Ray Burton (Figure A.6).
V. 144 Herrick's 'Moon Phases' covers carried by Worden (Figure A.7), printed with a cachet showing 15 phases of the Moon. On the USS Okinawa, the Apollo 15 recovery ship, the astronauts placed two 8¢ stamps on each of these covers, which Worden purchased, and then had the covers cancelled by the ship's own post office. The astronauts later autographed these covers while flying back from Hawaii to Houston. Sixteen covers were torn or damaged and were destroyed. Because of the furor created by the Apollo 15 covers incident, NASA confiscated 61 of these covers, even though they had been duly listed and authorized.
VI. 87 Apollo 12 covers (Figure A.8) that, for unknown reasons, did not fly on that mission. They were carried on Apollo 15 for Mrs. Barbara Gordon, a stamp collector and the wife of Richard 'Dick' Gordon (the Apollo 12 astronaut, who at that time was the backup pilot for Apollo 15).

Apollo 15: The First Lunar Post Office

Dave Scott was officially appointed lunar Postmaster and carried with him two postmarks, and two covers with non-perforated die-proofs of the twin 8¢ 1971 U.S. 'Decade of Space Achievement' stamps (still to be issued) affixed to

---

sion appeared in Scott's interview, quoted by Russel Still in his book. (Still [2001], p. 263.)

[4] As Irwin recalled in his book, when he was selected as an astronaut in 1966 he intended to take the shamrock on his missions. The shamrock is a symbol of Ireland and of Saint Patrick, who used its three leaves to explain the Christian doctrine of the Holy Trinity to the Irish people. *"Since I am Irish and was born on St. Patrick Day, I had planned, since the time I was first selected for the program, to take shamrocks to the moon"* (Irwin [1982], p. 95). Irwin had prepared hundreds of shamrock covers and flew eight of them on the Command Module *Endeavour* and an unspecified number on the LM *Falcon*. Unfortunately, after leaving the lunar surface, they realized only during the return that neither Irwin nor Scott had retrieved the PPKs from the LM, in which *"there were envelopes, medallions, stamps, medals, flags, shamrocks and coins. Dave and I had at least a hundred two-dollar bills that we were going to split after the flight"*.

318   Apollo 15: "The Problem We Brought Back From the Moon"

them (Figure A.3). He had one postmark bearing the text 'Moon Landing USA' (Figure A.1) and the other with 'United States on the Moon' (Figure A.2). The First Lunar Post Office was opened on August 2, 1971, when Scott postmarked only the official cover (and brought back the official backup cover uncancelled).

A handful of covers had been postmarked before launch, to test both cancelling devices. Dr. Matthew Radnofsky – who had already simulated cancelling in space in his laboratory for Apollo 11 – made the tests of both postmarks to ensure that the cancellation devices were in working order, especially the changeable date mechanism. Unlike what had been done for Apollo 11, covers without stamps were used for these tests. Apollo 15 lunar cancellation proofs are much scarcer than Apollo 11 ones. Collector Walter Hopferwieser, who provided the following list of the known tests done by Radnofsky before launch, stated: "*As of now, I am aware of only 16 items. Most of them had been sold a few years ago at Regency-Superior and Aurora auctions. However, I am pretty sure that more of them were done.*" [8]

- 3 covers – Moon Landing, USA, Jul 29, 1971
- 2 covers – Moon Landing, USA, Jul 30, 1971
- 3 covers – Moon Landing, USA, Jul 31, 1971 (Figure A.1)
- 5 covers – United States on the Moon, Aug 2, 1971
- 2 covers and 1 cut on brown paper – United States on the Moon, Aug 3, 1971

Figure A.1: Apollo 15 Proof Cover without stamps, used for a test of the 'Moon Landing USA' postmark. (From the collection of Walter Hopferwieser, Austria.)

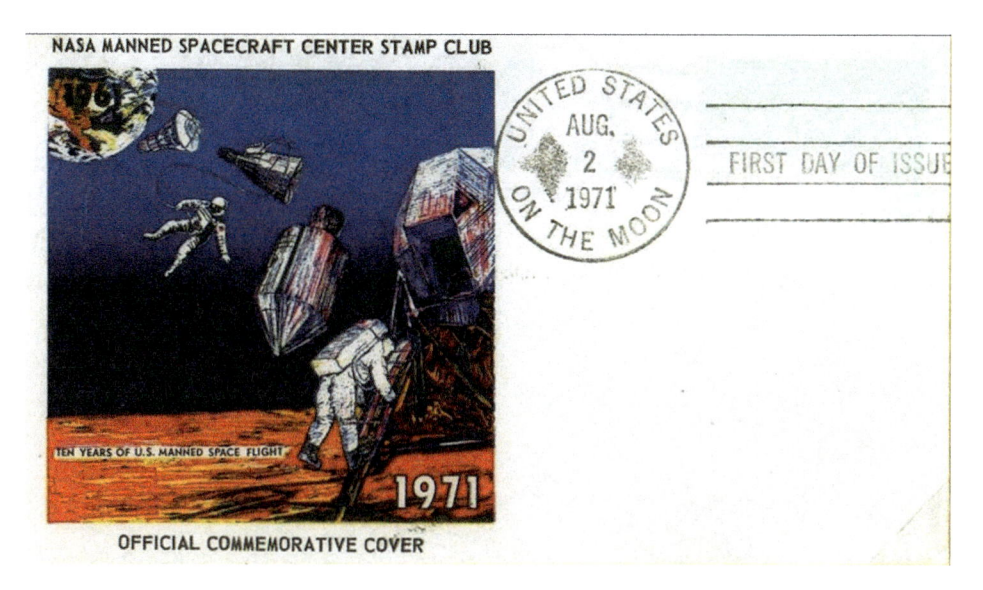

Apollo 15: "The Problem We Brought Back From the Moon" 319

Figure A.2: Apollo 15 Proof Cover with the 'United States on the Moon' postmark. Unfortunately, the actual usage on the Moon was almost unreadable. (From the collection of Walter Hopferwieser, Austria.)

The official Apollo 15 cover postmarked on the Moon (Figure A.3) is displayed at the U.S. National Postal Museum. A short address was given by Dave Scott while cancelling the cover on the Moon[5]: "*To show that our good Postal Service has deliveries any place in the universe, I have the pleasant task of cancelling, here on the Moon, the first stamp of a new issue dedicated to commemorate U.S. achievements in space. And I'm sure a lot of people have seen pictures of the stamp. I have the first one here on an envelope. At the bottom it says, 'United States in Space, a decade of achievement,' and I'm very proud to have the opportunity here to play postman. I pull out a cancellation device. Cancel this stamp. It says, 'August 2, 1971, first day of issue...' What could be a better place to cancel this stamp than right here at Hadley Rille...! By golly, it even works in a vacuum.*"

The first postmark Scott made was faint, so he made another below it. The smudges on the left side are thumbprints made by his spacesuit glove, soiled with lunar dust, as can be seen in the video. Immediately after this ceremony, the Postmaster General, who was waiting some 238,000 miles away at Mission Control in the Manned Spacecraft Center in Houston, Texas, gave the signal to

---

[5] The video of the cancellation, with transcript, can be seen at the Postal Museum website. https://postalmuseum.si.edu/stampstakeflight/moonmail.html (accessed April 2018).

Figure A.3: The official U.S. Postal Service Apollo 15 Lunar Mail cover postmarked on the Moon. Postmaster General's Collection. (Courtesy Smithsonian National Postal Museum, postalmuseum.si.edu/stampgallery/moonmail.html.)

start the 'terrestrial' First Day cancellations at three post offices, in Houston, at the Kennedy Space Center, and in Huntsville.

According to Walter Hopferwieser, the other postmark, "Moon Landing, USA" was believed to have been left aboard the Command Module with Al Worden, together with a stamp pad and the back-up cover, which was returned to Earth uncancelled.

The Souvenir Covers in the PPK List

Figures A.4 to A.8 show examples of the 241 souvenir covers that had been properly listed and authorized for the flight of Apollo 15

Controversial Envelopes Flown to the Moon

The idea of taking additional covers was suggested during a cocktail party by Walter Eiermann, a naturalized American citizen, who was a salesman for the heatshield contractor, had frequent business and social contacts with NASA personnel, and was well-acquainted with many in the astronaut corps. He suggested flying 100 lightweight covers for his friend Herman Sieger, a major European

Figure A.4: Vintage cover (December 17, 1928) commemorating the 25th anniversary of the Wright Brothers first aircraft flight, signed by Orville Wright.

Figure A.5: Barbara Baker flown cover (from the collection of Walter Hopferwieser, Austria).

stamp dealer based in Lorch, Germany. Eiermann offered each of the astronauts approximately $7,000 in the form of savings accounts, after agreeing that there was to be no commercialization or advertising of the covers and that nothing would be done with them until after the completion of the Apollo program. The

322   Apollo 15: "The Problem We Brought Back From the Moon"

Figure A.6: One of the eight shamrock covers carried by Jim Irwin on the CSM.

Figure A.7: One of the 144 Herrick's covers carried by Al Worden on the CSM.

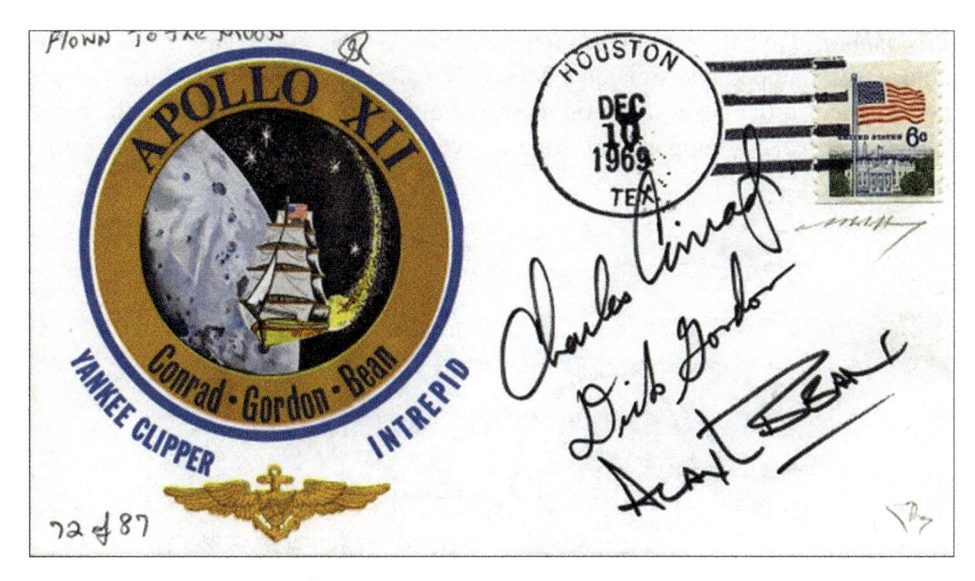

Apollo 15: "The Problem We Brought Back From the Moon" 323

Figure A.8: One of the 87 Apollo 12 covers carried by Apollo 15 for astronaut Dick Gordon's wife Barbara.

astronauts also decided among themselves to carry 300 more such covers for their personal use, although only 298 actually flew as two of them were damaged and discarded before being packaged.

The lightweight covers bore, as a cachet, a replica of the official Apollo 15 patch overprinted with an Air Force wing and propeller emblem. They were supplied by Al Bishop, a friend of many of the astronauts who was with the Howard Hughes organization. He had already provided covers for the previous flights, starting with Apollo 12. The crews usually returned a couple of flown covers to Bishop, signed with gratitude. In his book, Walt Cunningham wrote: *"To my knowledge, Bishop never sold any and never made a dime off his relationship. He was simply a fan. To Al, Apollo 15 was no different from any other flight, except for a phone call he received from Hal Collins, the Astronaut Office manager at the Cape. Collins told Bishop that the crew would like to know whether he could obtain some very lightweight envelopes for them. Al said that he'd be happy to do so. He was unaware then that many of them would be smuggled on board the next lunar flight. Al was trusted. That's why many of us imposed on him with our problems, special requests and, sometimes, matters which we would rather not share with NASA."* [9] This time, however, his trust was abused and he emerged as a scapegoat of the whole 'scandal', with the inference that he was the 'Mr. Big' of an international stamp conspiracy.

Although the unauthorized covers were similar lightweight envelopes with the same cachet (the official Apollo 15 emblem), the 100 Sieger covers can easily be identified (Figure A.9) as they have a handwritten inscription on their front upper left corner stating: *"Landed at Hadley moon July 30, 1971. Dave Scott, Jim Irwin."* On the reverse is a typed notarized inscription: *"This is to certify that this cover*

324 Apollo 15: "The Problem We Brought Back From the Moon"

was onboard the Falcon *at the Hadley-Apennine, Moon, Jul 30-Aug 2, 1971.*" The notary was stamped and signed by Mrs. C.B. Carsey and her notary raised seal is also applied to the cover. Also on the reverse, in the lower left corner, the name "*H.W. Sieger*" is stamped and signed, accompanied by a handwritten serial number.

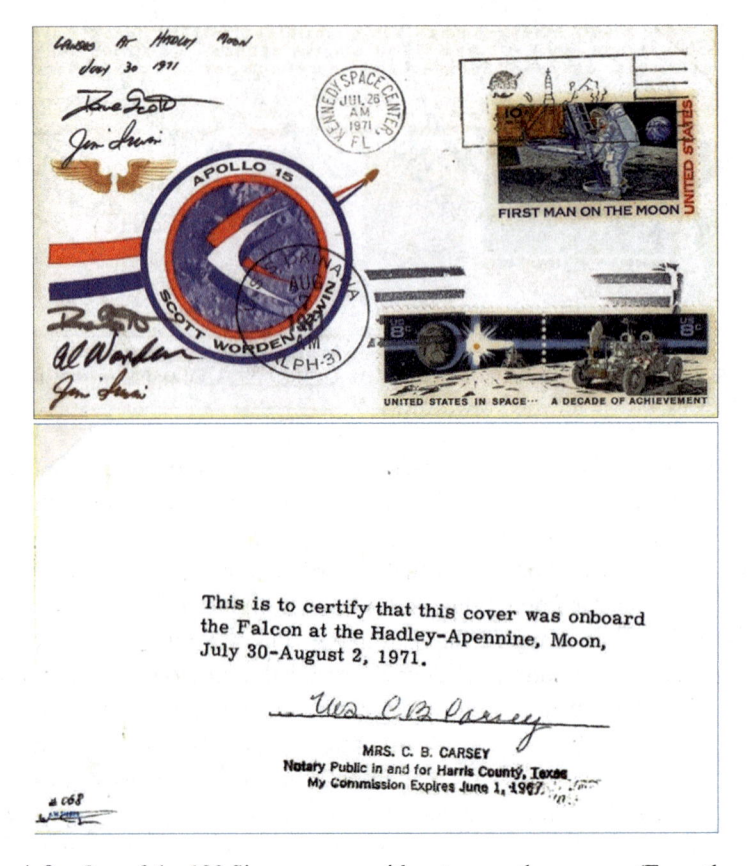

Figure A.9: One of the 100 Sieger covers with notary on the reverse. (From the collection of Walter Hopferwieser, Austria.)

In comparison, the 298 crew-owned covers (Figure A.10) have the following inscription printed on their front upper left corner:

> THIS ENVELOPE WAS
> CARRIED TO THE MOON
> ABOARD THE APOLLO 15
> #_____ OF 300 TO THE
> LUNAR SURFACE IN
> L.M. "FALCON"

Apollo 15: "The Problem We Brought Back From the Moon"   325

Figure A.10: One of the 298 crew owned covers, with the handwritten NASA serial number on the back.

These covers were also autographed by the crew on the front lower left corner, but instead of the notarized description on the back, each cover had a card inserted which reads:

> This cover is #____ of 300 postmarked just prior to the launch of Apollo 15 on July 26, 1971 at Kennedy Space Center; stowed aboard the spacecraft in a sealed fireproof packet; carried to the lunar surface in LM "FALCON"; returned to Earth in CM "ENDEAVOUR"; and postmarked immediately after splashdown on August 7, 1971 by the U. S. Navy Postal Station aboard the recovery ship USS OKINAWA.

326   Apollo 15: "The Problem We Brought Back From the Moon"

Some of these cards have been found to bear the signature of one of the crew. Additionally, on the reverse of each, there is a small, handwritten serial number. These were assigned by NASA at the time of their confiscation.

The covers were cancelled at Kennedy Space Center several days before July 26, 1971. The date of the post office machine canceller was moved forward[6]. Additionally, after their splashdown, the crew purchased the new twin 8¢ stamps honoring the 10th anniversary of the space program (the same as the one cancelled on the Moon for the U.S. Postal Service) aboard their recovery ship the USS Okinawa and affixed them to these covers. The covers were then cancelled and date-stamped 'Aug 7, 1971' in the ship's post office. The astronauts later autographed these covers while flying from Hawaii to Houston.

On August 31, the 100 Sieger covers, already carrying the handwritten notation on the front, had the additional notarized inscription typed on the back and were duly signed. On September 2, Scott mailed the 100 covers to Eiermann, who was in Stuttgart, Germany at the time. Eiermann then delivered these covers to Sieger who paid an unspecified sum to him. With their consent, Eiermann then opened a $7,000 savings account for each of the astronauts in a German bank. But it was not long before news began to circulate that the German dealer Sieger – in violation of the agreement made with the astronauts – was selling the covers in Europe for a reported average of 4,850 deutschmarks (about $1,500) each.

These reports infuriated the U.S. Congress, not least because they only found out about such controversial information by reading it in the press rather than being informed by NASA, which was obligated to keep Congress up to date about its activities. Recollections of the Apollo 14 medallions incident must have echoed in the minds of certain members of Congress, many of whom were not keen on NASA anyway. [5]

Scott telephoned Eiermann to request that sales be stopped and the covers returned, and took steps to ensure that the funds in the savings accounts were returned. Eiermann suggested an alternative to the accounts, however, offering each astronaut a commemorative stamp album for their families. This suggestion, initially accepted by the astronauts, was rejected in April 1972 after further consideration. [10]

NASA began an internal investigation. This was the most controversial development in the Apollo program and although most of the astronauts were involved to some degree, NASA made an example of the Apollo 15 crew. As Scott recalled: *"It was turning into a witch-hunt. Our bosses had abrogated their responsibilities and we were left alone on a very wet day."* [11] Jim Irwin recalled the irony of their situation: *"We had been back from the moon for less than a year, and during*

---

[6] Bulver [2001], p. 3.37. According to a different version reported by David S. Ball in his book, the covers carried onboard were unstamped and unaddressed and were then backdated after their return from space (Ball [2010], p. 99).

this brief period we had addressed a joint session of Congress as heroes, and now we were going back before these same senators in disgrace, because of this envelope scandal." [12]

After this 'incident', NASA prohibited carrying any covers or stamps during a space mission from Apollo 17 onwards (they were too late to block the covers carried on Apollo 16). The only exceptions since then have been the items flown in cooperation with the U.S. Postal Service, chiefly the 266,000 covers carried in 1983 in the payload bay of the Space Shuttle Challenger during the STS-8 mission[7] and the 500,000 $9.95 Express Mail Stamps created by Paul and Chris Calle in 1994 that were flown on Shuttle Endeavour's STS-68 mission, in recognition of the 25th anniversary of the Apollo 11 Moon landing. The huge number of such items carried in space discourages speculation and these items are still widely available for a few dollars.

So far, fewer than 1000 envelopes in total have flown to the Moon and there will be no other American covers going there for the foreseeable future. Perhaps the next 'Moon covers' will bear Chinese or Indian stamps?

References

1. **To Rule the Night**, James B. Irwin, Holman Bible Publishers, Nashville, 1982 (2nd ed.), p. 227.
2. NASA News Release 72-189, September 15, 1972, p. 1.
3. **The All-American Boys**, Walt Cunningham, iBooks, New York, 2003, p. 282.
4. **Two Sides of the Moon: Our Story of the Cold War Space Race**, David Scott and Alexei Leonov, St. Martin's Press, New York, 2004, p. 328. No covers were likely flown on Mercury or Gemini missions, but there were certainly postage stamps and other collectibles carried.
5. Reference 4, p. 329.
6. Reference 3, p. 291.
7. NASA News Release 72-189, September 15, 1972, p. 3; also, **Study of Suspect Space Covers**, Paul C. Bulver, Reuben A. Ramkissoon, and Lester E. Winick, 2nd Edition, ATA Space Unit, Dallas, Texas, 2001, pp. 3.35–3.45.
8. Communication from Walter Hopferwieser in April 2018.
9. Reference 3, p. 288.
10. Reference 2, p. 4.
11. Reference 4, p. 330.
12. Reference 1, p. 236.

---

[7] After this U.S. Postal Service initiative, Al Worden, seeing the similarities between his confiscated covers and those set to fly on STS-8, sued the government for the return of the crew's 298 flown covers (and his 61 Herrick's 'Moon Phases' covers). In an out-of-court settlement, the covers were returned to the crew (see *Apollo 15 'Sieger' Covers* at www.collectspace.com).

Appendix B
Timeline of Main Space Events: 1957–1969

Date	Mission name	Mission achievements	Country
1957, Oct 4	Sputnik 1	First artificial satellite	USSR
1957, Nov 3	Sputnik 2	First living passenger in orbit, the dog Laika	USSR
1958, Jan 31	Explorer 1	First U.S. artificial satellite	USA
1958, May 3	Sputnik 15	First scientific satellite in orbit, with a large array of instruments for geophysical research	USSR
1958, Jul 29	NASA	The U.S. Congress approves the Space Act (NASA's founding law)	USA
1959, Jan 2	Luna 1	First to reach Earth escape velocity or Trans Lunar Injection	USSR
1959 Feb 28	Discoverer 1	First U.S. spy satellite	USA
1959, Aug 7	Explorer 6	First photograph of Earth from orbit	USA
1959, Sep 13	Luna 2	First impact into another celestial body (the Moon)	USSR
1959, Oct 4	Luna 3	First photos of the far side of the Moon	USSR
1960, Mar 1	Pioneer 5	First solar probe	USA
1960, Aug 19	Sputnik 5	First plants and animals to return alive from Earth orbit	USSR
1960, Oct 10	Mars 1	First probe launched to Mars (failed to reach target)	USSR
1961, Jan 31	Mercury 2	First Mercury test with a chimpanzee (Ham).	USA
1961, Apr 12	Vostok 1	First human in space: Yuri Gagarin	USSR
1961, May 5	Mercury 3	First American in space: Alan Shepard	USA
1961, Jul 21	Mercury 4	Second U.S. suborbital flight: Gus Grissom	USA
1961, Aug 6	Vostok 2	First spacefarer to spend an entire day in space: Gherman Titov	USSR
1961, Oct 25	Saturn 1	First flight test of Saturn 1	USA
1962, Feb 20	Mercury 6	First U.S. manned orbital flight: John Glenn	USA

(continued)

© Springer Nature Switzerland AG 2018
U. Cavallaro, *The Race to the Moon Chronicled in Stamps, Postcards, and Postmarks*, Springer Praxis Books,
https://doi.org/10.1007/978-3-319-92153-2

(continued)

Date	Mission name	Mission achievements	Country
1962, May 24	Mercury 7	Second U.S. manned orbital flight: Scott Carpenter	USA
1962, Jun 14	ELDO & ESRO	Agreements are signed establishing the European Space Organization	Europe
1962, Aug 11	Vostok 3	First 4-day flight and first 'group' flight: Andrian Nikolayev	USSR
1962, Aug 12	Vostok 4	First 'group' flight, with Vostok 3: Pavel Popovich	USSR
1962, Sep 29	Alouette 1	First non-USSR, non-USA satellite	Canada
1962, Oct 3	Mercury 8	Walter Schirra orbited the Earth 6 times in a 9-hour mission	USA
1963, May 15	Mercury 9	Final and longest Mercury mission (34 hours in space): Gordon Cooper	USA
1963, Jun 14	Vostok 5	Solo flight endurance record of 5 days in space: Valery Bykovsky	USSR
1963, Jun 16	Vostok 6	First woman in space: Valentina Tereshkova	USSR
1964, Apr 8	Gemini 1	First two-seat spacecraft launched unmanned in space	USA
1964, Oct 12	Voskhod 1	First three-person crew, launched aboard a modified Vostok orbiter	USSR
1964, Dec 15	San Marco	First satellite implemented and launched by a non-USSR, non-USA team	Italy
1965, Mar 1	Gemini 3	First crewed Gemini, making Gus Grissom the first astronaut to travel to space twice	USA
1965, Mar 18	Voskhod 2	First extra-vehicular activity: Alexei Leonov	USSR
1965, Jun 3	Gemini 4	First American extra-vehicular activity: Ed White	USA
1965, Aug 21	Gemini 5	First Gemini eight-day mission	USA
1965, Nov 26	Astérix	First satellite launched by a non-USSR, non-USA rocket (Diamant)	France
1965, Dec 15	Gemini 6A & 7	First orbital rendezvous (parallel flight, no docking)	USA
1966, Feb 3	Luna 9	First soft landing of a probe on the Moon	USSR
1966, Mar 16	Gemini 8	First docking between two spacecraft in orbit	USA
1966, Apr 3	Luna 10	First artificial satellite to orbit the Moon	USSR
1966, Jun 2	Surveyor 1	First successful U.S. soft landing on the Moon	USA
1966, Jun 3	Gemini 9	Rendezvous with Agena and failed docking	USA
1966, Jul 18	Gemini 10	Record altitude, docked with two Agenas, Mike Collins performed two EVAs	USA
1966, Aug 1	Lunar Orbiter 1	First U.S. probe to map the Moon	USA
1966, Sep 12	Gemini 11	New record altitude. Dick Gordon performed two EVAs	USA
1966, Nov 11	Gemini 12	Last Gemini mission. Buzz Aldrin performed three EVAs	USA
1967, Jan 27	Apollo 1	A tragic fire causes the first 3 U.S. astronaut fatalities	USSR

(continued)

Timeline of Main Space Events: 1957–1969

(continued)

Date	Mission name	Mission achievements	Country
1967, Apr 23	Soyuz 1	Soyuz 1 launches with myriad problems. Komarov is killed during return	USSR
1967, Oct 30	Cosmos 186/188	First automated (crewless) docking	USSR
1968, Sep 1	Zond 5	First animals and plants to orbit Moon and return safely to Earth	USSR
1968, Oct 11	Apollo 7	First manned Apollo mission launches for an 11-day mission in Earth orbit	USA
1968, Dec 21	Apollo 8	First manned mission to orbit the Moon	USA
1969, Jan 14	Soyuz 4	Vladimir Shatalov launched. First docking between two spacecraft in Earth orbit. Alexei Yeliseyev and Yevgeny Khrunov took a spacewalk over to Soyuz 4 and returned to Earth with Shatalov	USSR
1969, Jan 15	Soyuz 5	Yeliseyev, Khrunov and Boris Volynov launched. Volynov returned to Earth alone	USSR
1969, Mar 3	Apollo 9	First test of lunar module in Earth orbit	USA
1969, May 22	Apollo 10	Dress rehearsal of the Moon landing	USA
1969, Jul 20	Apollo 11	First Men on the Moon	USA

Bibliography

Ball [2010]   David S. Ball, **American Astrophilately. The first 50 Years**, A&A Publishers, LLC, Charleston SC, 2010, 344 pages + DVD.

Bulver [2001]   Paul C. Bulver, Reuben A. Ramkissoon and Lester E. Winick, **Study of Suspect Space Covers**, 2nd Edition, ATA Space Unit, Dallas, TX, 2001, 258 pages (CD-Version)

Burgess-Hall [2009]   Colin Burgess, Rex Hall, **The First Soviet Cosmonaut Team: Their Lives and Legacies**, Springer-Praxis, Chichester, UK, 2009, 395 pages.

Cacka [2006]   Julius Cacka, *Typy a Padělky Ruskỳch Razítek Tématu Kosmos* (*Russian space postmarks and fakes* - in Czech), Prague 2006, (2nd Edition), 115 pages

Calle C [2009]   Chris Calle, **Celebrating Apollo 11 – The Artwork of Paul Calle**, AeroGraphics Inc., Bradenton, FL, 2009, 96 pages.

Calle P [1974]   Paul Calle, **The Pencil**, published by North Light Publishers, Westport, Conn, 1974; distributed by Watson-Guptill Publications, New York, 100 pages.

Cartier [1997]   Ray E. Cartier, *Fake Baikonur Cancel Story Uncovered*, in *Astrophile*, Vol. 42 No. 4, Jul. 1997, pp. 8–9.

Cavallaro [2017]   Umberto Cavallaro, **Women Spacefarers: Sixty Different Paths to Space**, Springer-Praxis Books, New York, 2017, 403 pages.

Cernan [1999]   Eugene Cernan, Donald A. Davis, **The Last Man on the Moon: Astronaut Eugene Cernan and America's Race in Space**, St Martin's Griffin, New York, 1999, 356 pages.

Chertok [2006]   Boris Chertok, **Rockets and People – Vol. 2: Creating a Rocket Industry**, NASA SP-4110, Washington D.C., 2006, 669 pages.

Chertok [2009]   Boris Chertok, **Rockets and People – Vol.3: Hot Days of the Cold War**, NASA SP-4110, Washington D.C., 2009, 831 pages.

Chertok [2011]   Boris Chertok, **Rockets and People – Vol.4: The Moon Race**, NASA SP-4110, Washington D.C., 2011, 663 pages.

Collins [2009]   Michael Collins, **Carrying the Fire**, Farrar, Strauss and Giroux, New York, 2009, 486 pages.

Bibliography

Courtney [1979]   Brooks Courtney G., Grimwood James M., Swenson Loyd S., **Chariots for Apollo: A History of Manned Lunar Spacecraft**, NASA SP-4205, Washington D.C., 1979, 540 pages.

Cunningham [2003]   Walt Cunningham, **The All-American Boys**, iBooks, New York 2003, 488 pages.

Doran-Bizony [2011]   Jamie Doran and Piers Bizony, **Starman: The Truth Behind the Legend of Yuri Gagarin**, Walker Publishing, 2011, 256 pages.

Evans [2009]   Ben Evans, **Escaping the Bonds of Earth: The Fifties and the Sixties**, Springer-Praxis Books, Berlin 2009, 492 pages.

Goebel [airvector.net]   Greg Goebel, *Race to the Moon 1957:1975*, in vc.airvectors.net

Grahn [svengrahn.pp.se]   Sven Grahn, *Sven's space place*, www.svengrahn.pp.se

Grassani [2003]   Enrico Grassani, **Yuri Gagarin e i primi voli spaziali sovietici – con documentazione filatelica** [Yuri Gagarin and the first Soviet space flight – with philatelic documentation - in Italian], Selecta, Pavia 2003, 228 pages.

Green [1970]   Constance Mclaughlin Green and Milton Lomask, **Vanguard: A History**, NASA SP-4202, Washington D.C., 1970, 810 pages.

Grimwood [1963]   James M. Grimwood, **Project Mercury: A chronology**, NASA, SP-4001, Washington D.C., 1963, 257 pages.

Hacker [1974]   Hacker, Barton C., Grimwood, James M., **On the Shoulders of Titans: A History of Project Gemini**, NASA SP-4203. Washington D.C., 1974, 628 pages.

Hall-Shayler [2001]   Rex Hall, David Shayler, **The Rocket Men: Vostok & Voskhod. The First Soviet Manned Spaceflights**, Springer-Verlag London, 2001 - 326 pages.

Hall-Shayler [2003]   Rex Hall, David Shayler, **Soyuz: A Universal Spacecraft**, Springer-Praxis Books, 2003, 461 pages.

Hansen [1995]   James R. Hansen, **Spaceflight Revolution**, NASA SP-4308, Washington D.C., 1995, 542 pages.

Harvey [2007]   Brian Harvey, **Soviet and Russian Lunar Exploration**, Praxis Publishing, Chichester, UK, 2007, 317 pages.

Hendrickx]1995]   Bart Hendrickx, *Soviet Lunar Dream that Faded*, in *Spaceflight* Vol. 37, April 1995, pp. 135–137.

Hendrickx]1996]   Bart Hendrickx, *Korolev Facts and Myths*, in *Spaceflight* Vol. 38, February 1996, p. 44–48.

Hillger [colostate.edu]   Don Hillger and Garry Toth *Soviet propaganda-design satellites*, in *rammb.cira.colostate.edu*, Colorado State University, 2001-2017.

Hoak [2008]   Frank M. Hoak, *Naval Cover Fakes, Forgeries and Frauds, Part VIII*, in *USCS Log*, September 2008, pp. 12–15.

Hopferwieser [2016]   Walter Hopferwieser **Pioneerraketenpost und kosmiche Post**, Austria Netto Katalog Verlag, Vienna, Austria 2016, 210 pages.

Irwin [1982]   James B. Irwin, **To Rule the Night**, Holman Bible Publishers Nashville, 1982 (2nd ed.), 279 pages.

Jenkins [2003]   Dennis R. Jenkins, Tony Landis, Jay Miller, **American X-Vehicles: An Inventory—X-1 to X-50**, NASA SP 4531, Washington D.C., 2003, 63 pages.

Klochko [2009]   Viacheslav Klochko, **Outer Space Mail of the USSR and Russia**, Zvezdnyi Gorodok, Moscow, 2009, 112 pages.

Kloman [1972]   Erasmus H. Kloman, **Unmanned Space Project Management - Surveyor and Lunar Orbiter**, NASA SP-4901, Washington D.C., 1972, 42 pages.

Kranz [2000]   Gene Kranz, **Failure is not an option**, Simon & Schuster, New York 2000, pp. 200–01.

Launius [1997] Roger D. Launius (Ed.), Logsdon, John M. (Ed.), Smith, Robert W. (Ed.), **Reconsidering Sputnik: Forty Years Since the Soviet Satellite**, NASA, Washington D.C. 1997, 464 pages.

Lebedev [1992] Daniel A. Lebedev, *The N1-L3 Programme* in *Spaceflight* Vol, 34 (September 1992), pp.288–90.

Lindroos [2011] Marcus Lindroos, *The Soviet Manned Lunar Program*, MIT, October 4, 2011, 14 pages.

Mishin [1990] Vasily Mishin (Interview), *Designer Mishin speaks on early space Soviet Programmes and the manned lunar project"*, in *Spaceflight* Vol. 32, March 1990, pp. 104–6.

Oberg [1981] James Oberg, **Red Star in Orbit**, Random House, New York, 1981.

Oberg [1988] James Oberg, **Uncovering Soviet Disasters: Exploring the Limits of Glasnost**, Random House, New York, 1988, 317 pages.

Oberg [1990] James Oberg, *Disaster at the Cosmodrome* in *Air & Space Magazine* (December 1990), pp. 74–77.

Pealer [1995] Donald Pealer, *Manned Orbiting Laboratory-Part 1*, in *Quest* (1995) Vol. 4 No. 2, pp. 4–16.

Pesavento [2003] Peter Pesavento, *Declassified American Government Documents show a broader and in-depth interest in Soviet Space Activities*, in *JBIS* Vol. 56, (2003) pp. 175–191.

Pitts [1985] John A. Pitts, **The Human Factor: Biomedicine in the Manned Space Program to 1980**, NASA SP-4213, Washington D.C., 1985, XII + 389 pages.

Portree [1997] David S. F. Portree and Robert C. Treviño, **Walking to Olympus: An EVA Chronology**, NASA Monographs in Aerospace History Series #7, Washington D.C., 1997, 131 pages.

Reichman [2013] James G. Reichman, **Philatelic Study Report 2013-1. Space related Soviet Special Postmarks 1958-1991**, American Astrophilately, Framingham, MA, 2013, 994 pages.

Scott-Leonov [2004] David Scott, Alexei Leonov, **Two Sides of the Moon. Our Story of the Cold War Space Race**, St. Martin's Press, New York 2004, 416 pages.

Siddiqi [1994-1] Asif Siddiqi, *Mourning Star*, in *Quest* Vol. 2, Winter 1994, pp. 38–47.

Siddiqi [1994-2] Asif Siddiqi, *Soyuz-1 revisited: From Myth to Reality*, in *Quest* Vol. 6, Fall 1998, pp. 5–16.

Siddiqi [2000] Asif Siddiqi, **Challenge to Apollo: The Soviet Union and the Space Race, 1945-1974**, NASA SP-4408, Washington D.C., 2000, 1010 pages.

Siddiqi [2004] Asif Siddiqi, *A secret uncovered. The Soviet decision to land cosmonauts on the Moon*, in *Spaceflight* Vol 46, May 2004, pp.205–213.

Still [2001] Russel Still, **Relics of the Space Race**, 3rd Edition, PR Products, Roswell, GA (USA) 2001, 342 pages.

Steiner [1994] Tom Steiner, *Riser Covers Revisited*, in *Astrophile*, November 1994, pp. 14–17.

Stoetzer [1953] Carlos Stoetzer, **Postage Stamps as Propaganda**, Public Affairs Press, Washington D.C., 1953, 27 pages.

Swanson [1999] Glen E. Swanson [Ed.], **Before This Decade Is Out...**, NASA SP-4223, Washington D.C., 1999, XVI + 408 pages.

Swenson [1966] Loyd S. Swenson [Editor] **This New Ocean: A History of Project Mercury**, NASA SP-4201, Washington D.C., 1966, XV + 672 pages.

Bibliography

Vick [2011]   Charles P. Vick, *Anti-Climatic End of the Lunar Race*, in globalsecurity.org *Unmasking N1-L3*.

Vladimirov [1973]   Leonid Vladimirov, **The Russian space bluff - The inside story of the Soviet drive to the moon**, Dial Press, London, 1973, 190 pages.

Wade [astronautix.com]   Mark Wade, *Encyclopedia Astronautica* in *www.astronautix.com*.

Wolfe [1979]   Tom Wolfe, **The Right Stuff**, Picador, New York, 1979, 353 pages.

Index

A

Agena (satellite), 57, 154, 187, 188, 191, 193, 195, 198–201, 205, 329
Aldrin, B., 161, 196, 205, 287, 291, 303, 312, 329
Antonchenko, A., 50, 52
Apollo
    Apollo 1, 73, 115, 181, 223, 225, 226, 228, 229, 232, 238, 242, 252, 298, 329
    Apollo 2, 181, 223–225
    Apollo 3, 181, 223, 224
    Apollo 4, 209, 224, 225
    Apollo 6, 209
    Apollo 7, 141, 225, 241, 242, 252–257, 260, 280, 330
    Apollo 8, 181, 201, 209, 242, 250, 253, 259–264, 275, 276, 283, 294, 308, 330
    Apollo 9, 178, 181, 242, 250, 265, 277–279, 330
    Apollo 10, 181, 242, 278–280, 288, 330
    Apollo 11, 115, 209, 211, 242, 278, 280–283, 285–287, 291–316, 327, 330
    Apollo 12, 181, 222, 316, 317, 323
    Apollo 13, 296, 316
    Apollo 14, 302, 316, 326
    Apollo 15, 293, 297, 300, 302, 305, 315–320, 323, 324, 326, 327
    Apollo 16, 296, 316, 327
    Apollo 17, 211, 297, 327
Apollo-Saturn mission, 223
Armstrong, N., 13, 65, 66, 108, 155, 191, 195, 258, 283, 286, 287, 291–293, 296–298, 303, 305–308, 311, 312

B

Baikonur (Cosmodrome), 20, 21, 23, 101, 103, 116, 257, 267, 268, 281
Ball, D., 196, 302, 326
Bassett, C., 161, 195, 196
Bean, A.L., 161, 222
Belyayev, P., 71, 72, 74, 171, 174–178
Beregovoy, G., 193, 234, 257, 258, 267
Berezovsky, B., 6, 94
Blount, W., 301, 307, 312
Bondarenko, V.V., 71, 73, 74, 174, 228
Borman, F., 155, 188–190, 237, 250, 260–262, 283, 291
Brezhnev, L., 82, 167, 170, 219, 232, 234, 248, 267
Briskin, V.M., 284
Bykovsky, V.F., 71, 74, 95, 96, 143–145, 149, 176, 233, 329

C

Cacka, J., 99, 267
Calle, C., 147, 183, 184, 186, 308–310, 327
Calle, P., 126, 183, 184, 186, 302, 307, 311
Carpenter, S.M., 69, 132, 136–138, 155, 329
Cernan, E., 161, 195–198, 205, 279, 280, 297
Chaffee, R.B., 161, 223–225, 227, 228
Chelomey, V., 4, 37, 38, 61, 124, 163, 205, 212, 214–216, 231, 289
Chertok, B., 4, 13, 70, 74, 82, 97, 107, 112, 117, 236, 238, 258, 277, 281, 288
Collins, M., 161, 193, 260, 298, 312

© Springer Nature Switzerland AG 2018
U. Cavallaro, *The Race to the Moon Chronicled in Stamps, Postcards, and Postmarks*, Springer Praxis Books,
https://doi.org/10.1007/978-3-319-92153-2

Conrad, P., 186, 199–201, 222
Cooper, G.L., 69, 137, 142–143, 155, 186, 187, 237, 329
Corona (spy satellites), 28, 55–59, 61, 282
Covers
    Apollo flown covers, 302, 303, 305, 306
    Apollo insurance covers, 286, 291–297, 302
    Apollo 15 scandal, 293, 297, 315
    Discoverer 17: first American space mail, 59, 60
    "Prisoner" Apollo VII covers, 255, 256
    Soyuz 4-5: first Soviet space mail, 265–275
Cunningham, W., 161, 225

D

Della Maddalena, P., 158
Dementiev, R., 118
Dillman, D., 45
Dobrovolsky, V., 94
Durst, S., 11, 30, 112, 156, 158, 162, 169, 182
Dyna-Soar, 159–160, 202

E

Eiermann, W., 320, 321, 326
Eisele, D., 161, 225, 252
Eisenhower, D.D., 10, 27, 28, 40–42, 45, 49, 53, 56, 59, 61, 62, 68, 88, 103, 153
Explorer
    Explorer 1, 10, 27–30, 328
    Explorer 2, 30
    Explorer 3, 30

F

Faget, M.A. (Max), 138, 244
Feoktistov, K., 164, 166, 167
Freeman, T., 161

G

Gagarin, Y., 15, 21, 22, 71–74, 76, 78, 85, 90–94, 96, 97, 99–102, 104, 107, 109, 111, 116–119, 140, 146, 148, 176, 232, 234, 247–249, 258, 286, 309, 328
Gazenko, O., 25, 85
Gemini
    Gemini 1, 162, 329
    Gemini 2, 169
    Gemini 3, 162, 178–180, 227, 278, 329
    Gemini 4, 181, 182, 206, 227, 329
    Gemini 5, 187, 192, 193, 199, 329
    Gemini 6, 141, 187–189, 206
    Gemini 7, 188, 189, 191, 205, 206, 329
    Gemini 8, 193, 198, 258, 329
    Gemini 9, 198, 206, 329
    Gemini 10, 193, 197, 198, 329
    Gemini 11, 199–202, 329
    Gemini 12, 205, 206, 329
    Gemini B (MOL), 131, 169, 202–205
Gemini-Titan 5 (GT-5) mission, 186
Gemini-Titan 8 (GT-8) mission, 191
Gemini-Titan 9 (GT-9) mission, 195
Gemini-Titan 10 (GT-10) mission, 198
Gemini-Titan 12 (GT-12) mission, 206
Gilruth, R. (Bob), 124, 155, 159, 193, 196, 250
Glenn, J., 69, 124–137, 155, 157, 266, 328
Glushko, V., 4, 8, 37–39, 111, 112, 124, 289
Golovanov, Y., 4, 7, 73, 174
Gordon, R., 161, 199, 317
Grissom, V.I. (Gus), 69, 114–117, 120, 155, 162, 177–179, 187, 223–225, 227, 228, 328, 329
Gugerell, A., 225

H

Ham's mission, 89
Healey, J., 241, 242
Hillger, D., viii
Hoffman, P., 92, 309
Hopferwieser, W., 60, 176, 193, 274, 299, 301, 318–321, 324
Houbolt, J., 121, 122
Hubble Servicing Missions, 201

I

Irwin, J. (Jim), 315, 317, 322, 323, 326
Istrebitel Sputnik A (IS-A), 212
Ivanov, V., 47

J

Johnson, L., 27, 39, 40, 107, 111

K

Kamanin, N., 72, 97, 116, 117, 138–140, 146, 176, 231, 234, 257, 270, 272
Karpov, Y.A., 70, 73, 74, 96
Keldysh, M.V., 4, 212, 275, 283
Kennedy, J.F., 10, 62, 91, 107, 108, 110–114, 120, 121, 159, 161, 209, 253, 287, 289
Khlebtsevich, Y., 112–114
Khrunov, Y.V., 71, 74, 233, 265, 269, 274, 330
Khrushchev, N., 3–5, 7, 9, 28, 49, 110, 119, 138, 163, 213, 287
Khrushchev, S., 4, 5, 9, 10, 166
Klimashin, V.S., 51
Klochko, V., 268

Index

Komarov, V.M., 71, 73, 74, 96, 97, 143, 164–167, 232–234, 236–238, 243, 286, 330
Kondratyuk, Y., 112–114, 121
Koroleva, N., 5, 215
Korolev, S., 4, 5, 7–9, 13, 23, 24, 31, 36–39, 42, 44, 70, 71, 73–76, 79, 84, 85, 91, 96, 112, 113, 116, 117, 120, 124, 138–140, 146, 163–166, 168, 170, 171, 175, 176, 194, 195, 212–217, 231, 232, 234, 267, 288, 289
Kraft, C., 89, 125, 137, 138, 180, 250
Kruchina, A., 94
Kruchina, E., 94
Kuznetsov, N., 282

L

Laika, 24, 25, 328
LANDSAT satellite, 282
Leonov, A.A., 37, 71, 73, 74, 94, 96, 103, 168, 171–178, 182, 228, 230, 233, 236, 238, 248, 249, 260, 267, 288, 329
Lesegri Collective, 94
Litvinov, N., 52
Lollini, 17, 20, 21
  Luna 15, 282, 283
  Luna 16, 283
  Luna 17, 283
Lovell, J., 188–191, 260–262
Lunar mission, 42
Lunik
  Luna 1, 46, 47, 53, 328
  Luna 2, 46, 48–52, 328
  Luna 3, 49, 53–56, 328
  Luna 9, 55, 217, 218, 220, 329
  Luna 10, 219, 329
Lunokhod, 114, 275, 276, 283

M

Makarov, O., 230, 249, 260
Manned Orbiting Laboratory (MOL, Gemini B), 131, 169, 202–205
Matteassi, S., 160
McDivitt, J., 181, 250, 277, 278
Mercury
  Mercury 1, 86
  Mercury 2, 328
  Mercury 3, 109, 111, 328
  Mercury 4, 187, 328
  Mercury 6, 126, 328
  Mercury 7, 128, 136, 329
  Mercury 8, 138, 329
  Mercury 9, 142, 329
Mercury-Atlas 2 (MA-2) mission, 88, 89
Mercury-Atlas 6 (MA-6) mission, 136
Mercury-Atlas 7 (MA-7) mission, 137
Mercury-Atlas 8 (MA-8) mission, 138
Mercury-Atlas 9 (MA-9) mission, 142, 143
Mercury-Redstone Booster Development (MR-BD) mission, 89, 90
Mercury-Redstone 2 (MR-2) mission, 87, 88
"Meteorological Survey" missions, 49
Mishin, V., 4, 39, 124, 166, 195, 214, 230, 232, 238, 250, 252, 257–260, 272
Monateri, R., 270, 273, 287
Mueller, G., 159, 193, 198, 209–211, 223, 242

N

Nedelin, M.I., 79–81, 83
Nelyubov, G.G., 71, 74, 96
Nikolayev, A.G., 71, 73, 74, 95, 96, 139–141, 143, 150, 151, 267, 329
Nixon, R.M., 59, 62, 285, 291–293, 307

P

Photoreconnaissance satellite, 59
Pioneer, 30, 38, 42–46, 54, 56, 59, 81, 88, 96, 159
Popovich, P.R., 71, 73, 74, 95, 96, 139, 140, 143, 329
Postmark
  Baikonur first special cancel, 267, 268
  "French" fakes, 17–21, 23, 35, 56
  Gagarin Kiev postmark, 97–99
  Moscow Gagarin special postmark, 99, 100
  Noa backdated, 133–135
  Plugged-9 cover, 181
  "Prisoner" recovery covers, 255
  Riser fakes, 130–132, 204
Powers, F.G., 56, 206
Project A119, 42–43
Prosteishy Sputnik ('P.S.') satellites, 8

R

Radnofsky, M., 298, 303, 318
Ranger (satellite), 220, 285
Rega, R., 136, 141, 299, 300
Reichman, J., 100, 103, 283
Rigo, A., 196

S

Safire, W., 291
Schirra, W.M. (Wally), 69, 136–138, 141–143, 155, 157, 162, 180, 188, 225, 252, 329
Schweickart, R.L. (Rusty), 161, 277, 278
Scott, D., 161
Sedov, L., 2–4, 7, 217

See, E., 155, 195, 196
Shepard, A., 69, 89–91, 108–111, 114, 115, 117, 120, 132, 143, 155, 211, 227, 316, 328
Shonin, G.S., 71, 74, 192
Sieger, H., 320, 323, 324, 326
Signal Communications by Orbiting Relay Equipment (SCORE) satellite, 45, 88
Slayton, D.K. (Deke), 69, 124, 136, 137, 142, 155, 180, 196, 211, 242, 250, 302, 316
Smirnov, L., 194
Soloviev, M., 50, 119
Soyuz
   Soyuz 1, 229–236, 238, 243, 245, 249, 252, 257, 265, 330
   Soyuz 2, 233, 236, 238, 245, 257, 258, 265
   Soyuz 3, 257, 258
   Soyuz 4, 265–275, 330
   Soyuz 5, 265–275, 330
Sputnik
   Sputnik 1, 7, 11–13, 15, 17, 18, 23–25, 31, 103, 328
   Sputnik 2, 17, 23–26, 29, 103, 328
   Sputnik 3, 17, 31–36
   Sputnik 4, 78
Sputnik-Korabl-3 mission, 84
Stafford, T., 162, 188, 195–197, 260, 279, 280, 295
Stamps
   Apollo 8, 294
   Apollo First Man on the Moon, 183, 287, 297, 307–314, 317
   Gemini, 183
   Mercury, 126–128, 130, 133–135, 183
STS-8 mission, 327
Surveyor (satellite), 217, 220–222, 285, 329

T

Tereshkova, V.V., 145–151, 176, 267, 329
Tikhonravov, M., 4, 8, 31, 42
Titov, G.S., 71, 73, 74, 91, 96, 97, 116–120, 140, 176, 232, 248, 328
Toidse, I.M., 95

U

Ustinov, A., 119
Ustinov, D., 38, 138, 212, 232

V

Vanguard (satellite), 2, 3, 7, 12, 27–30, 96
Viktorov, V., 31, 32, 47, 94, 284
Volynov, B.V., 71, 74, 192, 257, 258, 265–267, 330
Von Braun, W., 27, 28, 30, 40, 41, 89, 91, 107, 109, 111, 121, 159, 209, 242
Voskhod
   Voskhod 1, 163, 164, 166, 168–171, 176, 329
   Voskhod 2, 168, 170–172, 174–177, 180–182, 186, 188, 189, 191, 198, 232, 329
   Voskhod 3, 192–195, 258
Voskresensky, L., 4, 8, 124, 163, 171
Vostok
   Vostok 1, 73, 74, 91, 95, 328
   Vostok 2, 116, 118, 120, 140, 232, 328
   Vostok 3, 139–141, 329
   Vostok 4, 139–141, 329
   Vostok 5, 144, 145, 186, 233, 329
   Vostok 6, 144, 145, 147, 329

W

Warsaw Pact satellites, 14
Webb, J., 159, 201, 210
White, E., 181–183, 223–225, 227, 228, 329
Williams, C.C., 161, 229
Worden, A., 316, 317, 320, 322, 327

Y

Yangel, M., 38, 39, 80, 81, 212
Yegorov, B., 164, 166, 167
Young, J., 155, 162, 177–179, 198, 227, 279, 280

Z

Zenit spy satellites, 117, 120
Zond 5 mission, 251

Druck:
Canon Deutschland Business Services GmbH
im Auftrag der KNV-Gruppe
Ferdinand-Jühlke-Str. 7
99095 Erfurt